Introduction to
VISUAL COMPUTING
Core Concepts in Computer Vision, Graphics, and Image Processing

Introduction to
VISUAL COMPUTING
Core Concepts in Computer Vision, Graphics, and Image Processing

Aditi Majumder
University of California, Irvine, USA

M. Gopi
University of California, Irvine, USA

CRC Press
Taylor & Francis Group
Boca Raton London New York

CRC Press is an imprint of the
Taylor & Francis Group, an **informa** business

A CHAPMAN & HALL BOOK

CRC Press
Taylor & Francis Group
6000 Broken Sound Parkway NW, Suite 300
Boca Raton, FL 33487-2742

First issued in paperback 2020

© 2018 by Taylor & Francis Group, LLC
CRC Press is an imprint of Taylor & Francis Group, an Informa business

No claim to original U.S. Government works

ISBN-13: 978-0-367-57225-9 (pbk)
ISBN-13: 978-1-4822-4491-5 (hbk)

Visit the Taylor & Francis Web site at
http://www.taylorandfrancis.com

and the CRC Press Web site at
http://www.crcpress.com

Contents

Preface

This book is the culmination of over a decade of teaching of a newly designed umbrella course on *visual computing* that would provide students with fundamentals in the different areas of computer graphics, computer vision and image processing. Looking back, this was a very forward looking curriculum which became the launching pad for all computer graphics, computer vision and image processing students at UCI and helped future new faculty hires in this direction to count on this course to provide exposure to fundamentals that are common to all these domains. This course is a core entry-level course in the graduate curriculum providing students the opportunity to explore a larger breadth before moving on to more focused channels of computer graphics, computer vision and/or image processing. It is also being adopted as one of the core courses for our professional masters degree program which began in Fall 2017. Interestingly, the research community has also followed this trend since 2006 when we started to see researchers from one of the domains of computer graphics, computer vision and image processing having strong presence in others leading to a young and dynamic research sub-community that traverses all these domains with equal dexterity. Therefore, having a breadth of knowledge in the general area of visual computing is perceived today as a strength that helps students delve easily into inter-disciplinary domains both within CS and other domains where it is being extensively used.

The inspiration for writing this book came from many instructors and educators who inquired about our visual computing course at UCI, designed a similar course at their home institutions, and were requesting a standard single textbook to cover all the topics. The key exercises that we undertook prior to writing this book were (a) to carefully choose a *lean* set of topics that would provide adequate breadth for an introductory course in visual computing enabling the students to take one course instead of three different courses in CG, CV and IP before deciding on the direction they would like to pursue; (b) to carefully design the *depth* of material in each of these topics so that it can be dealt with nicely during the offering of a single course without being overwhelming; (c) to categorize the topics from the perspective of visual computing in such a manner that students are able to see the common threads that run through these different domains. This exercise led to the organization of the book into five different parts.

1. Part 1: Fundamentals provide an exposure to all kinds of different visual data (e.g. 2D images and videos and 3D geometry) and the core mathematical techniques that are required for their processing in any of the CG, CV or IP domains (e.g. interpolation and linear regression).

2. Part 2: Image Based Visual Computing deals with several fundamental techniques to process 2D images (e.g. convolution, spectral analysis and feature detection) and corresponds to the low level retinal image processing that happens in the eye in the human visual system pathway.

3. Part 3: Geometric Visual Computing deals with the fundamental techniques used to combine the geometric information from multiple eyes creating a 3D interpretation of the object and world around us (e.g. transformations, projective and epipolar geometry). This deals with the higher level processing that happens in the brain that combines information from both the eyes helping us to navigate through the 3D world around us.

4. Part 4: Radiometric Visual Computing deals with the fundamental techniques for processing information arising from the interaction of light with the objects around us. This topic covers both lower and higher level processing in the human visual system that deals with intensity of light (e.g. interpretation of shadows, reflectance, illumination and color properties).

5. Part 5: Visual Content Synthesis presents fundamentals of creating virtual computer generated worlds that mimic all the processing presented in the prior sections.

The book is written for a 16 week long semester course and can be used for both UG and graduate teaching. The recommended timeline for teaching would be to dedicate two weeks for Part 1, three weeks each for Parts 2 and 4, and three and half weeks each for Parts 3 and 5. The exercises following each chapter can be used to provide weekly or biweekly written assignments. The ideal way to provide hands-on implementation experience would be to have one programming assignment accompany each part of the course picking a subset of topics taught in each part based on the expertise level of the students. The decision of making this book independent of any programming language or platform is to enable each instructor to choose the most convenient topics, platforms, and programming language for their assignments based on the resources at hand and the skill set of the audience. Evaluation via two midterms at the end of the 6th and 12th week and a comprehensive final is probably most conducive.

Teaching the material in this book in a 10 week quarter usually poses a challenge. There can be multiple ways to handle this. The easiest way is to increase the number of credits for this course leading to more contact hours to compensate for the reduced number of weeks. The second way is to pare down or divide the content presented in a standard semester long offering of the course. For example, Visual Computing-I can focus on low level visual computing focusing on Chapters 1-5 and 9-10 and the first two sections of Chapter 11 while Visual Computing-II can focus on higher level visual processing and representation focusing on Chapters 6-8, the last section of Chapter 11 and Chapters 12-15. Alternatively, parts of a chapter or complete chapters can be skipped to created a pared down version of the course that avoids reducing the rigor of the concepts taught in the class. Such an approach has been explored in the past in UCI by removing Chapters 8,10,15, and most of Chapter 14 beyond texture mapping. The decision of what to present, what to shorten and what to completely remove resides best with the instructors. The book has been written carefully to minimize dependencies between chapters and sections so that they can be chosen independently by instructors without worrying overtly about dependencies on other parts of the book.

We hope that the material presented in this book and its non-traditional organization inspires instructors to design a visual computing course in their institutions, use this book as a textbook for its offering, and hopefully see an increased interest amongst the students towards the study of the general domain of visual computing. We would like to get feedback from instructors who are using this book as a textbook. Please feel free to write to us about anything you faced while using this book — desired additions, details, or organization. Such feedback will be instrumental towards more refined and better suited subsequent editions of this book.

We acknowledge our colleagues at the University of California at Irvine for their support in designing non-traditional courses leading to experimentation which provided the building blocks for this book. We would like to thank the numerous students who took the Visual Computing course at UCI and the teaching assistants who helped us execute and experiment during different offerings of this course which led to the development and organization of the material presented in this book. We also acknowledge the help rendered by our students, Nitin Agrawal and Zahra Montazeri, in designing and rendering to perfection the various figures used in this book. We deeply appreciate the special efforts of Prof. Shuang Zhao of the University of California, Irvine, Prof. Amy and Bruce Gooch of the University of British Columbia, Dr. David Kirk of nVidia, Prof. Chee Yap of New York University, and Prof. Jan Verschelde of the University of Illinois, Chicago, in providing some of the images in this book on physically based modeling, non-photorealistic rendering, geometric compression and GPU architecture respectively.

Aditi Majumder
Gopi Meenakshisundaram

Author Biographies

Aditi Majumder is Professor in the Department of Computer Science at the University of California, Irvine. She received her PhD from the Department of Computer Science, University of North Carolina at Chapel Hill in 2003. She is originally from Kolkata, India and came to the US in 1996 after completing her bachelors of engineering in Computer Science and Engineering from Jadavpur University, Kolkata.

Her research resides at the junction of computer graphics, vision, visualization and human-computer interaction. Her research focuses on novel displays and cameras exploring new degrees of freedom and quality while keeping them truly a commodity, easily accessible to the common man. She has more than 50 publications in top venues like ACM Siggraph, Eurographics, IEEE Visweek, IEEE Virtual Reality (VR), IEEE Computer Vision and Pattern Recognition (CVPR) including best paper awards in IEEE Visweek, IEEE VR and IEEE PROCAMS. She is the co-author of the book *Practical Multi-Projector Display Design*. She has served as the program or general chair and program committee in several top venues including IEEE Virtual Reality (VR), ACM Virtual Reality Software and Technology (VRST), Eurographics and IEEE Workshop on Projector Systems. She has served as Associate Editor in Computer and Graphics and IEEE Computer Graphics and Applications. She has played a key role in developing the first curved screen multi-projector display being marketed by NEC/Alienware currently and was an advisor at Disney Imagineering for advances in their projection based theme park rides. She received the Faculty Research Incentive Award in 2009 and Faculty Research Midcareer Award in 2011 in the School of Information and Computer Science in UCI. She is the recipient of the NSF CAREER award in 2009 for Ubiquitous Displays Via a Distributed Framework. She was a Givens Associate and was a student fellow at Argonne National Labs from 2001-2003, a Link Foundation Fellow from 2002-2003, and is currently a Senior Member of IEEE.

Gopi Meenakshisundaram is a Professor of Computer Science in the Department of Computer Science, and Associate Dean at the Bren School of Information and Computer Sciences at the University of California, Irvine. He received his BE from Thiagarajar College of Engineering, Madurai, MS from Indian Institute of Science, Bangalore, and PhD from University of North Carolina at Chapel

Hill. His research interests include geometry and topology in computer graphics, massive geometry data management for interactive rendering, and biomedical sensors, data processing, and visualization. His work on representation of manifolds using single triangle strip, hierarchyless simplification of triangulated manifolds, use of redundant representation for big data for interactive rendering, and biomedical image processing have received critical acclaim including best paper awards in two Eurographics conferences and in ICVGIP. He is a gold medalist for academic excellence at Thiagarajar College of Engineering, a recipient of the Excellence in Teaching Award at UCI and a Link Foundation Fellow. He served as the program co-chair and papers co-chair of ACM Interactive 3D Graphics conference in 2012 and 2013 respectively, area chair for ICVGIP in 2010 and 2012, program co-chair for the International Symposium on Visual Computing 2006, an associate editor of the Journal of Graphical Models, a guest editor of IEEE Transactions on Visualization and Computer Graphics and serves in the steering committee of ACM Interactive 3D Graphics.

Part I

Fundamentals

1

Data

In the context of visual computing, data can be thought of as a function that depends on one or more independent variables. For example, audio can be thought of as one dimensional (1D) data that is dependent on the variable time. Thus, it can be represented as $A(t)$ where t denotes time. An image is data that is two dimensional (2D) data dependent on two spatial coordinates x and y and can be denoted as $I(x, y)$. A video is three dimensional (3D) data that is dependent on three variables – two spatial coordinates (x, y) and one temporal coordinate t. It can therefore be denoted by $V(x, y, t)$.

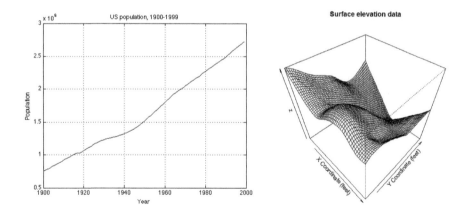

Figure 1.1. Most common visualization of 1D (left) and 2D (right) data. The 1D data shows the population of US (Y-axis) during the 20th century (specified by time in the X-axis) while the 2D data shows the surface elevation (Z-axis) of a geographical region (specified by X and Y-axes). This is often called *height field*.

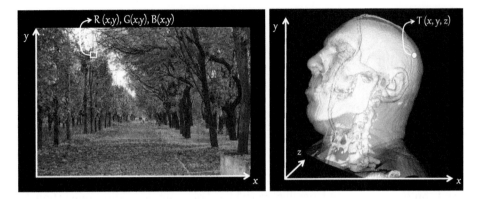

Figure 1.2. Conducive Visualizations: An image is represented as three 2D functions, $R(x,y)$, $G(x,y)$ and $B(x,y)$. But instead of three height fields, a more conducive visualization is where every pixel (x, y) is shown in RGB color (left). Similarly, volume data $T(x, y, z)$ is visualized by depicting the data at every 3D point by its transparency (right).

1.1 Visualization

The simplest visualization of a multi-dimensional data is a traditional plot of the dependent variable with respect to the independent ones, as illustrated in Figure 1.1. For example, such a visualization in 2D is called *height field*. However, as data becomes more complex, such visualization do not suffice due to the inherent inability of humans to visualize geometrical structures beyond three dimensions. Alternative perceptual modalities (e.g. color) are therefore used to encode data. For example, color image comprises of information of three color channels, usually red, green and blue, each dependent on two spatial coordinates (x, y) – $R(x, y)$, $G(x, y)$ and $B(x, y)$. However, often visualizing these three functions together is much more informative that visualizing them as three different height fields. Thus, the ideal visualization is an image where each spatial coordinate is visualized as a color which is also a 3D quantity. Similarly, a 3D volume data $T(x, y, z)$, providing scalar data at each 3D grid point, is visualized in 3D by assigning color or transparency to each grid point computed using a user defined *transfer function* $f(T(x, y, z))$ that is common to the entire data set (See Figure 1.2).

1.2 Discretization

Data exists in nature as a continuous function. For example, the sound we hear changes continuously over time; the dynamic scenes that we see around us also

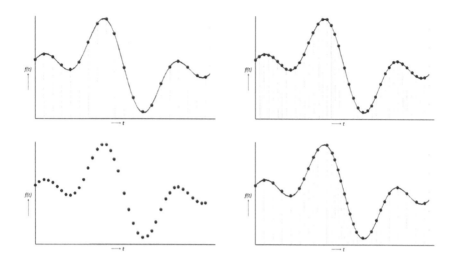

Figure 1.3. This figure illustrates the process of sampling. On top left, the function $f(t)$ (curve in blue) is sampled uniformly. The samples are shown with red dots and the values of t at which the function is sampled is shown by the vertical blue dotted lines. On top right, the same function is sampled at double the density. The corresponding discrete function is shown in the bottom left. On the bottom right, the same function is now sampled non-uniformly i.e. the interval between different values of t at which it is sampled varies.

change continuously with time and space. However, if we have to digitally represent this data, we need to change the continuous function to a discrete one, i.e. a function that is only defined at certain values of the independent variable. This process is called *discretization*. For example, when we discretize an image defined in continuous spatial coordinates (x, y), the values of the corresponding discrete function are only defined at integer locations of (x, y), i.e. pixels.

1.2.1 Sampling

A *sample* is a value (or a set of values) of a continuous function $f(t)$ at a specified value of the independent variable t. Sampling is a process by which one or more samples are extracted from a continuous signal $f(t)$ thereby reducing it to a discrete function $\hat{f}(t)$. The samples can be extracted at equal intervals of the independent variable. This is termed as *uniform* sampling. Note that the *density* of sampling can be changed by changing the interval at which the function is sampled. If the samples are extracted at unequal intervals, then it is termed as *non-uniform* sampling. These are illustrated in Figure 1.3.

The process of getting the continuous function $f(t)$ back from the discrete

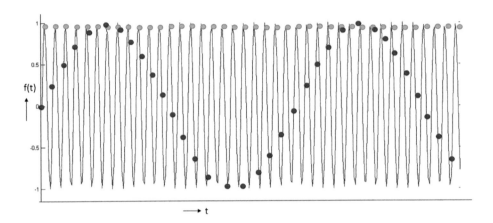

Figure 1.4. This figure illustrates the effect of sampling frequency on reconstruction. Consider the high frequency sine wave shown in blue. Consider two types of sampling shown by the blue and red samples respectively. Note that none of these sample the high frequency sine wave adequately and hence the samples represent sine waves of different frequencies.

function $\hat{f}(t)$ is called *reconstruction*. In order to get an accurate reconstruction, it is important to sample $f(t)$ *adequately* during discretization. For example, in Figure 1.4, a high frequency sine wave (in blue) is sampled in two different ways, both uniformly, shown by the red and blue samples. But in both cases the sampling frequency or rate is not adequate. Hence, a different frequency sine wave is reconstructed – for blue samples a zero frequency sine wave and for red samples a much lower frequency sine wave than the original wave. These incorrectly reconstructed functions are called aliases (for imposters) and the phenomenon is called *aliasing*.

This brings us to the question of *what is adequate sampling frequency?* As it turns out, for sine or cosine waves of frequency f, one has to sample them at a minimum of double the frequency, i.e. $2f$, to assure correct reconstruction. This rate is called the *Nyquist sampling rate*. However, note that the reconstruction is not a process of merely connecting the samples. The reconstruction process is discussed in details in later chapters.

We just discussed adequate sampling for sine and cosine waves. But, *what is adequate sampling for a general signal – not a sine or a cosine wave?* To answer this question, we have to turn to the operation complementary to reconstruction, called *decomposition*. Legendary 19th century mathematician, Fourier, showed that any periodic function $f(t)$ can be decomposed into a number of sine and cosine waves which when added together give the function back. We will revisit Fourier decomposition in greater detail at Chapter 4. For now, it is sufficient to

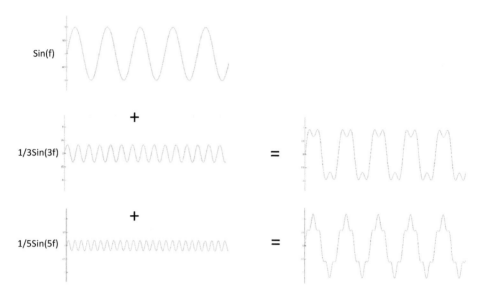

Figure 1.5. This figure illustrates how addition of different frequency sine waves results in the process of generation of general periodic signals.

understand that there is a way by which any general signal can be decomposed into a number of sine and cosine waves. An example is shown in Figure 1.5 where different frequency sine waves are added to create new signals. Hence, the adequate sampling rate of a general signal is guided by the highest frequency sine or cosine wave present in it. If the signal is sampled at a rate that is greater than twice the highest frequency sine or cosine wave present in the signal, sampling will be adequate and the signal can be reconstructed. Therefore, the signal in Figure 1.5 has to be sampled at least at a rate of $6f$ to assure a correct reconstruction.

1.2.2 Quantization

A analog or continuous signal can have any value of infinite precision. However, whenever it is converted to digital signal, it can only have a limited set of value. So a range of analog signal values is assigned to one digital value. This process is called *quantization*. The difference between the original value of a signal and its digital value is called the *quantization error*.

The discrete values can be placed at equal intervals resulting in uniform step size in the range of continuous values. Each continuous value is usually assigned the nearest discrete value. Hence, the maximum error is half the step size. This is illustrated in Figure 1.6.

Put a Face to the Name

Harry Theodore Nyquist is considered to be one of the founders of communication theory. He was born to Swedish parents in February 1886 and immigrated to the United States at the age of 18. He received his B.S. and M.S. in electrical engineering from the University of North Dakota in 1914 and 1915 respectively. He received his PhD in physics in 1917 from Yale University. He worked in the Department of Development and Research at AT&T from 1917 to 1934, and continued there when it became Bell Telephone Laboratories until his retirement in 1954. He died in April 1976.

However, human perception is usually not linear. For example, human perception of brightness of light is non-linear, i.e. if the brightness is increased by a factor of 2, its perception increases by less than a factor of 2. In fact, any modality of human perception (e.g. vision, audio, nervous) is known to be non-linear. It has been shown that most human perception modalities follow Steven's power law which says that for input I, the perception P is related by the equation $P \propto I^\gamma$. If $\gamma < 1$, as is the case of human response to brightness of light, the response is said to be sub-linear. If $\gamma > 1$, as is the case for human response to electric shock, the response is said to be super-linear.

Due to such non-linear response of the human system, in many cases, a non-uniform step size is desired when converting a continuous signal to digital. For example, in displays, the relationship of the input voltage to the produced brightness needs to be super-linear to compensate for the sub-linear response of the human eye. This function in displays (e.g. projectors, monitors) is commonly termed as the gamma function. When such non-uniform step size is used during the conversion of the continuous signal to digital, the maximum quantization error is half the maximum step size, as illustrated in Figure 1.6.

1.3 Representation

In this section we will discuss data representation in the context of visual computing – namely audio, images, videos and meshes. An analytical representation of data is in the form of a function of one or more independent variables. Audio data $A(t)$, where t denotes time, can be represented as $A(t) = sin(t) + \frac{1}{2}sin(2t)$. However, for digital representation of an arbitrary audio signal, we usually use a 1D array to represent the audio data. From now on, we will distinguish the digital representation from the analog by using $A[t]$ instead of $A(t)$. Note that

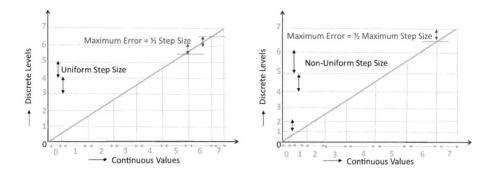

Figure 1.6. This figure illustrates the effect of step size on quantization error. The blue dotted lines show the eight discrete values. Note that these can be distributed at equal intervals resulting in uniform step size throughout the range of continuous values. The intervals can also change to create non-uniform step size. The range of continuous signal values that are assigned a particular discrete value is shown on the independent axis leading to maximum quantization error of half the maximum step size. Hence, for uniform step size, the maximum error is half the uniform step size.

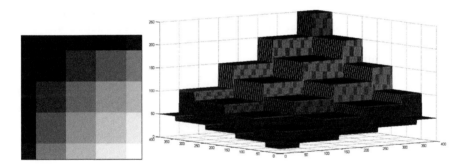

Figure 1.7. This figure illustrates the gray scale image (left) being represented as a height field (right).

representation using an 1D array follows an underlying assumption that the data is *structured*, which in this case means uniformly sampled.

Similarly, a 2D digital grayscale image I is denoted by the 2D array $I[x, y]$ where x, y stands for spatial coordinates. This also assumes structured data. This can be visualized as an image with a grayscale color assigned to every (x, y) coordinates. It can also be visualized as a *height field* in which the height (Z-value) is the grayscale value at every (x, y) coordinate forming a surface (Figure 1.7).

Color images also have multiple channels, typically red, green and blue.

Hence, they are represented by a three dimensional array $I[c, x, y]$ where c denotes the channel, $c \in \{R, G, B\}$. Video involves the additional dimension of time and hence is represented by a four dimensional array $V[t, c, x, y]$. Note that all of these data are structured, which assumes a uniform sampling in each dimension. All these aforementioned representations are called the time or spatial domain representation.

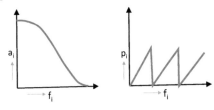

An alternate representation, called the *frequency domain representation*, considers the signal as a composition (e.g. linear combination) of a number of more fundamental signals (e.g. sine or cosine waves). Then the signal can be represented by the coefficients of these fundamental signals in the composition that would result in the original signal.. For example, the Fourier

Figure 1.8. Informal representation of the frequency domain response of a 1D signal

transform provides us with a way to find the weights of the sine and cosine waves that form the signal. Since the frequencies of these fundamental signals are predefined based on their sampling rate, the signal can then be represented by a set of coefficients for these waves. In this chapter we will briefly discuss the Fourier transformation, and will revisit this topic in greater detail in Chapter 4.

Let us consider a 1D signal $c(t)$ (e.g. audio). This can be represented as

$$c(t) = \sum_{i=1}^{\infty} a_i Cos(f_i + p_i)$$

where a_i and p_i denote respectively the amplitude and the phase of the constituting cosine waves. Therefore, the frequency domain representation of $c(t)$ is two plots – amplitude plot that shows

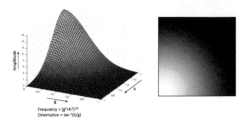

Figure 1.9. Left: Informal representation of the amplitude part of the frequency domain response of a 2D signal. Right: A grayscale representation of the same plot on the left.

a_i plotted with respect to f_i and phase plot that shows p_i plotted with respect to f_i. Together they show the amplitude and phase of each wave of frequency f_i. A typical 1D frequency response plot is shown in Figure 1.8. Since higher frequency waves only create the sharp features, they are usually present in very small amplitudes. Hence, most amplitude plots, especially for natural signals, taper away at higher frequencies as shown in 1.8.

Let us now try to extend this concept intuitively to 2D signals (e.g. grayscale image). Note that when considering these waves in 2D, they can now not only differ in frequency f but also in orientation o. A horizontal cosine wave is entirely different than a vertical one even if they have the same frequency. Therefore, the frequency response of 2D signals results in 2D plots where the amplitude/phase are functions of both frequency and orientation. However, understanding a 2D plot whose one axis is frequency and other orientation is very hard for us to comprehend. An easier way to plot these is to use polar coordinates g and h such that frequency f at coordinate (g, h) is given by the length $\sqrt{g^2 + h^2}$ and the orientation is given by the angle $\tan^{-1} \frac{h}{g}$. This means that a circle in (g, h) would provide cosine waves of the same frequency and different orientation and a ray from the origin will provide cosine waves of the same orientation and different frequencies. Figure 1.9 shows an example 2D amplitude plot. Note that here also the higher frequencies have much less amplitude than the lower frequencies given by the radially decreasing values of the plot. Alternatively, the same plot can be visualized as a grayscale image where the amplitude is normalized and plotted as a gray value between black and white (Figure 1.9).

1.3.1　Geometric Data

A geometric entity (e.g. lines, planes or surfaces) can be represented analytically. Alternatively, a discrete representation can also be used. Continuous representations can be implicit, explicit or parametric.

An explicit representation is one where one dependent variable is expressed as a function of all the independent variables and constants. The explicit equation of a 2D line is

$$y = mx + c$$

where m and c are the slope and y-intercept of the line. Similarly, the explicit representation of a 2D quadratic curve can be

$$y = ax^2 + bx + c$$

where a, b and c are the coefficients of the quadratic function representing the curve. Another popular explicit function occuring in physics and signal processing is

$$y = A sin(\omega t + \phi).$$

This represents a sine wave of amplitude A, frequency ω and phase ϕ. Note that an explicit representation allows easy evaluation of the function at different values of the independent variables.

However, more complex functions are sometimes not easy to represent using explicit form. Implicit representations consider a point $\mathbf{p}$ to be of interest if it satisfies an equation $F(\mathbf{p}) = c$, where c is a constant. The implicit equation of a 2D line is

$$ax + by + c = 0,$$

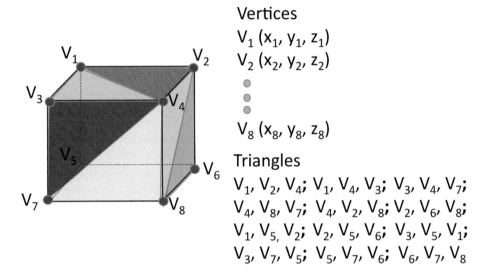

Figure 1.10. This figure shows the representation of a 3D mesh of a cube. It comprises of a list of vertices followed by a list of triangles. Each triangle is described by the indices of the vertices it comprises.

while that of a 3D plane is

$$ax + by + cz + d = 0.$$

Similarly, the implicit equation of a 2D circle is

$$(x - a)^2 + (y - b)^2 = r^2$$

where (a, b) is the center and r is the radius of the circle. The implicit equation of a 3D sphere is

$$(x - a)^2 + (y - b)^2 + (z - c)^2 = r^2$$

where (a, b, c) is the center and r is the radius of the sphere. In explicit function, sometimes dependent and independent variables have to be swapped to represent special cases. For example, it is not possible to represent a vertical line using explicit equation $y = mx + c$ since $m = \infty$. So we need to change x to be a dependent variable to represent this horizontal line $x = m'y + c'$ where $m' = 0$. On the other hand, there are no special cases in implicit function representation. The advantage of an implicit representation is an easy inside or outside test. If $F(\mathbf{p}) < 0$, the point is 'above' or 'outside' the surface and if $F(\mathbf{p}) < 0$, the point is 'below' or 'inside' the surface.

Finally, the parametric equation allows the representation of the function using one or more parameters. For example, a point $\mathbf{p} = L(t)$ on a line segment

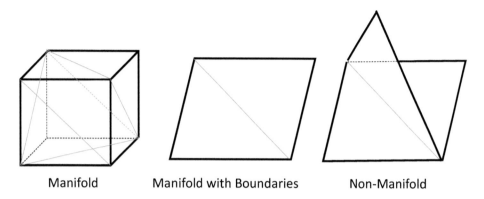

Manifold   Manifold with Boundaries   Non-Manifold

Figure 1.11. This figure illustrates manifold (closed objects), manifolds with boundaries (objects with holes) and non-manifolds (objects with folds and creases.

between two points and P and Q can be represented in the parametric form as

$$L(t) = P + t(Q - P),$$

where the parameter is t and $0 \leq t \leq 1$. Similarly, the parametric equation of a point inside the triangle formed by P, Q and R is given by the two parameter equation given by

$$\mathbf{p} = P + u(Q - P) + v(R - P).$$

where the parameters are u and v such that $0 \leq u, v \leq 1$ and $u + v \leq 1$. The parametric equation allows easy sampling of the parametric space and evaluating any function at these different sampled values.

In a discrete representation, a geometric entity is represented as a collection of other geometric entities as opposed to an analytical equation. For example, a 2D square can be defined by a set of lines embedded in the 2D space; a 3D cube can be defined by a set of quadrilaterals or traingles embedded in the 3D space. Such a representation is called a *mesh*. For example, when using triangles to define a 3D object, we call it a *triangular mesh*. The entities that make up the mesh (e.g. lines, triangles or quadrilaterals) are called the *primitives*.

Though there are many different ways to represent 3D geometry, the most common is a triangular mesh. So, we discuss some key elements of triangular mesh representation here. More details of other geometric representations and their use are presented in later chapters. A triangular mesh is defined by a set of vertices and a set of triangles formed by connecting those vertices. The representation therefore consists of two parts: (a) a list of vertices represented by their 3D coordinates; and (b) a list of triangles each defined by indices of the three vertices of its corners. Figure 1.10 shows an example of the mesh

Figure 1.12. Left: This shows how a genus 1 donut is transformed to a genus 1 cup by just changing the geometry. Right: This shows the diagram of a mobius strip.

representation of a simple 3D object, a cube. The coordinates of the vertices define the *geometry* of the mesh. In other words, changing these coordinates changes the geometry of the object. For example, if we want change the cube into a rectangular parallelepiped or a bigger cube, the 3D coordinates of the vertices will be changed. However, note that this will not change the triangle list since the connectivity of the vertices forming the triangles does not change. Hence, the latter is termed as the *topological* property of the mesh. Topology refers to connectivity that remains invariant to changes in geometric properties of the data.

Next, we will define certain geometric and topological properties of meshes, but not in a rigorous fashion. We will give you intuitions and informal definitions. Closed meshes (informally defined to have no holes) have several nice properties in the context of computer graphics operations like morphing, mesh simplification and editing. Such meshes are manifolds where every edge has exactly two incident triangles. A mesh where every edge has one or two incident triangles is called manifold with boundaries. For example, a piece of paper denoted by two triangles where the four edges forming the sides of the paper have only one incident triangle, is a manifold with boundaries. Note that manifold with boundaries are less restrictive than manifolds and hence a superset of manifolds. Meshes where edges can have more than two incident triangles are called non-manifolds. Note that non-manifolds are a superset of manifolds with boundaries. Figure 1.11 illustrate this.

Meshes can be defined with geometric properties or attributes. In Figure 1.10 each vertex has 3D spatial coordinates. In addition to this basic information, each vertex can have RGB color, normal vectors, or 2D coordinates of an image to be pasted on the mesh (formally known as texture coordinates), or any other vertex-based attribute that is useful for the given application. Topological properties are properties that do not change with change in geometric properties. For mesh processing, a few topological properties are very important. First is *Euler characteristics* e defined as $V - E + F$ where V is the number of vertices, E is the number of edges, and F is the number of faces (not necessarily triangular) of the mesh. Note that if you change the cube to a parallelepiped by changing the position of the vertices which is a geometric property, e does not change.

Essentially e may change only when the object undergoes some change in the mesh connectivity. *Genus* of a mesh is defined as the number of handles. For example, a sphere has a genus zero, a donut has a genus 1 and a double donut has a genus 2. One will need to change the topology of the mesh to change from one genus to another while only geometric changes are sufficient to change one object to another with same genus (Figure 1.12). Finally, a mesh is not orientable if you start walking on the top of the mesh and end up in its backside. An example of a non-orientable mesh is the mobius strip (Figure 1.12).

Fun Facts

A non-orientable surface that has been intriguing to topologists is the Klein bottle. Unlike a mobius strip, it does not have any boundary. It is what you get when you put two mobius strips together. The Klein bottle was first described in 1882 by the German mathematician Felix Klein. It cannot be embedded in 3D space, only in 4D space. It is hard to say how much water Klein bottles would hold, they contain themselves when embedded in 4D space! This has not stopped people from trying to embed them in 3D however, and there are some beautifully-made representations on display at the London Science Museum!

We have so far only considered triangular primitives for meshes. Though other primitives can be used (e.g. six quadrilaterals instead of 12 triangles for mesh representation of a cube), triangles are preferred for various reasons. First, triangles are always planar since three non-collinear points define a plane. Hence, modeling packages do not need to assure that a surface fits the vertices when they output the mesh representation. Second, as we will see in the next chapter, in computer graphics it is important to find out the attributes or properties of points lying inside a primitive from the properties at its vertices via techniques called interpolation where triangular primitives hold a great advantage.

1.4 Noise

Any discussion on data cannot be complete without discussing noise. Noise can be caused due to several factors like mechanical imprecision, sensor imprecision (e.g. occasional always-dead or always-live pixels) and so on. The origins of noise in different systems are different. It is best described as addition of random values as random locations in the data. In this chapter we will discuss some common types of noise.

Figure 1.13. This figure illustrates random noise in 1D audio data (left), 2D image data (middle), and 3D surface data (right). In each example, the clean data is shown on the left and the corresponding noisy data is shown on the right.

The most common and general kind of noise is what we call random noise i.e. addition of small random values at any location of the data. Figure 13 shows some examples. A common technique to reduce such noise in data is what we call low pass filtering and it will be dealt with in detail in Chapter 3.

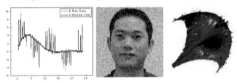

Another common type of noise originates from having outliers in the data i.e. samples which clearly cannot belong to the data. For example, in a camera some sensor pixels may be dead making thereby blocking or allowing all the light providing pixels that are always either black or white respectively. The lo-

Figure 1.14. This shows the outliers or salt and pepper noise in 1D (left), 2D (middle) and 3D (right) data. On the left, we show the effect of a median filter in removing the outliers in red.

cations of such pixels may be random. In the specific case of 2D images, this noise is called salt and pepper noise (see Figure 1.14). Such outliers are handled adequately by median filters or other order statistics filters. We will see some of these in Chapter 5.

Finally, some noise may look random in the spatial domain but can be isolated to a few frequencies in the spectral domain. An example of such noise is shown in Figure 1.15. Such noise can be removed by applying a filter in the frequency domain called the notch filter and we are going to talk about that in detail in Chapter 4.

1.5   Conclusion

In this chapter we discussed the fundamentals of representing and visualizing different kinds of visual data like images, 3D surfaces and point clouds. We also learned about two alternate representations of data in the spatial/time domain and frequency domain. We talked about practical issues involving noise in data

Figure 1.15. This figure shows the frequency domain noise that can be removed or reduced by notch filters.

and how it needs to be handled on a case by case basis. Here are some references for familiarizing yourself for some advanced concepts. [Ware 04] explores in details all about information visualization. [Goldstein 10] provides an excellent first reading for topics related to sensation and human perception. The chapter on Data Structures for 3D graphics in [Ferguson 01] provides a detailed description of representation of 3D models. The chapter on noise on [Gonzalez and Woods 06] provides a very detailed treatise on noise that is worth reading.

Bibliography

[Ferguson 01] R. Stuart Ferguson. *Practical Algorithms for 3D Computer Graphics.* A. K. Peters, 2001.

[Goldstein 10] Bruce E. Goldstein. *Sensation and Perception.* Thomas Wadsworth, 2010.

[Gonzalez and Woods 06] Rafael C. Gonzalez and Richard E. Woods. *Digital Image Processing (3rd Edition).* Prentice-Hall, Inc., 2006.

[Ware 04] Colin Ware. *Information Visualization: Perception for Design.* Morgan Kaufmann Publishers Inc., 2004.

Summary: Do you know these concepts?

- Height Field

- Discretization

- Sampling and Nyquist Sampling Theorem

- Decomposition and Reconstruction

- Aliasing

- Quantization

- Time and frequency domain representation

- Mesh - Geometry and Topology

- Manifold, manifold with boundaries and non-manifolds

- Euler characteristics, genus, orientability

- Random Noise

- Salt and Pepper Noise

Exercises

1. Consider an 8-bit grayscale image $I(x, y)$ whose size is 256×256. Each column of the image has the same gray value which starts from 0 for the left most column and increases by 1 as we sweep from left to right. What kind of shape does the height field of this image form? Find its equation?

2. Consider a height field $H(x, y)$ of size 256×256 given by the function $H(x, y) = (x \mod 16) * 16$. What kind of shape would this height field have? How many gray levels would this image have? Create a table to show the percentage of pixels that belong to each of these gray levels.

3. Consider a gray scale spatial function $A(x, y)$ which does not vary in the y-direction but form a sine wave as we go from left to right in x-direction making 50 cycles. What will be the minimum horizontal resolution of an digital image that can sample this function adequately? Consider another function $B(x, y)$ formed by rotating A about the axis perpendicular to the plane formed by x and y. Now consider the function formed by adding A and B. What is the minimum horizontal and vertical resolution of the image required to sample $A + B$ adequately?

4. Consider an object moving at 60 units per second. How many frames per second video is required to adequately capture this motion? What kind of artifact would you expect if the frame rate is lesser than this desired rate? What is this artifact more commonly known as?

5. The image of your TV looks washed out. The technician says that the intensity response curve of the TV is linear and hence the problem. To correct the problem, he has to make it non-linear. Why? What kind of non-linear response do you think he will put in?

6. If the number of bits used for representing the color of each pixel is increased quantization error is reduced. Justify this statement.

7. Can quantization be explained as an artifact of insufficient sampling? Justify your answer.

8. Your TV has three channels – R, G and B. However one of these channels is broken and now you can only see blacks and purples. Which channel is broken?

9. A 1D function contains all the harmonics of the sine wave that makes 1 cycle with a spatial span of 1 unit. Choose the correct answer.

   (a) The amplitude plot of the frequency domain response of this function is a (i) a sine wave; (ii) a horizontal line; (iii) a comb function.

(b) The phase plot of the frequency domain response of this function is a
(i) a sine wave; (ii) a horizontal line; (iii) a comb function.

10. What is the euler characteristics of a cube represented by six planar quadrilaterals. Euler characteristics of an object are related to its genus by the formula $e = 2 - 2g$. Can you derive the genus of a sphere from the Euler characteristics of a cube? If so, how?

11. Topologically, a cube is an approximation of a sphere using quadrilateral faces. In such a cube, all vertices have degree three. It is claimed that one can construct an approximation of a sphere using quadrilaterals where each vertex has degree 4. Prove or disprove this claim.

12. Objects like spheres are usually approximated in computer graphics by simpler objects made of flat polygons. Start with a regular tetrahedron constructed from four triangles. Derive one or more methods to obtain a close approximation of a sphere based on subdividing each face of the tetrahedron recursively using the same geometric operation. Does these constructions change the topological properties of the sphere? Can you think of some criteria to evaluate the quality of these constructions?

13. Match the noisy images in the top row with the filters that will remove the noise in the bottom row.

Notch Filter   Low Pass Filter   Median Filter

14. Consider the mesh representing a pyramid with a quadrilateral as base and four triangles attached to each of its sides to form the structure of the pyramid. Find its Euler characteristics and genus.

2

Techniques

We are familiar with different kinds of data. In this chapter, we will introduce two fundamental techniques that we will be using throughout this book: interpolation and computation of geometric intersections.

2.1 Interpolation

Consider a function (e.g. attributes or properties like color or position) sampled at certain parametric values. Interpolation is a process by which this function is estimated at parametric values at which it has not been measured or sampled. An image data can be considered as samples of a 2D function I(x,y) that provides color at each spatial location (x,y). Typically, image data points are sampled at integer values of x and y. Given the function values of I(x,y) at the integer grid points (x, y), we use interpolation to find the function value $I(x, y)$ at in-between, and possibly non-integer values of x, y. Or, consider a triangle. The position function (defined by 3D coordinates) is defined only at the vertices. We need to *interpolate* the positions of the vertices to the interior of the triangle to compute the position of any point lying inside the triangle.

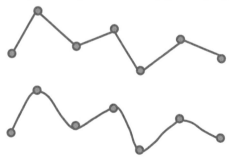

Interpolation is based on the assumption that the function changes smoothly between the different sampled values. However, interpolation techniques differ based on the degree of smoothness of change assumed between the samples. For example, consider the 1D function shown in Figure 2.1. The simplest assumption is that the 1D function changes linearly between two adjacent samples, i.e. two adjacent samples are connected by straight lines. Therefore, the function can be estimated at a parameter where it is not sampled, by considering the function values at its two nearest

Figure 2.1. This shows the assumption of smooth transition between samples in interpolation and how it can be modeled differently. Top: Linear; Bottom: Non-Linear.

neighbors which defines the straight

line. This kind of interpolation that uses a straight line to estimate function values at points where it is not sampled is called linear interpolation. However, it is evident from Figure 2.1 that at sample points, the function values can changes abruptly (also called C^0 continuity). In many applications such discontinuities in the derivative of the function values are not acceptable. Therefore, in more sophisticated interpolation techniques, it is assumed a smooth curve passes through multiple of these samples such that the tangent vector of the curve also smoothly changes. In order to compute the derivatives, we need a larger neighborhood of sample points rather than just two that we used for linear interpolation. For tangent continuity (aka first derivative continuous, C^1 continuous, quadratic interpolation), we use three sample points. Similarly, for second derivative continuous cubic curve (C^2 continuous), we use four sample points, and so on. In this book we will almost always use only linear interpolation and therefore we will explore this in detail. We will first describe linear interpolation in 1D (as shown in Figure 2.1) and then extend the concept to 2D data (e.g. images and meshes) in which case it is called a bilinear interpolation.

2.1.1   Linear Interpolation

Let us consider a straight line segment between the endpoints V_1 and V_2. Let the color at these two vertices be $C(V_1) = (r_1, g_1, b_1)$ and $C(V_2) = (r_2, g_2, b_2)$ respectively. Any point V on the line segment $V_1 V_2$ is given by $V = \alpha V_1 + (1-\alpha) V_2$ where $0 \leq \alpha \leq 1$.

We say a function f is linear if $f(aX + bY) = af(X) + bf(Y)$. Similarly, we say the color at the point V, $C(V)$, in the line segment $V_1 V_2$ is linearly interpolated when

$$C(V) = C(\alpha(V_1) + (1 - \alpha)(V_2)) = \alpha C(V_1) + (1 - \alpha)C(V_2). \qquad (2.1)$$

We can see that rate of change of color of a point between V_1 to V_2, with respect to the distance traveled between V_1 and V_2, is constant.

Technically, the above interpolation is much more specific than a general linear interpolation (or linear combination) – it is called a convex combination of $C(V_1)$ and $C(V_2)$ where the coefficients are positive and they add up to 1.0. It is said that the function value at V (in this case color) is interpolated from the function values at V_1 and V_2 linearly by weighting the function at those values using coefficients α and $(1 - \alpha)$. The coefficients α are typically computed as a relative function of the distance of the point V from V_1 and V_2. For example, we know that the parametric equation of a line is given by

$$(x, y, z) = \alpha(x_1, y_1, z_1) + (1 - \alpha)(x_2, y_2, z_2) \qquad (2.2)$$

where (x, y, z) are the 3D coordinates of any point on the line $V_1 V_2$. Note that, though the coordinates are 3D, the geometric entity is 1D line embedded in 3D

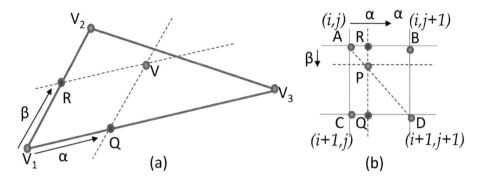

Figure 2.2. Left: Bilinear interpolation at V from V_1, V_2 and V_3 in a triangle. Right: Bilinear interpolation to find the value of F at P from the value of F at the integer pixels at (i, j), $(i + 1, j)$, $(i, j + 1)$ and $(i + 1, j + 1)$ in an image.

and therefore we are using linear interpolation. Given the locations of the points V_1, V_2 and V, we can find α by solving the equation

$$x = \alpha x_1 + (1 - \alpha) x_2 \qquad (2.3)$$

and use this α to find the value of $C(V)$ from $C(V_1)$ and $C(V_2)$. The coefficients for V_1 and V_2 are both between 0 and 1 and their sum is equal to 1.0. Therefore, the function at V is estimated by a weighted sum of functions at V_1 and V_2 where each of the weights are fraction between 0 to 1 and their sum is equal to 1. This is called a convex combination of V_1 and V_2.

If the constraint is only that the sum of the coefficients is 1.0, but the coefficients can be of any value, then it is called an affine combination. If there are no constraints on the coefficients, it is called a linear combination. Note that a linear combination does not always mean linear interpolation. For example, if the Equation 2.1 was $C(V) = \alpha^2 C(V_1) + (1 - \alpha^2) C(V_2)$, it would not be a linear interpolation in α, but would still be a linear combination of $C(V_1)$ and $C(V_2)$ because $alpha^2$ and $(1 - alpha^2)$ are still scalar values. In other words, for linear interpolation, the derivative of the interpolated function should be a constant.

2.1.2 Bilinear Interpolation

Instead of considering 1D data, let us now consider 2D data where the neighborhood of a sample extends in two different directions. Bilinear interpolation entails interpolating in one direction followed by interpolating in the second direction.

For this, let us consider a triangle with three vertices V_1, V_2 and V_3 (Figure 2.2a). To estimate the function C at a point V inside the triangle, we first estimate the function in the two directions $V_1 V_3$ and $V_1 V_2$. The point Q on $V_1 V_3$

is given by linear interpolation as

$$Q = (1 - \alpha)V_1 + \alpha V_3. \tag{2.4}$$

Similarly, the point R along V_2V_1 is given as

$$R = (1 - \beta)V_1 + \beta V_2. \tag{2.5}$$

where $0.0 \leq \alpha, \beta \leq 1.0$. Therefore V is given by the vector addition of V_1, R and Q as

$$V = V_1 + (1 - \alpha)V_1 + \alpha V_3 + (1 - \beta)V_1 + \beta V_2 \tag{2.6}$$
$$= (1 - \alpha - \beta)V_1 + \alpha V_3 + \beta V_2. \tag{2.7}$$

Therefore C at V can be estimated as

$$C(V) = (1 - \alpha - \beta)C(V_1) + \alpha C(V_3) + \beta C(V_2). \tag{2.8}$$

Bilinear interpolation also results in a convex combination and the values of α and β can be recovered by solving two equations formed by the coordinates of V, V_1, V_2 and V_3 in Equation 2.7. Further, you can verify for yourself that the coefficients for finding V does not change if you consider any two different directions like V_3V_2 and V_3V_1 or V_1V_2 and V_2V_3 (See exercise for problems on this).

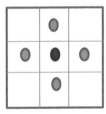 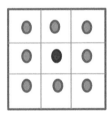

4-connected   8-connected

Figure 2.3. This shows the 4 and 8 connected neighbors in blue for the red pixel in the center.

Now let us consider bilinear interpolation in another scenario of an image (Figure 2.2b). Any non-integer spatial location may be considered 4-connected when its neighborhood is defined by four nearest neighbors, two in horizontal direction and two in vertical direction. A neighborhood can also be 8-connected when the diagonal neighbors are also included. The distance of the neighbors in the 8-connected neighborhood can be different (for example, diagonal neighbors are $\sqrt{2}$ distance away while the horizontal and vertical neighbors are unit distance away). This is illustrated in Figure 2.3.

Let us consider a function F that defines the color at integer pixels at (i, j), $(i, j + 1)$, $(i + 1, j)$ and $(i + 1, j + 1)$ denoted by A, B, C and D respectively. Let us consider a point P where C has to be estimated as shown in Figure 2.2b. Note that in this case, the location of each of these pixels is defined using 2D coordinates. Therefore, the distance between pixel (i, j) and P in horizontal and vertical direction can be found from their location. Let this be α and β respectively where $0 \leq \alpha, \beta \leq 1$. Therefore, the value of C at the pixel Q is given by linear interpolation in the horizontal direction as

$$F(Q) = (1 - \alpha)C + \alpha D \tag{2.9}$$

Similarly the value of F at R is given by

$$F(R) = (1 - \alpha)A + \alpha B. \tag{2.10}$$

Now, the value of C at P is found by interpolating between R and Q linearly in the vertical direction as

$$
\begin{aligned}
F(P) &= F(Q)\beta + F(R)(1 - \beta) & \text{(2.11)} \\
&= \beta(1 - \alpha)C + \beta\alpha D & \text{(2.12)} \\
&\quad + (1 - \beta)(1 - \alpha)A + (1 - \beta)\alpha B & \text{(2.13)} \\
&= \beta(1 - \alpha)F(i + 1, j) + \beta\alpha F(i + 1, j + 1) & \text{(2.14)} \\
&\quad + (1 - \beta)(1 - \alpha)F(i, j) + (1 - \beta)\alpha F(i, j + 1) & \text{(2.15)}
\end{aligned}
$$

Now, consider the case where P happens to be on the straight line connecting A and D. In this case, P can be expressed as a linear combination of these two points. The distance AP is given by $\sqrt{\alpha^2 + \beta^2}$ and the distance PD is given by $\sqrt{(1 - \alpha)^2 + (1 - \beta)^2}$. Therefore,

$$F(P) = \sqrt{\alpha^2 + \beta^2}D + \sqrt{(1 - \alpha)^2 + (1 - \beta)^2}A \tag{2.16}$$

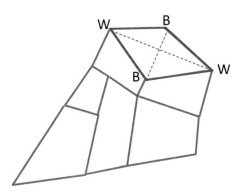

Therefore, there are multiple ways to interpolate F at P, using either equation 2.14 and 2.16 and many more can be found. For example, we may want to interpolate the point at the intersection of AD and BC using A, B and C followed by an interpolation of this point and A to get P. Each of these will result in different coefficients. This non-uniqueness of interpolation in an image data which is sampled uniformly in the same two directions, horizontal and vertical, is avoided by always interpolating along these two directions. You can verify

Figure 2.4. This shows a mesh made of quadrilaterals.

that interpolating in the vertical direction first and then in the horizontal direction will yield the same result as Equation 2.14.

However, unlike this case of a uniform planar grid, if this 2D surface happens to be a mesh as in Figure 2.4, the result of the interpolation will completely depend on the particular intermediate points you use. Consider the quadrilateral highlighted in red in Figure 2.4. Let two opposite vertices have color black denoted by B and the other two vertices have color white denoted by W. A point at the intersection of the two diagonals shown by dotted red lines can be interpolated to have two completely different colors. If interpolated from the black vertices, it will be black. If interpolated from the white vertices it will be white.

Therefore, there is something special about a triangle, where a point inside it will be interpolated uniquely irrespective of how you do it. This is because a triangle is a simplex. A simplex is a geometric construct achieved by connecting n points to each other in $(n-1)$ dimensions. A straight line is the 1D simplex. A triangle is the 2D simplex (encloses a surface). Similarly, a tetrahedron is the 3D simplex.

Figure 2.5. From left to right: The smallest 1D (line), 2D (triangle) and 3D (tetrahedron) simplex.

These are illustrated in Figure 2.5. The linear interpolation on these simplices (called bilinear for 2D, trilinear for 3D) yields unique interpolation coefficients which are also called *barycentric coordinates* of a point inside the simplex with respect to the vertices forming the simplex. Therefore in Equation 2.8 provides the barycentric coordinates of the point V with respect to V_1, V_2 and V_3 respectively in Figure 2.2a. This is the reason triangles and not any other polygons are chosen for representing geometric meshes. The advantages will be even more evident when we cover computer graphics later in the book.

2.2 Geometric Intersections

Linear equations represent lines (when using 2 variables) or planes (when using three variables) and we would often need to compute the intersection of such geometric entities. Such intersections are computed by solving a set of linear equations. In order to solve equations with n unknowns, we need at least n equations. First, let us derive a matrix formulation for this problem. Let us consider n linear equations with n unknowns, $x_1, x_2, \ldots x_n$, where the ith equation is given by

$$a_{i1}x_1 + a_{i2}x_2 + \ldots + a_{in}x_n = b_i \tag{2.17}$$

Now, this can be written as

$$Ax = b \tag{2.18}$$

where A is a $n \times n$ matrix given by

$$A = \begin{pmatrix} a_{11} & a_{12} & \ldots & a_{1n} \\ a_{21} & a_{22} & \ldots & a_{2n} \\ & & \ldots & \\ & & \ldots & \\ & & \ldots & \\ a_{n1} & a_{n2} & \ldots & a_{nn} \end{pmatrix} \tag{2.19}$$

and x and b are $n \times 1$ column vectors given by $x^T = (x_1 \ x_2 \ \ldots \ x_n)$ and $b^T = (b_1 \ b_2 \ \ldots \ b_n)$ respectively. The solution to this system of linear equations is

$$X = A^{-1}B \tag{2.20}$$

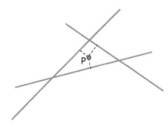

Figure 2.6. This shows three non-parallel lines in 2D which may not meet at a single point. The red lines show the perpendicular distance of each of these lines from the point P.

Note that A^{-1} exists only when all the equations are linearly independent of each other. Geometrically, you can imagine this problem in an n dimensional space where each of the n equations defines a hyperplane and x will give you the intersection point in n dimensions of all these n hyperplanes. Full rank A indicates non-parallel hyperplanes which will have a unique intersection point. For example, when $n = 2$, two non-parallel lines will always intersect at a point.

Now consider the case when you have the same number of unknowns but have a much larger number of equations, m, where $m > n$. Such a system of equations is called an over-constrained system of equations. In this case, note that there may not be one common intersection point. For example, consider $m = 3$ and $n = 2$. Therefore, we are considering three lines which may not intersect at a point as shown in Figure 2.6. Therefore, one (and probably the most widely used) way to geometrically solve this set of over-constrained equations is equivalent to finding a point P such that the sum of the squares of the distance from P to the lines defined by the set of linear equations is minimized. The squaring is done to make sure that negative and positive distances do not cancel each other out. This process is called *linear regression*. Since the square of the distances is used, this is also often referred to as linear least square optimization.

An over-constrained system of linear equations can be expressed by the same Equation 2.18. However, the dimension of A is now $m \times n$ and that of b is $m \times 1$. Now, since A is no longer a square matrix, its inverse is not-defined. Therefore, let us consider the following.

$$Ax = b \tag{2.21}$$

$$Or, A^T A x = A^T b \tag{2.22}$$

$$Or, x = (A^T A)^{-1} A^T b \tag{2.23}$$

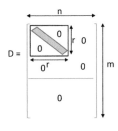

Figure 2.7. This shows the structure of D matrix in singular value decomposition.

Note that A^T is a $n \times m$ matrix. Therefore, $A^T A$ is a $n \times n$ square matrix whose inverse is defined and is called the pseudo-inverse of A. Therefore, x can be solved now using this pseudo-inverse. However, it only works well when $A^T A$ is a full-ranked matrix and not singular (i.e. determinant is not zero). Often when $m >> n$, it is very hard to assure that the pseudo inverse is full-ranked.

Singular value decomposition (SVD) is a technique that helps us to solve x in such situations. It decomposes A into three matrices U, D and V such that

$$A = UDV^T \tag{2.24}$$

where U is a $m \times m$ square matrix, D is a $m \times n$ diagonal matrix and V is a $n \times n$ square matrix.

Put a Face to the Name

Linear regression, one of the most used optimization techniques, was first conceptualized in 1894 by Sir Francis Galton, who was a cousin of Charles Darwin. This started with the then vexing problem of heredity – understanding how strongly the characteristics of one generation of living things manifested in the following generation. Galton initially approached this prob-

Figure 2.8. Left: Sir Francis Galton; Right: Karl Pearson

lem by examining characteristics of the sweet pea plant. Galton's first insights about regression sprang from a two-dimensional diagram plotting the sizes of daughter peas against the sizes of mother peas. Galton realized that the median weights of daughter seeds from a particular size of mother seed approximately described a straight line with positive slope less than 1.0. Later on, Galton's colleague and researcher from his own lab, Karl Pearson, formalized this concept mathematically in 1922 after Galton's death in 1911.

If the rank of A is $r < n$, then only the first r of the diagonal entries of D are non-zero as shown in Figure 2.7. Also, U and V are orthonormal matrices, i.e. they represent unit vectors that are orthogonal to each other. One property of orthonormal matrices that is important here is that their inverse is their transpose. Therefore, $U^{-1} = U^T$ and $V^{-1} = V^T$. Now, let us consider a matrix $D^\star$ given by inverting the $r \times r$ submatrix of D while the rest of the entries remain zero as in D. Note that $D^\star$ is again a diagonal matrix whose top left $r \times r$ submatrix has diagonal elements that are reciprocal of those of D. It can be shown that $AVD^\star U^T b = b$. Therefore, $x = VD^\star U^T b$ is the solution of $Ax = b$.

2.3 Conclusion

We have given enough details in this chapter to take you through the book. However, such mathematical fundamentals and their geometric interpretations are an interesting area of study by itself. To learn more about this, refer to [Lengyel 02]. Matrices inherently represent geometry and analysis of matrices and in fact analysis of the underlying geometry they represent. To know more in this direction, refer to [Saff and Snider 15, Nielsen and Bhatia 13].

Bibliography

[Lengyel 02] Eric Lengyel. *Mathematics for 3D Game Programming and Computer Graphics*. Charles River Media Inc., 2002.

[Nielsen and Bhatia 13] Frank Nielsen and Rajendra Bhatia. *Matrix Information Geometry*. Springer Verlag Berlin Heidelberg, 2013.

[Saff and Snider 15] Edward Barry Saff and Arthur David Snider. *Fundamentals of Matrix Analysis with Applications*. John Wiley and Sons, 2015.

Summary: Do you know these concepts?

- Interpolation

- Bilinear Interpolation

- Linear Regression

- Singular Value Decomposition

Exercises

1. Consider the triangle $P_1P_2P_3$ where $P_1 = (100, 100)$, $P_2 = (300, 150)$, $P_3 = (200, 200)$. Consider a function whose values at P_1, P_2 and P_3 are $\frac{1}{2}$, $\frac{3}{4}$ and $\frac{1}{4}$ respectively. Find the interpolation coefficients at $P = (220, 160)$. Compute them considering two different directions to verify getting the same interpolation coefficients. What is the interpolated value of the function at P?

2. Consider two planes given by $4x+y+2z = 10$ and $3x+2y+3z = 8$. Consider a line given by $2x+y = 2$. Solve the equations to find the intersection points of these planes and the line. Next verify your result using matrix based on solution of the equation $Ax = b$.

3. Consider the set of linear equations given by $x - y = 0$, $2x + 5y = 10$, $4x - 3y = 12$ and $x = 5$. Solve these using SVD.

Part II

Image Based Visual Computing

3

Convolution

3.1 Linear Systems

A *system* is defined as a method that modifies a signal. An audio amplifier that modifies the 1D audio signal to make it louder, or an image processing method that modifies the 2D image signal to detect some features are a few examples of a system. Systems can be very complex. But we will be mostly dealing with a specific class of systems that are simpler, namely *linear systems.*

Linear systems satisfy some conducive properties of linearity. We will denote the ystem by S, input to the systems by x and output by y and z. For the sake of simplicity of explanation, we assume 1D signals which depend on the single parameter t. However, the following properties hold for a linear signal in any dimension.

1. *Homogeneity:* If the input to a linear system is scaled, the output would also be scaled by the same factor.

   If $\xrightarrow{\;x(t)\;}\boxed{S}\xrightarrow{\;y(t)\;}$ then $\xrightarrow{\;Kx(t)\;}\boxed{S}\xrightarrow{\;Ky(t)\;}$

2. *Additivity:* The independent responses (output) of multiple different input signals are added when the inputs are added. This implies that each signal is passed through the system independently without interacting with others.

   If $\xrightarrow{\;x_1(t)\;}\boxed{S}\xrightarrow{\;y_1(t)\;}$ and $\xrightarrow{\;x_2(t)\;}\boxed{S}\xrightarrow{\;y_2(t)\;}$ then $\xrightarrow{\;x_1(t)+x_2(t)\;}\boxed{S}\xrightarrow{\;y_1(t)+y_2(t)\;}$

3. *Shift Invariance:* Finally, the output of a shifted input is also shifted.

If $\xrightarrow{x(t)}$ S $\xrightarrow{y(t)}$ then $\xrightarrow{x(t+s)}$ S $\xrightarrow{y(t+s)}$

When that data passes through multiple systems, the properties of linearity assures the following.

1. *Commutative:* If two linear systems are applied to a signal in a cascaded manner (i.e. in series), the order of their application does not matter. Given two linear systems S_A and S_B,

If $\xrightarrow{x(t)}$ S_A $\longrightarrow$ S_B $\xrightarrow{y(t)}$ then $\xrightarrow{x(t)}$ S_B $\longrightarrow$ S_A $\xrightarrow{y(t)}$

2. *Superposition:* If each input generates multiple outputs in a linear system, the addition of the inputs will generate an additions of the outputs.

If $\xrightarrow{x_1(t)}$ S $\Big\langle \begin{matrix} y_1(t) \\ y_2(t) \end{matrix}$ and $\xrightarrow{x_1(t)}$ S $\Big\langle \begin{matrix} z_1(t) \\ z_2(t) \end{matrix}$ then $\xrightarrow{x_1(t)+x_2(t)}$ S $\Big\langle \begin{matrix} y_1(t)+y_2(t) \\ z_1(t)+z_2(t) \end{matrix}$

This superposition property is especially important for finding the response of a complex signal when passing through a linear system. A complex input signal $x(t)$ can be broken into a bunch of simpler input signals $x_1(t), x_2(t), \ldots, x_n(t)$, via different processes of *decomposition*. It is usually easier to find the outputs $y_i(t)$ of the simpler input signals $x_i(t)$ when passing through the system. The y_is are then combined or added via the process of *synthesis* to create the output $y(t)$ for the complex signal. This is illustrated in Figure 3.1. We will study many different ways to decompose and synthesize in the following sections and chapters.

3.1.1   Response of a Linear System

An *impulse*, $i[t]$, is a discrete signal with only one non-zero sample. Therefore, it is a signal with a sharp spike at one location and zero elsewhere. *Delta*, $\delta[t]$,

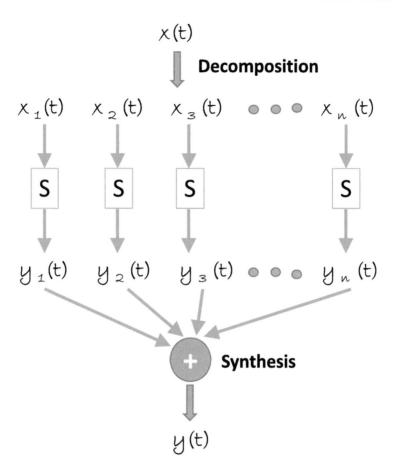

Figure 3.1. This figure illustrates the process of decomposing a complex signal to simpler signals which are then passed through the linear system and combined to generate the output of the complex signal.

is a special kind of impulse whose non-zero sample is at $t = 0$ and has a value $\delta[0] = 1$. Therefore, $\delta[t]$ has a normalized spike at 0. Considering each sample to be of unit width and height proportional to its value, the area covered by a delta is therefore 1. $\delta[t]$ is considered the simplest signal.

Consider an impulse with value $i[2] = 3$ and zero elsewhere. This impulse can be represented as a scaled and shifted δ as $3\delta[t-2]$. Therefore, any impulse with a non-zero value of k at $t = s$ can be represented in general as a scaled and shifted δ as

$$i[t] = k\delta[t - s] \tag{3.1}$$

The *impulse response* (also called kernel or filter), $h[t]$, of a linear system is

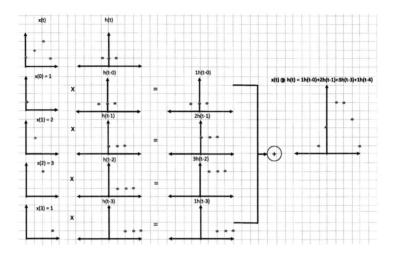

Figure 3.2. This figure illustrates the input side algorithm for convolution.

defined as the output of the system to the input $\delta[t]$. The size of h (width in case of 1D h) is also called its *support*. Due to the properties of shift invariance and homogeneity, the response of the same linear system to a general impulse function is given by a scaled and shifted h as $kh[t-s]$.

Convolution is the method to find the response of a linear system with impulse response, h, to a general signal or function. It is rather evident that convolution is hence a pretty powerful function.

Let us consider a discrete signal $x[t]$ where $t = 1, 2, \ldots, n$. Note that x can be decomposed as the sum of n impulse functions $i_1[t], i_2[t], \ldots, i_n[t]$ where $i_l[t]$ has a non-zero value of $x[l]$ at $t = l$ and $x[t] = \sum_{l=1}^{n} i_l[t]$. The response of the system to each $i_l[t]$ is given by $x[l]h[t-l]$. Due to the additivity property of the linear system, the response $R[t]$ of the linear system to $x[t]$ is given by

$$R = \sum_{l=1}^{n} x[l]h[t-l] = x[t] \star h[t]. \tag{3.2}$$

$x[t] \star h[t]$ is the convolution of $x[t]$ with the impulse response $h[t]$. We will use the symbol $\star$ for convolution. This is illustrated in Figure 3.2.

Now that we have defined convolution, let us ponder for a while on Figure 3.2. First, note that when the first few or last few samples of x are being multiplied by h, h extends beyond x in left and right where x is not defined (for e.g. we do not know the value of x at $t = 0$ or $t = -1$ or $t = n+1$). In such cases, we assume some arbitrary values for x. The most common assumption is to consider x to be 0 at these indices. Sometimes, x is reflected about the left and right ends. Either way, this brings in two important issues. First, the size of R is larger than that of x. If the support of h is m, then the size of R is $n + m - 1$. This is illustrated in Figure 3.2 where $n = 4$ and $m = 3$ and therefore the size of the output is $5 + 3 - 1 = 6$. Second, some of the values of R are not accurate since they involve calculations from assumed information. For example in Figure 3.2, the output of convolution at $t = -1, 0, 3, 4$ depends on the values of x at $t = -1, 4, 5$ and x is not defined at these locations. Therefore, only a subset of samples of R, in fact only $n - m + 1$ of them, are obtained from precisely defined information and these samples are called *fully immersed samples*. In Figure 3.2, the only fully immersed samples are samples at $t = 1$ and 2.

Another point to note here is that in using Equation 3.2, each sample of the input x corresponds to a scaled and shifted impulse response to create an intermediate function $x[l]h[t - l]$. All these intermediate functions are added up to create R (Figure 3.2). Hence, any single output sample of R is generated by accumulating the samples at the same location from all of these intermediate functions. Since a single sample at l from input x contributes to multiple output samples via the corresponding intermediate function $x[l]h[t - l]$, this is called the *input side* algorithm

Now take a careful look at one of the fully immersed samples at $t = 1$ in Figure 3.2. Note that it is given by

$$R[1] = h[1]x[0] + h[0]x[1] + h[-1]x[2]$$

This is the same as flipping h and weighting the neighborhood of x at $t = 1$ with this flipped h. Therefore, another method to find the convolution is to generate each output sample at t by shifting a flipped h to align its center with l and find the weighted sum of the underlying x and the flipped shifted h. Therefore, the sample at l, $R[l]$, is generated by the dot product of $x[t]$ with flipped h shifted at

$$R[l] = x \cdot h[-(t - l)]. \tag{3.3}$$

Note that for all indices at which the value of h is not defined, it is assumed to be 0. In this method, each output sample of R gets constructed in each step by gathering contributions from multiple samples of h. Hence, this is called the *output side algorithm*. This is a more efficient algorithm since no intermediate functions needs to be maintained. Each output can be generated directly from the input and h. Further, if h is symmetric (as in Figure 3.2), then the flipping can also be avoided.

Extending convolution to 2D is really trivial. x, R, and h are now two dimensional functions. In the case of images, these two dimensions are due to two spatial coordinates s and t. However, the support of h is usually much smaller than x. $R[s, t]$ is now obtained by moving h that is flipped in both dimensions to the desired location (s, t) and finding the dot product of this flipped h with x. However, note that in most cases in image processing, we use symmetric filters or h which deems the flipping unnecessary.

In the rest of the chapter we will study the properties of convolution using 1D signal since they are simple to understand. The concepts can be easily extended to 2D and we will mention them at the end of each treatise on 1D signals.

3.1.2   Properties of Convolution

In this section we are going to discuss the properties of convolution. Consider a signal x convolved with δ. This means that if the impulse response of a system is the impulse itself, what will be its response to an arbitrary function $x[t]$. It is rather intuitive that this system outputs the impulse unchanged and hence it will also output the signal x unchanged. Therefore,

$$x[t] \star \delta[t] = x[t]. \tag{3.4}$$

This is called an *all pass system.*

Let us now consider a system which simply scales the impulse response. Therefore,

$$x[t] \star k\delta[t] = kx[t]. \tag{3.5}$$

If $k > 1$, then this system is called an *amplifier* since it increases the strength of the signal $x[t]$. On the other hand, if $k < 1$, it is called an *attenuator* since it reduces the strength of $x[t]$.

Finally, if we consider a system whose impulse response is to shift the signal, then

$$x[t] \star \delta[t + s] = x[t + s]. \tag{3.6}$$

This is called a *delay* system.

As a mathematical operation, convolution has the following conducive properties.

First, it is commutative, i.e.

$$a[t] \star b[t] = b[t] \star a[t]. \tag{3.7}$$

This indicates that when convolving two functions, the order does not matter. This is why we always use the smaller sized function as the kernel for more efficient processing.

Second, convolution is associative, i.e.

$$(a[t] \star b[t]) \star c[t] = a[t] \star (b[t] \star c[t]). \tag{3.8}$$

This means that if an function x undergoes cascading convolutions using two different kernels b and c, the same operation can be achieved by first designing a new kernel by $d = b \star c$ and convolving x with this new d to provide $x \star d$ (Figure 3.3.

Third, convolution is distributive, i.e.

$$a[t] \star b[t] + a[t] \star c[t] = a[t] \star (b[t] + c[t]).$$

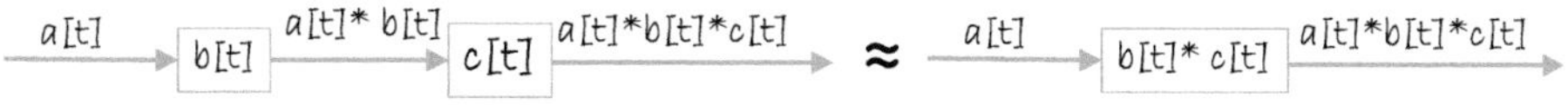

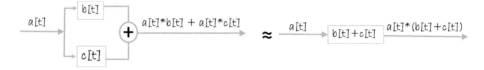

Figure 3.3. This figure illustrates the effect of cascading convolutions.

Figure 3.4. This figure illustrates the effect of combination of parallel convolutions.

This means that if a function x undergoes two different convolutions in parallel with b and c which are then combined, the same effect can be obtained by designing a new kernel d by adding b and c ($d = b + c$) and then undergoing a single convolution with this new kernel d (Figure 3.4).

3.2 Linear Filters

The next question to ask is how does knowing about convolution help us in any way? In fact, convolution can help us greatly in designing systems since instead of worrying about complex signals, we need to only worry about the simple δ function. If we can design the impulse response of a system, we know that convolving the input signal with impulse response would provide us with the correct answer for any general function.

Let us take the case of designing a filter (or impulse response) that will blur any general signal. To design this, we have to first think intuitively about what a blurred delta signal would look like. In other words, what would a linear system that blurs a signal produce when a delta is provided as its input. For this consider Figure 3.5. Delta is a function that has a single sample of value 1 at 0 which is essentially a sharp spike. Therefore, intuitively, blurring a delta function can be expected to produce a spike with a broader base and a smaller height. This is represented by a function that has multiple samples centered around 0 whose values are smaller than 1. Now the next question is, how much broader should the base be and how much shorter should the spike be? In fact, there can be many answers to this question. For example, it can be three samples centered at 0, each of value 0.7. Or, it can be five samples centered around 0, each of value 0.5. So, how do we constrain this problem to find an appropriate answer to these questions?

One way to constrain the problem may be to first fix the base of the spike

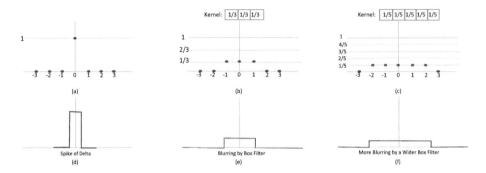

Figure 3.5. This figure shows how to design a blur filter by considering just the simple δ function. The top row shows the concept at work in a digital representation while the bottom row shows the analog counterpart. (a) shows a δ. (b)shows the impulse response of a blurring system, i.e. the fate of δ when passed through a blurring system. Instead of having a spike of 1 at 0, it now has a value of $1/3$ at each of -1, 0 and 1. The width of the filterv is also often referred to as the support of the kernel or filter. If the width of the blur is wider, it is said that the support of the filter is now higher.

as a parameter for the amount of blurring. Therefore, a width of seven pixels indicates more blurring than a width of five. And a width of five pixels indicates more blurring than a width of three pixels. Now the question that remains is once the base width is fixed, how should we decide the height of the blurred spike? To decide this, we can apply the constraint that the energy (defined by the area under the curve depicting a function) of the delta function will not be changed by blurring. This constraint is that the delta function spike has a height of 1 and a width of 1. Therefore, its energy is given by the multiplication of its width and height, i.e. 1. Now, if a three pixel wide blurred spike needs to have the area 1, then its height should be $\frac{1}{3}$. Therefore, this additional energy constraint has now allowed us to define a blurred delta to be a spike of three pixel width and value of $\frac{1}{3}$ centered around 0. Therefore, if delta is given as input to the blurring system we want to design, we would expect its output to be the aforementioned shorter and wider spike. Therefore, the impulse response of this blurring system is given by a function that is a three pixel wide spike, centered at 0 and has a constant value of $\frac{1}{3}$ at those three pixels. The advantage of defining this impulse function is that if we now convolve any other general function with this impulse response, it will now result in the blurring of this function.

Now, if we want to design a system that blurs the signal even more, we have to again design an impulse response for this system. Since we know more blurring implies widening the base and shortening the height of the spike further, one possible impulse response will be a five pixel wide spike centered around zero with height $\frac{1}{5}$. Hence, the impulse response of this system would be five pixels in

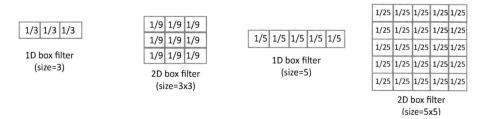

Figure 3.6. This figure shows how the extension of 1D box filters to 2D.

size with a value of $\frac{1}{5}$ at every pixel as shown in Figure 3.5(c). Notice the analog representations of these filters in (d) and (e). Since they look like boxes, these are most commonly referred to as *box filters*.

However, there are multiple ways to maintain the same energy as the δ of which only one is to assign same value at every pixel. Therefore, conceptually, the shape of the filter can change and still remain a blurring filter as long as the support of the kernel increases. It is indeed true that many such blurring filters exist and we will revisit this issue in the next chapter where we will find that though box filter is the easiest to implement, it is not the best for blurring.

Now the next question is how do we extend this blurring filter to 2D. In this case, the energy of the δ should be spread around a 2D box around the surrounding of origin. So, the extension of the three pixel 1D box filter to 2D will be a 3×3 filter with each value as $\frac{1}{9}$ and that of the wider box filter will be a 5×5 array with each value as $\frac{1}{25}$ (Figure 3.6).

$$\text{If } a[t] \xrightarrow{\;F\;} A[f] \quad \text{and} \quad b[t] \xrightarrow{\;F\;} B[f]$$

$$\text{then } a[t]*b[t] \xrightarrow{\;F\;} A[f]B[f] \quad \text{and} \quad a[t]b[t] \xrightarrow{\;F\;} A[f]*B[f]$$

Figure 3.7. The duality in convolution is given by the fact that multiplication of two functions in the spatial domain is a multiplication of their frequency responses in the frequency domain and vice versa. Here the F denotes an operation that converts the time domain function to the frequency domain.

3.2.1   All, Low, Band and High Pass Filters

Interestingly you will find that a box filter is often referred to as a *low pass filter*. In order to understand why it may be worthwhile to go back and review frequency domain representations discussed in Chapter 1. In this chapter we will understand this concept very informally and intuitively, and will revisit it more formally in the next chapter.

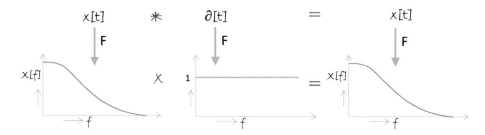

Figure 3.8. This shows the concept of an all pass filter. The frequency response of a δ function is a constant. So, the frequency response of a general function when multiplied by a constant does not cut away any frequencies. Since, it passes all the frequencies it is called an all pass filter.

Let us start again from the delta function in 1D. This can be thought of as a very sharp spike at origin. Now, intuitively, what would you expect the frequency domain representation of the δ function to be? Remember, that to form sharp features very high frequency signals are needed. This fact may convince you that a function like δ, which is the sharpest possible digital function, would involve all high frequencies. In fact, it can be shown formally that equal strength of all frequencies is needed to create δ. Therefore the frequency domain response (we only consider amplitude here) of the δ function is a constant. Now, lets us ask what would the frequency domain response of a constant function be i.e. a function which remains at a constant value in the time domain. It is probably easy to see that such a function is represented by a zero frequency cosine wave. Therefore, the frequency domain response of this constant function is a single value at the origin in the frequency domain, i.e. a δ function in frequency domain.

You probably notice an interesting pattern here. A δ is the time domain is a constant in the frequency domain while a constant in the time domain is a δ in the frequency domain. Is this really a coincidence? As it turns out, it is not! This is termed as the *duality* and we explore it more formally in the next chapter. But we will use this concept to understand a few things in this chapter.

The concept of duality gives rise to an important property of convolution which is as follows. If the frequency domain response of two functions in spatial domain $a[t]$ and $b[t]$ are $A[f]$ and $B[f]$ respectively, the frequency domain response of their convolution in time domain is given by the multiplication of their frequency domain responses and vice versa (Figure 3.7).

This provides us the background to understand the all, low, high and band pass filters. Let us first revisit Equation 3.4. We mentioned that a convolution of any function $x[t]$ with delta is termed as an *all pass system*. Let us see if we can explain this using what we just learned about duality. First, the frequency response of a δ is a constant. Therefore, the frequency response of $x[t] \star \delta[t]$

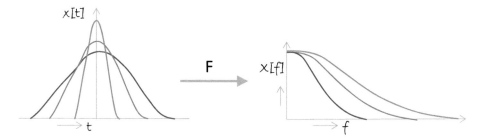

Figure 3.9. This shows the concept that as a function becomes narrower or wider, its frequency response gets wider and narrower respectively.

will be $X[f]$ multiplied by a constant. This says that by this convolution, no frequency will be blocked (or eliminated). Since this will *pass* all the frequencies, convolution with a delta is an *all pass filter*. This is illustrated in Figure 3.8.

Next, let us take a look at *low pass filter*. For this, we need to turn our attention to another intuitive consequence of duality. Consider a function which is compressed in space to create another function that undergoes similar changes but in a much smaller space. Therefore, intuitively, the latter function is similar to the former but has much sharper changes. How are the frequency responses of these two signals related? It is evident that we will need more higher frequency signals to create a "sharper" function and fewer higher frequency signals to create a "flatter" function. Therefore, we can probably infer that the frequency response of the sharper function will be wider than the other. The inverse is also true. If the function becomes smoother, the frequency response gets narrower (Figure 3.9). Now, assuming that delta is the sharpest of all functions, you can see how as the function gets sharper and sharper, its frequency gets wider and wider and finally comes to span all possible frequencies for δ.

Note that as the size of the filter (also called the support of the kernel) increases the cut-off frequency beyond which the frequencies are blocked reduces, i.e. the filter achieves more blurriness due to greater loss of higher frequencies. Thus, convolving with kernels with progressively increasing size creates a blurrier and blurrier function. The final stage would be a filter of the same size as the function which would provide the average value of the entire function.

Let us apply this concept to the three filters seen in the bottom row of Figure 3.5 where the filter gets wider to create a three-pixel-wide and subsequently, a five-pixel-wide box filter. So, intuitively, their frequency responses will be narrowing as shown in Figure 3.10(a). Therefore, the frequency response of these blur filters will have cut-off frequencies that will reduce as the width of the filter increases. Therefore, the cut-off frequency for the three pixel filter is f_1 and that of the five pixel is f_2 such that $f_2 < f_1$. Now let us consider a general function $x[t]$ as in Figure 3.10(b) and its frequency response. When we convolve $x[t]$ with

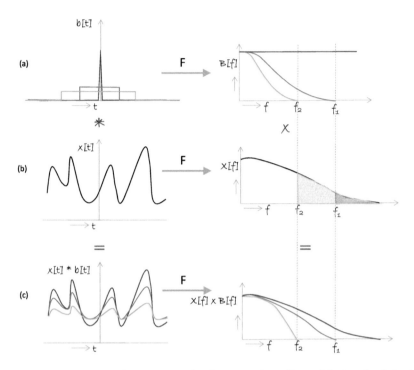

Figure 3.10. This shows the concept of a low pass box filter. (a) On the left the δ (red), three pixel (blue) and five pixel (green) box filters are shown in time domain. On the right, their frequency components are shown with cut-off frequencies at f_1 and f_2 respectively. (b) On the left, a general function $x[t]$ is shown in black. On the right, its frequency response is shown in black. A convolution between x and b on the left indicates a multiplication between $X[f]$ and $B[f]$ on the right. When this multiplication happens, the frequencies that are thrown away (blocked) by each filter are shown by their corresponding colors - all the frequencies above f_1 for the three pixel filter and those above f_2 for the five pixel filter are blocked. (c) This shows the result of the convolution and multiplication in spatial and frequency domain respectively. Note that all higher frequencies are removed to different degrees (based on the width of the filters) when convolved with the box filters while the lower frequencies are passed, hence the name *low pass filter*. The strength of the low frequency components is also changed, and in general reduced, after this filtering process. The signal in the time domain gets progressively smoother or blurrier.

the blur filter $b[t]$, their frequency responses $X[f]$ and $B[f]$ are multiplied. The consequence of this is that all frequencies beyond f_1 are multiplied by zero when convolving with the three pixel filter i.e. all the frequencies above f_1 are thrown away creating the blurry or smoother signal in Figure 3.10(c). Similarly, when multiplied by the frequency response of the five pixel filter, even more frequen-

Figure 3.11. This image shows the effect of increasing the size of the box filter and convolving it with an image. From left: Original image and the same image convolved with a 3×3, 5×5 and 15×15 box filter.

cies – essentially everything above f_2 where $f_2 < f_1$ – are thrown away thereby creating an even blurrier signal in Figure 3.10(c). Since higher frequencies are thrown away as a result of convolution with box filters while the lower frequencies are passed, these are called *low pass filters*. However, note that box filters are only one kind of low-pass filter. The shape of the filter need not be an exact box to spread the energy of the delta from a single pixel to multiple ones. For example, a triangular shaped filter can also be used for this purpose. The difference between these filters will be in the way they spread the energy which is defined by their shape. In a box filter, the energy is spread equally to all the pixels while in a triangular filter the amount of the energy reduces from the center towards the periphery. Another important thing to note in this context is that all the frequencies that are passed by the low-pass filter are not passed unchanged. In fact, different frequencies are attenuated differently, with the higher of the passed frequencies getting more severely attenuated. The figures in this chapter focuses on the frequency content (the range of frequencies passed) rather than the exact shape of the frequency response of the filters so that your attention is not distracted from the most important aspect of frequency domain analysis, the frequency content. The exact shape of the frequency response of the filter will only affect the amount of attenuation of the passed frequencies and will be discussed in details in the next chapter.

Now consider this same situation in 2D. Consider an image which is being progressively convolved with filters of larger and larger size (3, 5, 7 and so forth). The images that would be created would be progressively blurrier (Figure 3.11) with the final one being a flat gray colored image where the gray color is given by the average of all the pixels in the original image. In the frequency domain, we know that the cut-off frequency for each of these images will be progressively reducing. In 2D, different frequencies are represented as concentric circles with the length of the radius representing the frequency. Figure 3.12 illustrates the cut-off frequency of low pass filters beyond which all frequencies are blocked. With larger kernel size, this cut-off frequency gets smaller.

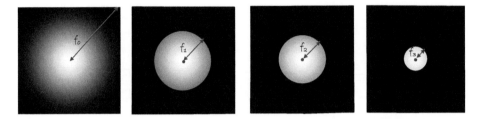

Figure 3.12. This image shows the effect of increasing the size of the box filter when convolving with an image in the frequency domain. The frequency response is visualized as a gray scale image. From left to right: Frequency response of the original image and the same image convolved with filters of increasingly bigger size. Note that the cut-off frequency is denoted by a circle beyond which every frequency has zero contribution and is hence black. The radius of this circle reduces as the size of the filter increases to show that more high frequencies are getting chopped off.

Issue of Sampling Let us now consider the sampling consequences of low pass filtering an image. Nyquist sampling criteria says that the minimum samples required to sample an image are double the highest frequency contained in the image. As an image undergoes low pass filtering, its frequency content decreases ($f_3 < f_2 < f_1$ in Figure 3.12). This means that the minimum number of samples required to adequately sample the low pass filtered image goes down too. This says that the low pass filtered image can be at a smaller size than the original image. Or, as we progressively increase the size of the low pass filter, we do not need to have the image at its original size, but we can resample and store them at a much smaller size, just adequately large to sample the highest frequencies in them.

This property is used to build a pyramid of progressively low pass filtered images called the *Gaussian pyramid*. For this, we resample the original image to the size of $2^n \times 2^n$. This forms level 0 of the pyramid. 2×2 pixel blocks of this image are low pass filtered to create a single pixel of the image at the next level of the hierarchy providing a $2^{n-1} \times 2^{n-1}$ image. When using a box filter, this amounts to just averaging every 2×2 blocks of pixels in level n to create each pixel of level $(n + 1) \times (n + 1)$. Note that since the image at level $i + 1$ is a low pass filtered from the image at level i, the lower resolution is adequate to sample this image with lower frequency content. This process if progressively continued creates n levels of the pyramid with the last level being a single pixel which can be considered to be an image of $2^{n-n} \times 2^{n-n} = 1 \times 1$. This is illustrated in Figure 3.13.

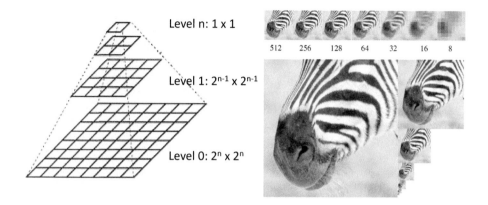

Figure 3.13. This image illustrates the concept of Gaussian Pyramid. On the left, it shows a 4-image pyramid with n=3. On the right, we show the example of a Gaussian pyramid starting with a 512×512 image where $n = 9$. Note how the images halve in size and it is very difficult to see the content of the smaller images. Therefore, on the top each image is shown resampled at the same size to show the reduction in the frequency content.

Put a Face to the Name

Johann Carl Friedrich Gauss (30 April 1777 to 23 February 1855) was a German mathematician who contributed significantly to the fields of number theory, algebra, statistics, analysis, differential geometry, geodesy, geophysics, mechanics, electrostatics, astronomy, matrix theory, and optics. He came from poor working-class parents. His mother was illiterate and never recorded the date of his birth, remembering only that he had been born on a Wednesday, eight days before the Feast of the Ascension, which itself occurs 40 days after Easter. Gauss would later solve this puzzle about his birthdate, deriving methods to compute the date in both past and future years. At the age of three, Gauss corrected an arithmetical error in a complicated payroll calculation for his father.

Gauss made his first ground-breaking mathematical discoveries while still a teenager. At age 19, he demonstrated a method which had eluded the Greeks for constructing a heptadecagon using only a straightedge and compass. Gauss's intellectual abilities attracted the attention of the Duke of

Brunswick, who sent him to the Collegium Carolinum (now Braunschweig University of Technology) from 1792 to 1795, and to the University of Gottingen from 1795 to 1798. He completed Disquisitiones Arithmeticae, his magnum opus, in 1798 at the age of 21, though it was not published until 1801. This work was fundamental in consolidating number theory as a discipline and has shaped the field to the present day. Unfortunately for mathematics, Gauss reworked and improved papers incessantly, therefore publishing only a fraction of his work, in keeping with his motto "pauca sed matura" (few but ripe). He kept a terse diary, just 19 pages long, which later confirmed his precedence on many results he had not published. Gauss wanted a heptadecagon placed on his gravestone, but the carver refused, saying it would be indistinguishable from a circle. The heptadecagon appears, however, as the shape of a pedestal with a statue erected in his honor in his home town of Braunschweig.

Let us assume that we are working with a box filter. Therefore, every 2×2 block of pixels in level 0 is averaged to create a single pixel in level 1. 2×2 pixels in level 1 are in turn averaged to create a single pixel in level 2. When 2×2 pixels in level 1 are averaged i.e. a weighted sum with equal weights of $\frac{1}{4}$, it is equivalent to averaging 4×4 pixels in level 0 i.e. a weighted sum with equal weights of $\frac{1}{16}$. In other words, applying a 2×2 box filter to level 1 is equivalent to applying a $4 \times 4 = 2^2 \times 2^2$ box filter to level 0 to create the image at level 2. This concept can be generalized to show that level i is equivalent to applying a $2^i \times 2^i$ filter to level 0. Therefore, as we are going up in the levels, each image is a low pass filtered version of the original image, but using filters of progressively larger sizes and therefore of progressively lesser frequency content as shown in Figure 3.10. But creating it from level $i-1$ using a 2×2 filter is computationally more efficient. Generalizing this concept, you can see that level n is created by applying a box filter of size $2^n \times 2^n$ to level 0 which is essentially averaging all the values of the image. Therefore, the level n of the pyramid is a single gray value. Note that this concept generalizes to any low pass filter, not necessarily a box filter. In case of other filters, the weights used for filtering are not equal for every pixel, but the notion of filter size increasing as the levels increase still remains the same.

From the aforementioned explanation, you may think that images in a Gaussian pyramid should progressively reduce in size. This is not true. The reducing size only defines the minimum sampling requirement at each level and it can be proved mathematically. However, having a size larger than $2^{n-i} \times 2^{n-i}$ for level i only provides a higher sampling density that the minimum sampling requirement. Therefore, an alternate way to create the pyramid is to simply convolve the image in level i with 2×2 box filter to create level $i+1$ creating an image of the same size of level i for level $i + 1$. In this case, though the images at all levels of the Gaussian pyramid will be the same size, the content slowly loses details as we go higher up in the pyramid. However, since the size is unchanged, it is much easier to perceive this removal of details. In both cases, the pyramid is still called a Gaussian pyramid since the important concept here is the progressively reducing frequency content as you go higher in the pyramid. The sampling is inconsequential as long as the minimum sampling criteria is met. Figure 3.13 shows the representation with reducing image size while Figure 3.14 shows the representation where the image size is kept unchanged.

In fact, this can also be shown mathematically from the properties of convolution. Let us denote the image at i level of the Gaussian pyramid by G_i and the low pass filter of size 2×2 as l. Therefore,

$$G_1 \star l = (G_0 \star l) \star l = G_0 \star (l \star l) \tag{3.9}$$

Note that $l \star l$ is a kernel of greater size than l. Similarly, for ith level, l is convolved multiple times with itself to create a much wider kernel and hence a much lower frequency content of the filtered image. These kinds of operation are also called multi-scale operations. This is due to the fact that the scale of the objects appearing in each level of the pyramid differs. For example, at the lowest level of the Gaussian pyramid, all the minute edges are present. But as we go up the pyramid, only the bigger changes show up as edges, the details are lost.

At this point, one question remain: what is the use of Gaussian pyramids? Here is a very common application. Suppose we want to reduce the size of an image to half to display the image in a smaller mobile device. The first instinct in this situation is to *subsample* the image which is essentially throwing away every other pixel in the horizontal and vertical direction. However, this may lead to a sampling that fall below the Nyquist rate for the highest frequencies in the image leading to aliasing artifacts. Therefore, a better way to achieve this is to first low pass filter the image and then subsample it. This way the low pass filtering first reduces the Nyquist sampling criterion by removing the high frequency content following which the subsampling resolution provides adequate sampling. This is called pre-filtering and then subsampling. Reducing the image size to half is equivalent to the next highest level of the Gaussian pyramid. This is illustrated in Figure 3.14. A more drastic example of aliasing in such situations is given in Figure 3.15.

3.2.2 Designing New Filters

Now that we know the concepts of low pass filters, let us see how we can use this and the knowledge of the mathematical properties of convolution to design new filters. Being able to design new filters arms us with a entirely new set of tools that we can start using in several contexts.

Figure 3.14. This figure illustrates the difference between simple subsampling (top) and pre-filtering and subsampling (bottom). The resolution is halved as we go from left to right – but the images are generated in the alternative way to keep their size unchanged throughout the pyramid.

A low pass filter is a filter that allows the lower frequencies of an image to be retained. Now let us consider a filter which is complementary to this – a filter that will throw away the frequencies passed by the low pass filter and retain the higher frequencies that are thrown away by the low pass filters. Such a filter, as you can probably guess, is called an *high pass filter*. The question is how we can design a high pass filter? One way that probably has come to the mind of most of you is to subtract the low pass filtered image from the original image. And this is a perfect route to take.

Let us consider I to be the image and l be the low pass filter. Let the low

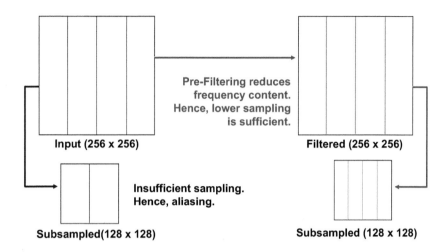

Figure 3.15. This figure illustrates severe aliasing artifacts that can occur due to subsampling without prefiltering.

pass filtered image be I_l and the high passed image be I_h. Therefore

$$I_h = I - I \star l \tag{3.10}$$
$$= I \star \delta - I \star l \tag{3.11}$$
$$= I \star (\delta - l) \tag{3.12}$$

In the second line of the above algebra, we consider the image to be an all pass filtered version of itself. Using this fact and the mathematical properties of convolution, we see that the high pass filtered image I_h can be expressed as the convolution of the original image I with a single filter given by $\delta - l$. This gives us the design of an *high pass filter* resulting from the subtraction of any low pass filter from δ, as shown in Figure 3.16. This figure also shows the general shape of any high pass filter with its charactaristic positive spike near the center and the negative lobes adjacent to it. For example, a Gaussian function is another good low pass filter which has a much smoother response than the box filter. In this case, the high pass filter will still look the same having smoother and deeper negative lobes. The image formed due to high pass filtering will have the complementary frequencies that will give the details of the image. This is illustrated in Figure 3.17. Now, do not confuse a Gaussian filter with a Gaussian pyramid. Both are named after the same person, but are entirely different concepts. A Gaussian filter is a kind of low pass filter, while a Gaussian pyramid is a pyramid of image formed by progressively applying any low pass filter on an image, not necessarily a Gaussian filter.

Now, let us consider the Gaussian pyramid $G_0, G_1, \ldots, G_n$. Let us now build

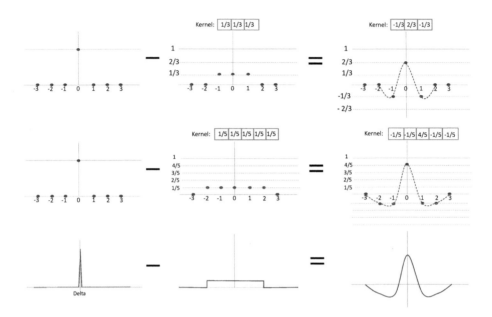

Figure 3.16. This figure illustrates the creation of high pass filters (bottom) in 1D from box filters of size 3 (top) and 5 (middle)

another pyramid $L_0, L_2, \ldots L_{n-1}$ where $L_i = G_i - G_{i+1}$. Note that to construct this pyramid, G_{i+1} has to be resampled at double the resolution since G_i and G_{i+1} are not at the same resolution to perform an image subtraction (which is essentially a pixel by pixel subtraction). Now note what happens from the frequency standpoint in this pyramid. If $f_0, f_1, \ldots, f_n$ are considered to be the cut-off frequency of $G_0, G_1, \ldots, G_n$, then each of L_i consists of only a band or range of frequencies $f_i - f_{i+1}$. Therefore, these images are created by passing a band of frequencies. This pyramid is called a *Laplacian Pyramid* as illustrated in Figure 3.18.

Now, let us consider the single filter that we will use on G_0 to create the ith level of the Laplacian pyramid L_i. This single filter is given by convolving l with itself for $i + 1$ times and subtracting from it l convolved for i times. This filter passes the band of frequencies between f_i and f_{i+1} and is the *band pass filter*.

3.2.3   2D Filter Separability

From the above discussions, we are now capable of visualizing or generating 2D filters like a 2D box filter or a 2D high pass filter (Figure 3.19).. Now, when considering 2D filters, there is another important property to be aware of. This is called *separability*. Let us consider a $p \times q$ 2D filter given by $h[i][j]$ where

Figure 3.17. This figure shows the original image (left), the corresponding low pass filtered image (middle) that provides the basic shape of the object, and the high pass filtered image (right) created by subtracting the low pass filter from a delta that provides the details required to identify the face. The frequency contents of the right 3two images are complementary to each other and together create the entire range of frequencies present in the original image on the left.

$1 \leq i \leq p$ and $1 \leq j \leq q$. If h can be separated into two 1D filters, a and b of size p and q respectively, such that $h[i][j] = a[i] \times b[j]$, then h is a separable filter.

As an example, let us consider a 3×3 box filter where $p = q = 3$. We know that h is a constant function where $h[i][j] = \frac{1}{9}$. Now, consider two filters a and b, each of size 3 such that $a[i] = \frac{1}{3}, 1 \leq i \leq p$ and $b[j] = \frac{1}{3}, 1 \leq j \leq q$. You can think of a and b as two 1D box filters, one in horizontal direction and the other in vertical direction. Note that in this case, $h[i][j]$ is indeed equal to $a[i]b[j], \forall(i,j)$. Therefore, a 2D box filter is separable.

The advantage of this is that it can be shown that the result of convolving an image with h is equivalent to convolving its rows with a and then its columns with b. This is because for any image I,

$$(I \star a) \star b = I \star (a \star b) = I \star h \qquad (3.13)$$

You can verify that $a \star b$ is indeed h for the box filter.

Let us now discuss the advantage of convolving the rows with a and the columns with b. Let us consider an image with N pixels. The convolution for each pixel with h of size pq will need pqN multiplications and pqN additions, i.e. a total of $2pqN$ floating point operations. If we instead apply a first, we will need $2pN$ operations. Next, with b, we will need $2qN$ operations. This leads to a total of $2(p + q)N$ operations that is much less than $2pqN$. In other words, separable filters can be implemented a lot more efficiently.

Let us now consider the 1D Gaussian function given by

$$\phi(x) = \frac{1}{\sigma\sqrt{2\pi}}e^{-\frac{x^2}{2\sigma^2}}, \qquad (3.14)$$

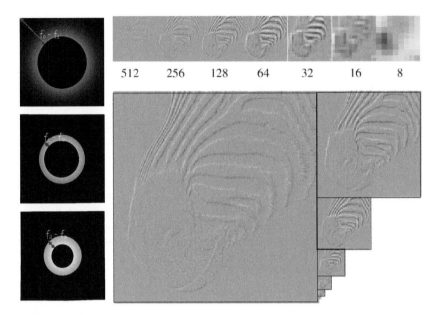

Figure 3.18. On the left we show the frequency response of the band pass filters which can act on the original picture to create the first three levels of the Laplacian pyramid. On the right we show the Laplacian pyramid itself for the same image shown in Figure 3.13. The images are shown in the same size on the top for better perception.

We will see in the next chapter that this Gaussian function is a very good low pass filter (Figure 3.19). Now, let us consider the 2D Gaussian filter given by

$$\phi(x,y) = \frac{1}{2\pi\sigma^2} e^{-(\frac{x^2+y^2}{2\sigma^2})} \tag{3.15}$$

$$= \phi(x) \times \phi(y). \tag{3.16}$$

Since the 2D Gaussian can be expressed as above, it is a separable filter. Therefore, it can be implemented efficiently using

$$I \star \phi(x,y) = (I \star \phi(x)) \star \phi(y)$$

3.2.4　Correlation and Pattern Matching

In this section, we will introduce yet another application of convolution. Consider a picture which is a checkerboard pattern and a filter which looks like a corner of the checker board (Fig 3.20). Consider the image and template both to be in grayscale with values between 0 and 1 (0 for black and 1 for white).

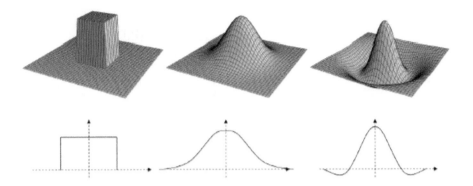

Figure 3.19. Visualizing the 2D filters from their 1D counter part: Box filter (left), Gaussian filter (middle) which is a very good low pass filter and High-Pass Filter (right) obtained by subtracting the Gaussian from a delta.

On the left in Fig 3.20 the value of the convolution is shown for sample colored windows. Note that the value of the convolution is highest where the template matches the image at the blue window. It is also the lowest where the image is exactly complementary to the template as in the red window. Further it is something in between at the purple window which is also showing a corner but with a different distribution of black and white around it. But the problem is, other areas like the green window also shows the highest value and the yellow window shows the lowest value. Intuitively, correlation should provide a measure of the match between the template and the subregion of the image being considered. From that perspective, the subregions defined by the green and yellow windows are similar despite their vastly different colors (one black and one white). The justification for this is based on the variation of colors present in the template which can be considered maximal. Since correlation is a measure of similarity, one would expect the blue window to have the highest value since it has identical spatial distribution as the template (white on top right and bottom left quadrants and black elsewhere). Using the same measure of similarity, the red window should have the lowest value though it has the same color variation due to the complementary spatial distribution (black on top right and bottom left quadrants and white elsewhere). Also, one would expect the yellow and green windows to have some value right in the middle of these lowest and highest values since the template has to go through either of these to go to the complementary pattern in the red window. Similarly, the pattern in the purple window should also have a value between the lowest and the highest, but it should be closer to the lowest value (red square) than the highest value (blue square) since the

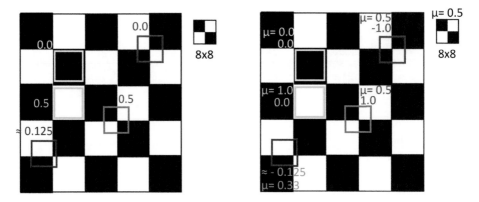

Figure 3.20. This image shows the convolution of an 8 × 8 template that needs to be matched with an identical subregion to the checkboard image. On the left image this is achieved by a simple convolution of the template with the image and on the right image this is achieved by first subtracting the average value of all the pixels from the template and the region to be matched before applying the convolution. On the left image, window of different colors show the result of convolution in those windows. On the right image, the information is also augmented with the value of the mean (μ) at those windows.

spatial distribution of the black and white pixels in the purple window is more similar more similar to the former than the latter.

Put a Face to the Name

 Pierre-Simon Laplace (23 March 1749 to 5 March 1827) was an influential French scholar whose work was important to the development of mathematics, statistics, physics, and astronomy. He summarized and extended the work of his predecessors in his five-volume Mecanique Celeste (Celestial Mechanics) (1799–1825). This work translated the geometric study of classical mechanics to one based on calculus, opening up a broader range of problems. He formulated the Laplacian differential operator widely used in mathematics, pioneered the Laplace transform that forms the cornerstone of many branches of mathematical physics, and was one the first scientists to postulate the existence of black holes and the notion of gravitational collapse.

Laplace was the son of a farmer and cider merchant who intended that he be ordained in the Roman Catholic Church and sent to the University of Caen to read theology at the age of 16. At the university, two enthusiastic teachers of mathematics, Christophe Gadbled and Pierre Le Canu, inspired his zeal for the subject. His brilliance as a mathematician was quickly recognized and while still at Caen he wrote his first paper in a journal founded by Lagrange. Recognizing that he did not have any inclination towards priesthood, Laplace became an atheist, broke away from the church and left for Paris to become the student of Jean le Rond d'Alembert (famous today for Lambertian reflectance models). Lambert tried to get rid of Laplace initially by giving him impossible assignments of reading tough math books and solving unsolved math problems. But when Laplace's brilliance allowed him to complete such tasks in much less time than was provided, he took him under his wings and recommended a teaching position in the Royal Military Academy of Belgium where Laplace devoted his time to path breaking research for the next seventeen years. Laplace became a count of the First French Empire in 1806 and was named a marquis (nobleman) in 1817. He died in 1827. Intrigued by the magnitude of his brilliance, his physician removed his brain and preserved it before it was displayed in an anatomical museum in Britain. It was reportedly much smaller than the average human brain.

To arrive at this intuitive result, we will still do the convolution, but first we will subtract the mean of all the pixel colors from the template and the region of the image it is overlapping with before performing the convolution. This is shown in the right image of Fig 3.20 where the mean is also indicated by μ. Note that now the convolution provides us with what we expected from the correlation operation intuitively. The blue window has the highest value of 1, the red window has the lowest value of -1 and the yellow and green windows have a value of 0 which is halfway between -1 and 1. Also, note that the value at the purple window is negative making it closer to the red window than the blue window. Therefore, convolution when used on functions offset by their mean, can be used to find the extent of similarity between a template and a region in the image. This process is called *cross-correlation*.

Recall that convolution usually involves flipping the impulse response which we did not need to do in this chapter (and in many image processing applications) since we deal with symmetric filters most of the time. The way cross-correlation differs from convolution is that the flipping of the kernel is not required.

Next, let us take another perspective to this cross correlation. You can think of it as an element wise multiplication of the template elements with the underlying image elements with which the template overlaps and then adding them up. Does this remind you of anything? Well, this is a dot product of two vectors, one vector consisting of the elements of the template and the other made up of

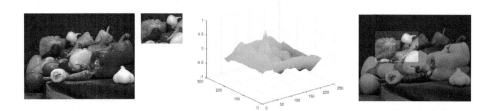

Figure 3.21. On the left is an image and the template to be matched in the image. In the middle is the result of the process of normalized cross correlation. On the right is the template (in color) superimposed at the location given by the highest value of the normalized cross-correlation finding an exact match.

the elements of the region of the image with which the template overlaps. The value of the dot product gives an estimate of how close is one vector to another. The dot product of 1 signifies identical vectors, the dot product of zero signifies orthogonal vectors and the dot product of −1 signifies opposite vectors. Notice that the same thing holds for cross correlation.

Now, you probably remember that while performing dot products for vectors, they needed to be normalized. This was to make sure that we are only considering the directions of the vectors and not their magnitudes. The values of this dot product range between −1 to +1 only when such normalization is applied. Our subtraction of the mean, as shown in Figure 3.21, was an attempting to do approximate this normalization. But a subtraction does not affect the magnitude of the colors to assure unit vectors. Therefore, ideally, we should subtract the mean and divide with the standard deviation, given by the square root of the sum of squared differences of each value from the mean. This step would achieve the desired normalization.

What does this normalization mean in the context of image processing. Fundamentally, cross-correlation is a way to examine if any part of the image matches with a template image. Ideally, we should be able to do this matching irrespective of three factors changing between the two images: (a) scene illumination; (b) the camera exposure that decides the brightness of the image; and the (c) the camera gain that decides the contrast of the image. The normalization allows us to make the cross-correlation robust against these three changes and is often called *normalized cross-correlation*.

3.3 Implementation Details

This brings us to the end of the fundamental concepts behind the linear system and convolution. However, you may still find some challenges when implementing convolution. So, following are the things to be remembered for this purpose.

1. When performing convolution of a filter with an image, it is important to align the filter with every pixel and then apply the convolution. Note that the result of convolution at each pixel should be stored in a different image. Otherwise it will affect the convolutions at other pixels performed later.

2. If the filter size is an even number, you will not be able to find a central pixel to align with the image pixel at which convolution is being performed. The common thing to do in this case is to align the image pixel with the top left pixel of the filter. This will shift the image by half the size of the kernel in each direction and should be shifted back after the operation.

3. When the filter overlaps an area outside the edge of the image, the pixel values of those pixels are undefined. What should we do then? Usually you can pad the image with 0 or 1 or by reflecting the image about the edge or any other way you choose. What is chosen does not matter since these will only contribute to the samples that are not fully immersed and hence should be ignored from a data accuracy perspective.

4. Convolution is made of many floating point operations while images are usually stored as 8-bit integers. Performing floating point operations on integers results in accumulation of errors at every step of the operation (e.g. every multiplication and addition). The best way to handle this is to first convert the image and the filter into a floating point representation, perform the filtering and then round it back to the integer representation.

5. Finally, sometimes convolution can lead to out-of-range values in the resulting image (beyond 0 to 1 or beyond 0 to 256. The best way to handle this is to find the minimum and maximum values after the operation and scale the image back to be within range.

3.4   Conclusion

In this chapter you were introduced to one of the most fundamental concepts of visual computing — systems and their responses and how to find responses of arbitrary inputs to the systems. We have tried to give you a less mathematical and more engineering view of convolution which is directly applicable to digital image processing. To get a more mathematical treatise of the subject, especially considering general multi-dimensional continuous signals, please refer to [Pratt 07, Gonzalez and Woods 06].

Bibliography

[Gonzalez and Woods 06] Rafael C. Gonzalez and Richard E. Woods. *Digital Image Processing (3rd Edition)*. Prentice-Hall, Inc., 2006.

[Pratt 07] William K. Pratt. *Digital Image Processing*. John Wiley and Sons, 2007.

Summary: Do you know these concepts?

- Linear System and its Properties
- Convolution and its Properties
- Low Pass Filters
- Gaussian Filter
- Gaussian Pyramid
- High Pass Filters
- Band Pass Filters
- Laplacian Pyramid
- 2D Filter Separability
- Normalized Cross Correlation

Exercises

1. Consider a signal blurring system. Every sample of the output signal is generated by averaging the values of the sample itself, and its left and right neighbors in the input signal. (Assume that the samples at the boundary of the input signal are zero).

   (a) Is this system linear? Prove your answer.

   (b) What is the impulse response of the system?

   (c) How would this impulse response change if a larger neighborhood of five samples is considered?

2. Calculate the convolution of the following signals (your answer will be in the form of an equation).

   (a) $h[t] = \delta[t-1] + \delta[t+1], x[t] = \delta[t-a] + \delta[t+b]$
   (b) $h[t] = \delta[t+2], x[t] = e^t$
   (c) $h[t] = e^{-t}, x[t] = \delta[t-2]$
   (d) $h[t] = \delta[t] - \delta[t-1], x[t] = e^{-t}$

3. $g[t]$ is a 1D discrete signal defined for $-3t4$. The impulse response $h[t]$ of a linear system is another discrete signal defined for $2t6$. The response of $g[t]$ when passed through this system is given by the convolution of $g[t]$ with $h[t]$ and denoted by $y[t]$. What is the length of $y[t]$? What is the range of t for which $y[t]$ is generated? What is the range of n for which the input $g[t]$ is completely immersed in the output $y[t]$?

4. The low pass filter is a linear operation. Given this prove that the high pass filter is also a linear operation.

5. System A is an "all pass" system, i.e. its output is identical to its input. System B is a low-pass filter that passes all frequencies below the cutoff frequency without change, and blocks all frequencies above. Call the impulse response of system B, $b[t]$.

   (a) What is the impulse response of system A?

   (b) How would the impulse response of system B need to be changed to make the system have an inverted output (i.e., the same output, just changed in sign)?

   (c) If the two systems are arranged in parallel with added outputs, what is the impulse response of the combination?

   (d) If the two systems are arranged in parallel, with the output of system B subtracted from the output of system A, what is the impulse response of the combination?

(e) What kind of filter is the system in (d)?

(f) In this problem, system B has the ideal characteristic of passing certain frequencies "without change." How would the outputs of the systems described in (c) and (d) be affected if the low-pass filter delayed (i.e. shifted) the output signal by a small amount, relative to the input signal?

6. Design a three pixel 1D kernel for ghosting the image where the ghost appears two pixels to the right of the image and has half its intensity. Extend this concept to design a 2D filter where two ghosts appear in horizontal and vertical direction in the same manner.

7. Let $f(x, y)$ denote an image and $f_G(x, y)$ denote the image obtained by applying a Gaussian filter $g(x, y)$ to $f(x, y)$. In the photography industry an operation called high boost filtering generates an image $f_B(x, y) = af(x, y) - fG(x, y)$, where $a1$.

   (a) You are asked to achieve high boost filtering by using a single filter. Derive an expression, $h(x, y)$, for such a filter.

   (b) How would the frequency response, H(u,v), of this filter look like?

8. You are asked to boost the edges of an image. How would you achieve this operation at multiple scales using the Gaussian pyramid? Can you design a single filter to be applied at every level of the pyramid to achieve the same?

9. Consider a 1D signal that has a repeatable pattern of width n pixels. Provide a method to find the value of n using correlation?

10. Given the Laplacian pyramid of an image, how would you reconstruct the original image?

11. Consider the frequency domain response of the box filter and the Gaussian filter, both of the same size. Which one these do you think will be smoother? Justify your answer.

12. Consider two image taken by a stereo pair of cameras (two cameras placed closed to each other like the two human eyes). We have marked some features in the first image. Provide a method to find accurately the location of these features in the second image.

4

Spectral Analysis

In this chapter we will learn a new way of decomposing a signal into simpler signals, each of which is a sinusoid. Studying the nature of each of these sinusoids and their relationship with each other can give us important insights about the signal and help us alleviate problems during common filtering techniques. This kind of analysis of a signal is often called *spectral analysis* – since it decomposes a complex signal into a bunch of signals that span a spectrum of frequencies, phases and orientations. Such a study will also provide us a way to synthesize signals and we will study some of that as well.

In this chapter, we will focus on the most fundamental and popular technique for spectral analysis — the discrete Fourier transform or DFT. However, it is important to be aware that several different ways exist for such spectral analysis. They mostly differ in the kind of simpler signals (or basis functions) that are used to decompose the complex signal into. Radial basis functions or wavelets can be used to provide a different kind of spectral analysis using data dependent basis functions. However, DFT provides one of the most popular tool for spectral analysis of visual signals using data independent or standard basis functions. We will first study DFT for 1D signals. Usually this provides us key insights which are applicable to higher dimensional data. Later in the chapter we will study its interpretation for 2D data like images.

4.1   Discrete Fourier Transform

Discrete Fourier Transform or DFT is a technique that takes as input a periodic signal of infinite length and decomposes it into a set of sine and cosine waves which when combined (via a process called inverse DFT) would provide the periodic signal itself. As soon as we define DFT likewise, the first question that comes to mind maybe how we would use DFT on digital signals? Digital signals are hardly periodic and never infinite. In order to make a digital signal periodic and infinite, we assume that the span of the signal is its period and the signal span is repeated infinite number of times. For example, if we have an audio signal with 100 samples, we will consider a periodic signal where each period

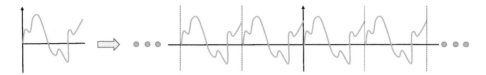

Figure 4.1. This figure shows how the finite 1D signal on the left is repeated to be considered a periodic, infinite signal in order to make it appropriate for applying DFT.

of the signal is 100 samples long and is the function defined by the given 100 samples, as illustrated in Figure 4.1. This assumption has its consequences which we will discuss later in the chapter.

Let the input to the DFT be a signal x with N samples denoted by $x[0, 1, \ldots, N-1]$. DFT is a process that generates two arrays from x, denoted by x_c and x_s each with $\frac{N}{2} + 1$ samples denoted by $x_c[0, 1, \ldots, \frac{N}{2}]$ and $x_s[0, 1, \ldots, \frac{N}{2}]$ respectively. x is said to be in time or spatial domain while x_c and x_s is the representation of the same function/signal in the frequency domain, as illustrated in Figure 4.2.

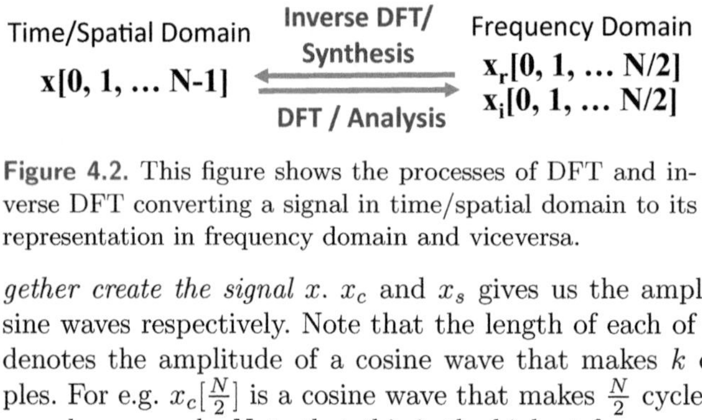

Time/Spatial Domain	Inverse DFT/ Synthesis	Frequency Domain
x[0, 1, ... N-1]	⟵⟶	**x$_r$[0, 1, ... N/2]** **x$_i$[0, 1, ... N/2]**
	DFT / Analysis	

Figure 4.2. This figure shows the processes of DFT and inverse DFT converting a signal in time/spatial domain to its representation in frequency domain and viceversa.

So, what does x_c and x_s mean? *DFT decomposes x into $\frac{N}{2} + 1$ cosine and sine waves, each of a different frequency. These cosine and sine waves when added together create the signal x.* x_c and x_s gives us the amplitude of the cosine and sine waves respectively. Note that the length of each of these waves is N. $x_c[k]$ denotes the amplitude of a cosine wave that makes k cycles over the N samples. For e.g. $x_c[\frac{N}{2}]$ is a cosine wave that makes $\frac{N}{2}$ cycles over N samples, i.e. 2 samples per cycle. Note that this is the highest frequency wave possible given N samples and abiding by the Nyquist sampling criteria. Similarly, $x_s[k]$ denotes the amplitude of a sine wave that makes k cycles over N samples.

Next, we will discuss alternate ways of expressing the frequency of these waves. We have already seen that the wave with index k makes k cycles over the N pixels (samples). This means that each pixel constitutes $f = \frac{k}{N}$ cycles. This is another way to represent the frequency in terms of cycles per pixel. Note that as k spans from 0 to $\frac{N}{2}$, f ranges from 0 to 0.5. Finally, the frequency can be expressed as $\omega = f \times 2\pi$, its natural frequency. ω ranges from 0 to π. You may come across any of these representations when you work with spectral data. Examining the range of the independent axes of such data will immediately tell

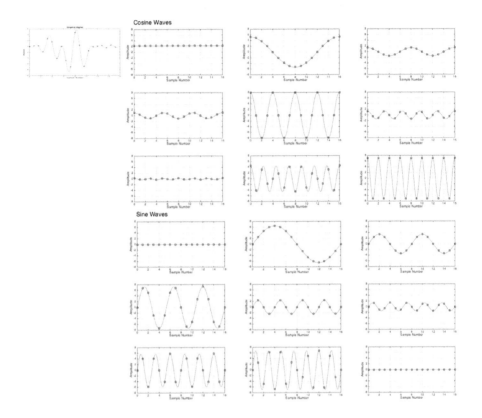

Figure 4.3. This figure shows how the sine and cosine waves are scaled by the factors produced in x_c and x_s. These waves when added will create x, the original signal shown on the left.

you the particular representation used.

Now, consider the cosine wave c_k whose amplitude is given by $x_c[k]$. Note that c_k is a signal with N samples. Further, it is a cosine wave that makes k cycles over these N samples, and has an amplitude of $x_c[k]$. This cosine wave can be described as

$$c_k[i] = cos\left(\frac{2\pi ki}{N}\right) = cos(2\pi fi) = cos(\omega i) \qquad (4.1)$$

where i denotes the ith sample of c_k, $0 \le i \le N-1$. Similarly, the kth sine wave, s_k, is given by

$$s_k[i] = sin\left(\frac{2\pi ki}{N}\right) = sin(2\pi fi) = sin(\omega i) \qquad (4.2)$$

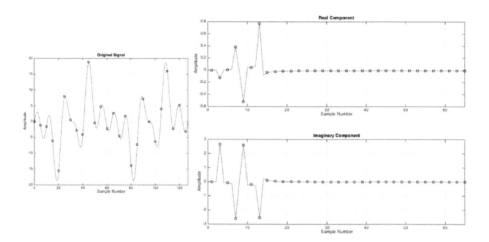

Figure 4.4. This figure shows decomposition of the time domain signal x to the frequency domain x_c and x_s.

Now, recall that the signal x is formed by the addition of all the cosine and sine waves weighted by their amplitudes given by x_c and x_s respectively. Therefore,

$$x[i] = \sum_{k=0}^{\frac{N}{2}} x_c[k] cos\left(\frac{2\pi ki}{N}\right) + \sum_{k=0}^{\frac{N}{2}} x_s[k] sin\left(\frac{2\pi ki}{N}\right) \tag{4.3}$$

Note that this provides us the equation for using x_c and x_s and combining the different sine and cosine waves to create the signal x. Therefore, this is the equation behind the inverse DFT or synthesis. It is simpler to understand this and hence we derived this first. Figure 4.3 shows weighted sine and cosine waves that when combined create the original signal.

However, the actual equations for synthesis are slightly different than Equation 4.3. There are some scale factors associated with each term. All the terms except for $x_c[0]$ and $x_s[\frac{N}{2}]$ are scaled by $\frac{2}{N}$. $x_c[0]$ and $x_s[\frac{N}{2}]$ are scaled by $\frac{1}{N}$. Therefore the actual amplitudes $\hat{x}_c$ and $\hat{x}_s$ are given by

$$\hat{x}_c[k] = \frac{2}{N} x_c[k] \tag{4.4}$$

$$\hat{x}_s[k] = \frac{2}{N} x_s[k] \tag{4.5}$$

for all k except for

$$\hat{x}_c[0] = \frac{1}{N} x_c[0] \tag{4.6}$$

$$\hat{x}_s[\frac{N}{2}] = \frac{1}{N} x_s[\frac{N}{2}] \tag{4.7}$$

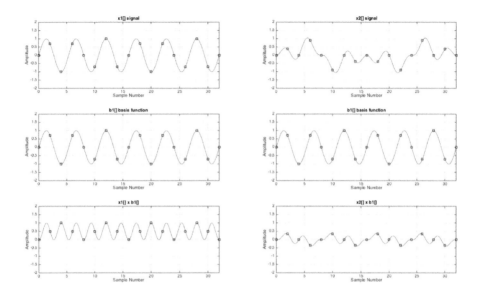

Figure 4.5. Two example signals are shown to be correlated with the same basis function. The first leads to the value of 0.5 indicating an amplitude of 1.0 for that basis function in the synthesis equation. The second one leads to a value of 0 indicating that this basis function has no contribution in synthesizing the function.

and the actual synthesis equation is

$$x[i] = \sum_{k=0}^{\frac{N}{2}} \hat{x_c}[k] cos\left(\frac{2\pi ki}{N}\right) + \sum_{k=0}^{\frac{N}{2}} \hat{x_s}[k] sin\left(\frac{2\pi ki}{N}\right) \tag{4.8}$$

These scale factors are related to the underlying process of discretization of the analog Fourier transform during its digital processing and we will come back to it soon.

The next question is how we really find x_c and x_s. Note that this decomposition is trying to figure out how much of each of these sine and cosine waves is contained in the signal x. What does this remind you of? Of course, this is best computed by correlation. If x is a cosine or sine wave, it will be completely correlated with the sine and cosine waves of the same frequency resulting in only one element of x_c and x_s to be non-zero. Thus, we can write the equation for

DFT as a correlation as

$$x_c[k] = \sum_{i=0}^{N-1} x[i]cos\left(\frac{2\pi ki}{N}\right) \tag{4.9}$$

$$x_s[k] = \sum_{i=0}^{N-1} x[i]sin\left(\frac{2\pi ki}{N}\right) \tag{4.10}$$

Here too, due to the same scale factors, the real coefficients are given by $\hat{x}_c$ and $\hat{x}_s$. In fact, these are the exact weights that are used in Figure 4.3. The arrays x_c and x_s thus generated from x are shown in Figure 4.4. An example of the process of correlation is shown in Figure 4.5.

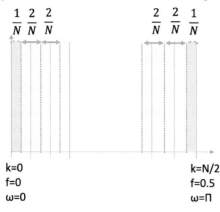

Figure 4.6. This figure shows the origin of the scale factors in Equation 4.8. The blue lines indicate the frequencies generated by DFT. The dotted red lines denote the boundaries of the ranges of frequencies represented by each of these frequencies.

Let us briefly examine Equation 4.10 first. Note that $x_c[0]$ is given by correlation with $cos(0) = 1$. This means

$$\hat{x}_c[0] = \frac{1}{N}\sum_{i=0}^{N-1} x[i], \tag{4.11}$$

i.e. the first coefficient for cosine waves is the average of all the samples of x. This is often called the DC component. Second, note that $sin[0] = sin[\pi i] = 0$. Therefore, $\hat{x}_s[0] = x\hat{_\frac{N}{2}} = 0$. Some of you may have been wondering before how we generate $N + 2$ samples in the frequency domain from N samples in the spatial domain. Here you can see that we actually do not generate more information since two of the $N + 2$ samples are zero.

Let us now revisit the issue of the scale factors introduced to generate Equation 4.8. In this book, we do not discuss analog Fourier transform. However, mathematically, discrete Fourier transform is derived from analog Fourier transform. In DFT, when we move from time/spatial domain to frequency domain, we generate a few discrete frequencies, with uniform distance between them. In other words, DFT only creates frequencies that makes k cycles over the N sample, where k is limited to an interger. However, when we perform the same Fourier transform in the analog domain, it can generate many different frequencies since conceptually both k and N can take any value.

Therefore, the set of frequencies generated by the DFT can be thought of as the sampling of the frequencies in continuous domain. Each of these discrete frequencies can be considered to be a representative of a range of frequencies. All the frequencies with $k = 1, 2, \ldots \frac{N}{2}$ can be thought of representative of a range of frequencies with width $\frac{2}{N}$ (Figure 4.6). However, the range of frequencies represented by the first and last frequency, i.e. $k = 0, \frac{N}{2}$ is half of this, i.e. $\frac{1}{N}$. These are exactly the scale factors used for the different frequencies in Equation 4.8. Therefore, these scale factors originate from the width of the spectral band represented by each of the discrete frequencies.

Put a Face to the Name

Jean-Baptiste Joseph Fourier was a French mathematician (21 March 1768 – 16 May 1830) who played an important role during the French Revolution for which he was briefly imprsioned. He accompanied Napoleon Bonaparte in his Egyptian expedition in 1798 and contributed heavily to the Egyptian institute in Cairo that Napoleon founded. Fourier's biggest contribution is in the investigation of Fourier series and their applications to problems of heat transfer and vibrations. Though Fourier transform is probably the most fundamental mainstay of image processing, it was not easy for Fourier to publish this work. Fourier first tried to publish this work in 1807 when he claimed that "any continuous periodic signal can be expressed as a sum of properly chosen sinusoids". Two stalwart mathematicians of those times, Lagrange and Laplace, reviewed the paper. While Laplace wanted to publish the paper, Lagrange vehemently opposed it saying that sinusoids cannot be sufficient to represent signals with corners. The paper was published 15 years later after Lagrange's death. As to the question of who was right, – well, both were. It is true that you cannot represent signals with corners with sinusoids, but you can get extremely close. In fact, you can get so close that the energy difference is zero which was shown by Gibbs later on and is famously called the Gibbs effect. Also, it was shown that this is true only in the analog domain. In the digital domain, the representation is exact, even for signals with corners.

4.1.1 Why Sine and Cosine Waves?

A question that is probably hovering in everyone's mind by this time, is what is so special about sine and cosine waves and why are we decomposing general signals into sine and cosine waves? In fact, sine and cosine waves are indeed special. You can show the sine and cosine waves of different frequencies are completely independent of each other. In other words, none of these waves can be given by a

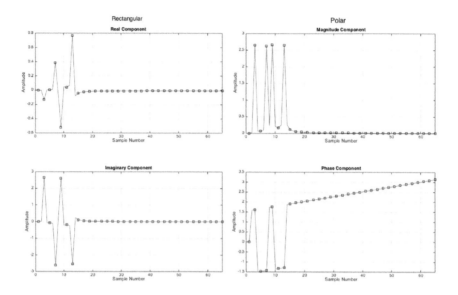

Figure 4.7. This shows both the rectangular and the polar representation for the same function.

linear combination of the others. You can verify this by correlating any of these waves with any of the other. The answer will always be zero confirming that each wave is completely uncorrelated and hence independent of another. Therefore, these waves form a basis for representing other functions. It can be shown that these set of waves are both necessary and sufficient to represent any general 1D periodic function.

4.2 Polar Notation

Now, the question is, how do we represent this frequency domain for us to interpret it for our needs. Of course, one obvious way is to plot x_c and x_s (as shown in Figure 4.4). This is referred to as the *rectangular* representation. But note that we are dealing with sine and cosine waves in the rectangular representation which when drawn out as in Figure 4.3 will cause constructive and destructive interference (yes, this is the concept in physics that you learned in high school!). Therefore, some parts of one wave will cancel out some parts of another wave while others will reinforce it. Therefore, it is very difficult to really understand anything useful from this plot of x_c and x_s.

However, there is respite from this. When considering a pair of scaled sine and cosine waves of the same frequency, they can be represented by a single cosine wave of certain amplitude and phase. This is due to the fact that sine and cosine waves are phase shifted versions of each other. Therefore, for each pair of same frequency sine and cosine waves, we can do the following.

$$x_c[k]cos(\omega i) + x_s[k]sin(\omega i) = M_k cos(\omega i + \theta_k) \tag{4.12}$$

where $M_k = \sqrt{x_c[k]^2 + x_s[k]^2}$ and $\theta_i = tan^{-1}\left(\frac{x_s[k]}{x_c[k]}\right)$ are the amplitude and phase respectively of the k cosine wave. Therefore, instead of two arrays x_c and x_s which makes it a little complicated to analyze, we can represent x by a set of only the cosine waves, each with a different amplitude and phase. $\frac{N}{2}$ such cosine waves each of which scaled by the right amplitude and shifted by the right phase is added together to create the original signal. This amplitude and phase generates two 1D plots which constitutes the frequency domain representation of the signal x. This representation is called the *polar representation*. Now note here, that the phase can be expressed either in degrees or in radians. If expressed in degrees, the independent axis will range from -180 to 180. If expressed in radians it will range from $-\pi$ tp π. Another aspect to keep in mind when coding this up is computing θ involves a division with $x_c[k]$. Therefore, sometimes this can lead to a division zero. The case of $x_c[k] = 0$ should be treated as an exception and appropriate phase assigned to it. Most of these are more pertinent if you have to code up DFT which will be rare since multiple mathematical packages (e.g. Matlab) are available today that do this for you.

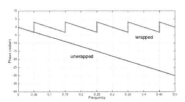

Figure 4.8. This figure shows the the process of unwrapping of phase.

Figure 4.7 shows the rectangular representation and the corresponding polar representation of the DFT of a 1D signal. Now, let us take a moment to study this polar representation a little more. Interpreting this plot is absolutely essential to internalize concepts in image analysis. While multiple computer programs will generate this plot for you in no time, none will tell you how to interpret it. First, note that the independent axis of the plot can be either k ranging from 0 to $\frac{N}{2}$ or f ranging from 0.0 to 0.5 or ω ranging from 0 to π (in Figure 4.7 we have used f). You should be prepared to see any of these representations. The ranges will tell you which particular parameter is being used. Second, note that phase provides us information about how synchronous the rise and fall of the different waves are, indicating whether they result in features (e.g. edges, corners). Such features are best studied in the spatial/time domain and not in the frequency domain. Therefore, the amplitude plot is of most importance to us as we saw in the previous chapter.

However, the polar notation is not devoid of problems. Consider the two cases of $x_s[k] = x_c[k] = 1$ and $x_s[k] = x_c[k] = -1$. In both these cases $M[k] = 1.414$. But the phase for one of them is $45°$ and the other is $-135°$. This is due to the phase ambiguity. Similar problem arises because, for example, a phase shift of θ is equal to that of $\theta + 2\pi$ or $\theta + 4\pi$ and so on. This ambiguity in representation needs to be taken care of after θ is computed. This process is called unwrapping of the phase and is illustrated in Figure 4.8.

4.2.1 Properties

It is now time to explore some basic properties of DFT as follows.

1. *Homogeneity:* If the DFT of a signal $x[t]$ is given by $M[f]$, the DFT of the signal scaled by a factor k is also scaled by k. In other words, homogeneity implies that if a signal is scaled in the spatial domain, its amplitude in frequency domain is also scaled similarly. Assuming $\rightarrow$ stands for DFT, this can be expressed as,

$$x[t] \rightarrow (M[f], \theta[f]) \implies kx[i] \rightarrow (kM[f], \theta[f]) \tag{4.13}$$

   Note that the phase $\theta[f]$ remains unchanged with scaling.

2. *Additivity:* Addition of signals in the spatial domain results in an addition of its responses in the frequency domain. This can be expressed as

$$x[t] \rightarrow (x_c[f], x_s[f]), y[t] \rightarrow (y_r[f], y_i[f]) \tag{4.14}$$
$$\implies x[t] + y[t] \rightarrow (x_c[f] + y_r[f], x_s[f] + y_i[f]) \tag{4.15}$$

   Addition of two sine or cosine waves makes sense only when they are of the same phase. Therefore, this addition cannot be performed in the polar notation (where we express x as sum of different cosine waves of different phase and amplitude). Hence, we resort to the rectangular notation (where we express the signal as a sum of sine and cosine waves of similar phase) to achieve this addition.

3. *Linear Phase Shift:* A shift in the signal in the spatial domain results in a linear phase shift proportional to the spatial shift in the frequency domain as follows.

$$x[t] \rightarrow (M[f], \theta[f]) \implies x[t+s] \rightarrow (M[f], \theta[f] + 2\pi f s) \tag{4.16}$$

   This can be intuitively explained also. Think of $x[t]$ getting shifted by s in the spatial domain. This means that all the waves comprising $x[t]$ will be shifted by s. Note that the same shift s makes up a larger part of a cycle for a lower frequency wave than a higher frequency wave which corresponds to a lower phase shift. Therefore the phase shift for each wave is proportional to its frequency.

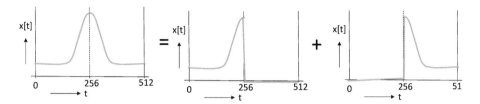

Figure 4.9. This shows how a symmetric signal can be decomposed to two signals which are complex conjugates of each other.

4.2.2 Example Analysis of Signals

Let us see if we can explain some of the phenomena using what we have learned so far. This will help you understand how to use these properties for analyzing signals. First let us consider a symmetric signal as seen in Figure 4.10. What can you say about the phase of a symmetric signal?

For this we need to understand something called the *complex conjugate* of a signal. A signal whose frequency domain response has the same magnitude but negative phase is called the complex conjugate of x and is denoted by $x^\star$. It can also be shown that if the frequency response of a signal and its complex conjugate are added, the resulting phase is zero at all frequencies. Further, if the phase response in the frequency domain is negated as above, the signal is flipped in the spatial domain.

Now considering this, let us look at a symmetric signal as shown in Figure 4.9. A symmetric signal can be decomposed into two signals, each of which is a flipped version of the other i.e. complex conjugate of the other. Therefore, the addition of these two signals provides a signal whose phase response is zero. Therefore a symmetric signal always has zero phase as shown in Figure 4.10.

Now, let us next consider what happens when we consider shifting of such a symmetric signal. For this, please refer to Figure 4.10. Here we show circular shifts of the symmetric signal. Note that from the property of linear phase shift due to spatial shift in the signal we know that any such shift will cause the phase to be shifted linearly. The slope of this shift will be positive or negative based on if the shift is to the right or left respectively. Finally, when the circular shift results in another symmetric signal as shown in Figure 4.10, it leads to a zero phase signal again.

This provides us with an example of how these properties can be used to analyze signals. However, the issue of phase of symmetric signals also brings in the question of what does it meant by a non-linear phase? Usually non-linear phase means non-linear features superimposed on linear phase as shown in Figure 4.11. It is typical of more general non-symmetric signals.

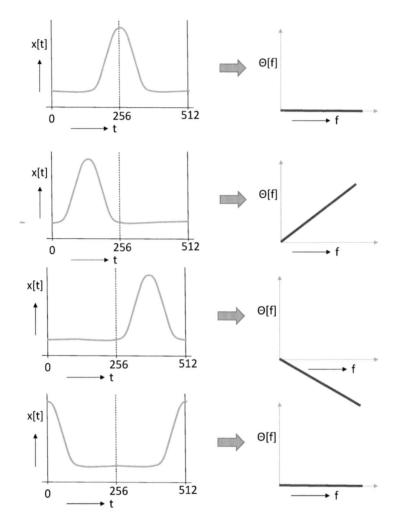

Figure 4.10. This shows the frequency responses of symmetric signals and their circular movement.

Let us now see another example of signal analysis called amplitude modulation. Lets us consider an audio signal $x[t]$ whose response $X[f]$ in frequency domain is bandlimited by b. This signal is multiplied by a very high frequency cosine wave $y[t]$ called the carrier wave or carrier frequency. The frequency of this wave is called the carrier frequency and denoted by c. Note that since this is a single cosine wave, its frequency response $Y[f]$ is a shifted delta at c.

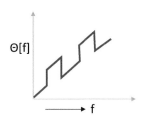

Figure 4.11. This shows the non linear phase of a typical non-symmetric signal.

A multiplication in spatial domain is a convolution is frequency domain which results in creating two mirror copies of $X[f]$ centered around c. To recover the signal in frequency domain a filter that extracts a region of bandwidth b to the left or right of c, often called a *notch filter*, would be sufficient. This is exactly how AM (standing for amplitude modulation) radio works with each station having their own carrier frequency transmitting a signal modulated by the carrier wave. When we tune in the radio, we are applying the notch filter to recover the signal back. Note that $c \gg b$ for this to work and also the carrier frequencies of the different channels needs to be at least $2b$ apart to ensure no mingling of signals. We will see in Section 4.4 that this mingling of signals have a name and distinct feature.

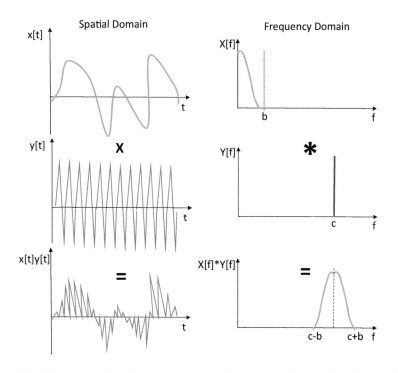

Figure 4.12. This shows the phenomenon of amplitude modulation. A audio signal $x[t]$ in spatial domain, bandlimited by b in frequency domain, is multiplied by a carrier sine wave $y[t]$ of very high frequency c. This multiplication results in a convolution of a shifted delta with the frequency response $X[f]$ of the audio signal. This results in a shifted copy of $X[f]$ centered at frequency c.

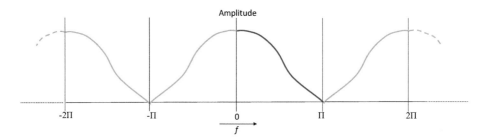

Figure 4.13. The amplitude plot is frequency domain is periodic and repeats itself as an even function. The red plot signifies the $[0, \pi]$ range that we have seen so far.

This constraint that carrier frequencies should be $2b$ apart did not allow enough carrier frequencies as were demanded by the fast growing number of stations. So, frequency modulation or FM came later. Here, the frequency, and not the amplitude, of the signal to be transmitted is modulated based on the amplitude of the spatial signal. For example, the frequency can be modulated from 55KHz to 65KHz for one station and between 40-50KHz in another station. Here also a frequency domain notch filter extracts the signal on the receiver end. However, the range of frequencies to be used for modulation can be much smaller allowing much larger number of stations to be packed in.

4.3   Periodicity of Frequency Domain

As explained in Figure 4.1, DFT consider the signal to be repeating itself periodically infinitely. However, we have not discussed yet, the consequence of such an assumption. A question may be hovering in your minds as to how does an infinitely periodic function have a finite DFT. In fact, it doesn't! We will now discuss how this periodicity in spatial domain induces a periodicity in the frequency domain. Let us explore now how this periodicity looks.

DFT decomposes a signal into cosine waves of different magnitude and phase. If we consider a frequency f, we know from trigonometry that

$$Acos(f) = Acos(-f) = Acos(2\pi - f) = Acos(n2\pi - f) \qquad (4.17)$$

where n is any integer. From this, you can see that if the value of the amplitude at f in the DFT magnitude response is A, then the same value would repeat at $-f$ and also at $2\pi - f$ and so on. Therefore the amplitude repeats as

$$M[f] = M[-f] \qquad (4.18)$$

as shown in Figure 4.13. Thus it repeats as an even function, a function whose

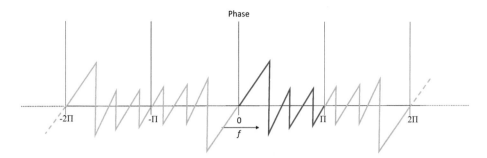

Figure 4.14. The phase plot is frequency domain is periodic and repeats itself as an odd function. The red plot signifies the $[0, \pi]$ range that we have seen so far.

value at a negated parameter is the same as its value at the corresponding positive parameter.

Let us now see what happens to the phase. We know from trigonometry that

$$cos(f + \theta) = cos(-f - \theta). \tag{4.19}$$

Therefore, the negative of the value of the phase θ repeats at $-f$. Therefore, the phase repeats itself as

$$\theta[f] = -\theta[-f] \tag{4.20}$$

as shown in Figure 4.14. This is an odd function, a function whose value at the negated parameter is the negative of its value at the corresponding positive parameter. Also, note that hence forth most of the times when you will be shown a 1D frequency plot you will be shown the plot for the entire range of $[-\pi, \pi]$.

4.4 Aliasing

Now that we have understood periodicity, this brings us to a very important phenomenon called aliasing. Let us start with the amplitude modulation we did in Section 4.2.2. Let us assume that the carrier frequencies of different stations are less than $2b$ apart. Let us see what happens in that case.

Check out Figure 4.15. We see here two signals, blue and green transmitted with carrier frequencies c and d respectively. On the left when c and d are $2b$ apart, the blue and green signals do not overlap with each other after convolution. Therefore, when the blue signal is reconstructed using a notch filter with frequencies from c to $c + b$, the original signal is reconstructed via inverse DFT. But now note what happens when c and b are not $2b$ apart as shown in the right. Here the blue and green signals overlap each other and now during reconstruction part of the green signals higher frequencies get added to the blue

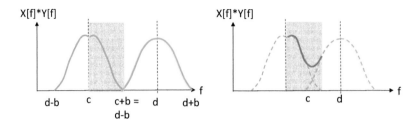

Figure 4.15. Left: The Y and X denote the frequency domain response of the carrier wave and the signal to be transmitted respectively. Since carrier waves are a cosine or sine wave of a single frequency, their response is a single spike at the carrier frequency. Two such carrier frequencies, c and d, are exactly $2b$ apart and the reconstructed signal by notch filter of frequency c to $c + b$ would give the correct signal back. Right: c and d are less than $2b$ apart. Now during reconstruction, part of the green signal gets added to the blue signal giving a signal which has some extra high frequencies. This reconstructed signal is highlighted in dark blue.

signal thereby amplifying its higher frequency. This would create high frequency artifacts in the reconstructed signal. The phenomena of ghost frequencies from other signals contaminating a signal is called aliasing.

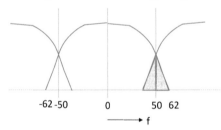

Figure 4.16. This figure shows the aliasing stemming from the periodicity of the DFT.

To start with this, let us start with the simplest case. Let us consider a discrete 1D signal $x[t]$ of size $n = 100$ samples being convolved with a filter $h[t]$ of size $m = 25$. We know that the size of the resulting signal will have $n + m - 1 = 125$ samples. Now let us consider performing the same convolution by first finding the DFT of x and h given by X and H of size 50 and 13 respectively. Then, we multiply X and H to get the response of size 50 and then find the inverse DFT that results in a signal with 100 samples. So, using one approach we get an answer which is 125 samples long and using another we get to a signal with 100 samples. So, what is the problem here? In fact, since the convolved signal should be 124 samples, it would need sinusoids that make 0 to 62 cycles to represent the convolved signal as per the Nyquist criterion. However, when going by the frequency domain multiplication, we are not sampling the frequency domain adequately by using only 50 samples. Hence, the convolved signal achieved via that route will show an aliasing artifact. Here the aliasing is purely due to inadequate sampling. To do this correctly via the frequency domain, you should first pad both x and h adequately to make them 125 samples in size. Next, we find the DFT of x and h

yielding X and H, each of which is 62 samples in size. These after inverse DFT will create the correct 124 sample sized convolved signal.

Note that this aliasing essentially stems from the periodicity of the DFT as shown in Figure 4.16. The size of the frequency response array is supposed to be 62 and when 50 samples are used the frequency response overlaps with each other. It is almost as if the last 12 samples have flipped over to be part of the frequency response when it is not supposed to be like that. This leads to leak-in high frequency that shows up as aliasing artifacts.

4.5 Extension for 2D Interpretation

In this chapter, we have so far been considering only 1D signals. Now lets us consider 2D data. As you already know, grayscale images are considered as 2D data. Therefore, they can undergo DFT too. A multitude of software can perform this DFT, but it is important to interpret the results of DFT. So, let us see how would the frequency domain representation of the DFT of 2D data be interpreted. We had already touched upon this briefly in Section 1.3 of

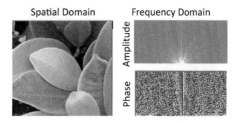

Figure 4.17. This figure shows the amplitude and phase plot on the right corresponding to the frequency domain representation of the image in the left.

Chapter 1. It may be useful if you want to go back and brush up on this section. We will get into lot more details in this chapter.

As we saw in Chapter 1, the frequency domain response of a 2D data will result in a 2D plot. Therefore, M and θ are now a 2D functions representing the amplitude and phase of cosine waves of different frequencies (as in the case of 1D) and different orientation. Let M and θ depend on two variables g and h such that $M[g, h]$ and $\theta[g, h]$ give the amplitude and phase respectively of a cosine wave of frequency $\sqrt{g^2 + h^2}$ and orientation $tan^{-1}(\frac{h}{g})$. The most common way of visualizing M and θ is to plot them as a grayscale image where the values of $M[g, h]$ and $\theta[g, h]$ is visualized as a gray value between black and white.

Let us now understand this representation thoroughly. Let us consider a grayscale image which is 2D data – note that an RGB image will be considered as a 3D data due to multiple channels. However, each channel of RGB image is usually considered as 2D data. The frequency domain representation of an example image is shown in Figure 4.17. There are a few things to note here in the amplitude and phase plots. First, the zero frequency (or the DC component) is at the bottom center of the image. Second, the orientation of the cosine wave

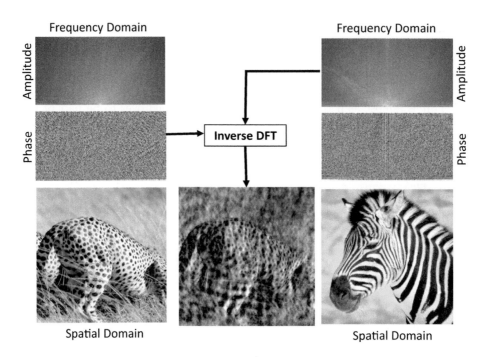

Figure 4.18. This figure shows that when the phase information of one image is combined with the amplitude of another to create a new image via inverse DFT, the perception of the image whose phase was used prevails.

can be from 0 to 180. Beyond that the orientation can be mapped back to 0 to 180. Therefore the bottom horizontal line denotes frequencies of 0 degree on the right and 180 degree on the left. Third, note that as expected, high gray values are at lower radius from the center denoting concentration of the most of the energy of the signal at lower frequencies. In this particular plot, note that the the vertical frequencies (90 degrees) have high frequencies. Finally, note that it is almost impossible to make any sense out of the phase plot. This is expected since this plot is not unwrapped. Also, as mentioned earlier, the phase plot shows how synchronous the rise and fall of the cosine waves are. This information is pertinent for detecting features (e.g. edges). However, feature detection is usually done in the spatial domain and hence the phase plot is typically not interpreted for spectral analysis.

This may give you an impression that the phase plot is not that important. But this is a grave misconception as is illustrated by Figure 4.18. Here we have taken the phase response of the cheetah image and the amplitude response of the zebra image and combined them using the inverse DFT to create a new image back. Note that the predominant perception of this image is that of cheetah whose phase plot has been used. This shows that phase information is very important, just that it is easier to study this in the spatial domain rather then the frequency domain.

4.5.1   Effect of Periodicity

Now let us see how this periodicity extends to 2D. We have seen that the amplitude and phase plots repeat themselves to infinity in positive and negative directions as an even and odd function respectively. The same will be true in 2D. But now since we are dealing with 2D functions, the plot will repeat itself in each of the four quadrant directions. We will be working with amplitude plots mostly and to show one period of the plot, we will be showing four quadrants where the bottom two quadrants will be the reflection of the top two quadrants.

Let us now look at a few frequency response plots (amplitude only) in Figure 4.19 to understand these better. (a) is the image of a sine wave making 8 cycles in the horizontal direction. Note that horizontal direction sinusoids result in vertical stripes and vice versa. Therefore, you see the plot with two high points on the x axis denoting the symmetric position for the horizontal sinusoid which can be considered to be of orientation 0 or 180. The bright spot in the center is due to the averaging of all the pixels in the image. (b) contains only one frequency in the vertical direction. Therefore, again we find two highlights in the vertical axis at 90 and 370 degrees. The next two images, (c) and (d), contains both vertical and horizontal frequencies and therefore you see four highlights instead of two in the amplitude graph in the frequency domain. Finally, (e) shows an image which is just a rotation of a single sinusoid. Ideally, you would see two highlights in the direction perpendicular to the direction of the stripes as shown in (g). However, instead we see strong horizontal and vertical patterns in Figure 4.19. This may look surprising but this is exactly what is due to the periodicity. Due to periodicity we are finding the DFT of the image in (e) repeated in both directions multiple times as shown in (h). The edges thus created causes the horizontal and vertical highlights in the frequency domain. To alleviate this effect, we perform a windowing operation on this image as shown in (f). This is essentially pixel-wise multiplication of an image with a gaussian image that has the peak brightness at the center and the brightness falls off smoothly to a medium gray near the fringes based on a Gaussian function. Though this brings in a new effect getting rid of the edges when repeated in horizontal and vertical direction, we can see a better approximation of the original non-repeated sinusoid where the two highlights can at least be deciphered.

4.5.2   Notch Filter

At this point, you may possibly be wondering about the possible uses of frequency domain computation and how it can be more effective than spatial domain computation. Towards this, let us see an example of a *notch filter*. Consider an image

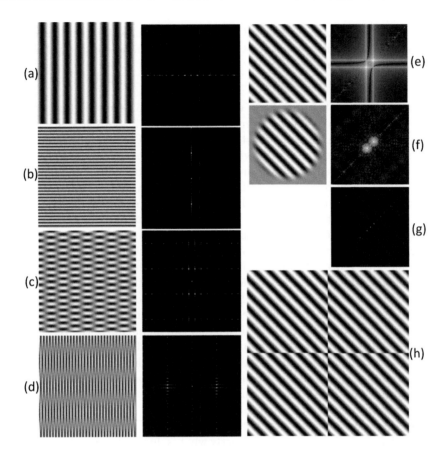

Figure 4.19. The top row shows the spatial domain image and the bottom row shows its amplitude response in the frequency domain. (a) A cosine of 8 horizontal cycles. (b) A cosine of 32 vertical cycles. (c) A cosine of 4 cycles horizontally and 16 cycles vertically. (d) A cosine of 32 cycles horizontally and 2 cycles vertically. (e) This is (a) rotated by 45 degrees. The ideal frequency domain response of an infinite image of this pattern is (g). But this is suppressed due to the extra pattern and therefore frequencies created by tiling as shown in (h). (f) This image is that of (e) but with a "windowing" that slowly tapers off to a medium gray at the edge and therefore the frequency response is closer to (g).

which has some periodic high frequency pattern superimposed on it — this can be due to the technology of its creation (e.g. newsprint, tapestry). Removing this superimposed pattern in spatial domain is not obvious. But if you now see the spectral response of the image you will see that its magnitude plot will clearly show this pattern as a outlier high frequency. It is easy to isolate this high frequency in the magnitude plot, remove it and then apply the inverse DFT to get

Figure 4.20. A: The original image with a high frequency pattern superimposed on it;
B: The DFT of the image in A - note the white high frequency regions corresponding to
the high frequency pattern; C: Removal of the outlier frequency in the spectral domain;
D: Inverse DFT is applied on C to get a new image back. This is devoid of the high
frequency pattern.

a new image back. And voila! This high frequency pattern is removed from this
image. This example is illustrated in Figure 4.20.

4.5.3   Example of Aliasing

To discuss aliasing in the context of 2D images, a very good example is provided
by the process of digital image generation and display. We will discuss this using
1D images, but you will see that it will be adequate. An analog image is sampled
to create a digital image. This process is called *sampling*. The process of using
a light spot (called pixel) of a particular size and intensity profile to display
these samples is called is called *reconstruction*. We will now provide a frequency
domain analysis of these two processes and show how aliasing artifacts can result
during these processes.

Let us now consider an analog signal with bandwidth f_s which means that
the highest frequency wave present in the signal f_s. Now sampling this image
to convert it to digital domain can be considered as a multiplication by a comb
function in the spatial domain. A comb function is a periodic function where
a scaled impulse repeats itself periodically (Figure 4.21). Since the frequency
domain response of a comb function is another comb function, the sampling
process becomes a convolution with a comb function in the frequency domain.
Since the largest frequency present in the function is f_s, it has to be sampled at
least at the rate of $2f_s$. This means that the interval between the combs in the
spatial domain is $\frac{1}{2f_s}$. Therefore, by duality, the interval between the combs in
the frequency domain is $2f_s$ as shown in Figure 4.21. Note that if the distance
between the combs is greater than $2f_s$ in frequency domain (by duality, if the
interval between the combs is greater than $\frac{1}{2f_s}$ in the spatial domain), then the
copies achieved due to convolution in the frequency domain would overlap with
each other leading to aliasing. This is also what Nyquist criterion states.

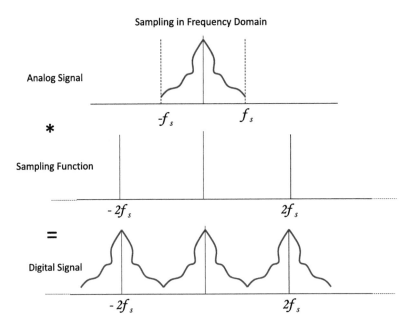

Figure 4.21. This figure shows the process of sampling an analog signal to its digital counterpart in frequency domain by convolution with a comb function. The bandwidth of the analog image is f_s.

Now, let us consider the process of displaying a digital signal on a display. Now note that a pixel, though thought of as a dot of light, practically is formed by illuminating a finite area. Though ideally we would like this area to be uniformly illuminated and the light to be cut-off sharply away from the boundaries of the pixel, practically this is hardly possible. The illumination of a pixel is usually highest at the center and then drops of smoothly towards its boundary and hopefully dies down before reaching the other pixel. Let us consider the illumination profile of the pixel to be a signal is spatial domain. We call this function as the *point spread function* or PSF. The process of image reconstruction is to convolve the sampled signal in the spatial domain with the kernel of PSF. Therefore, it is a multiplication in the frequency domain as shown in Figure 4.22. Now note that if the bandwidth of the PSF (or highest frequency in the PSF) is exactly f_s, a correct reconstruction will be assured. However, if the bandwidth is a little higher (i.e. pixels are sharper and become dark before reaching the boundary), as shown by the orange function, the reconstructed signal will have leakage high-frequencies from the other copies leading to aliasing artifacts. This artifact is commonly called *pixelization*. If the bandwidth is a little smaller (i.e. pixels are larger and bleeds into the next pixel), only a smaller band of frequencies will be

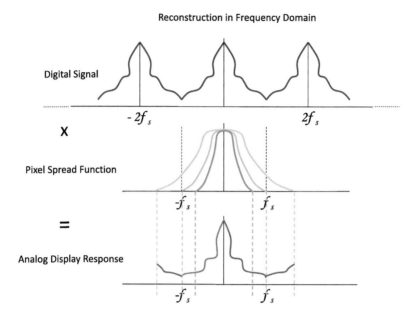

Figure 4.22. This figure shows the process of reconstruction of a digital signal to its analog counterpart in frequency domain. Note that if the PSF does not have the perfect bandwidth of f_s, artifacts will result. If the PSF is too wide, higher frequencies will leak in (shown in orange) causing a pixelization artifact. If the PSF is too narrow, higher frequencies will be cut off causing a blurring artifact.

recovered i.e. the image will be blurred as in low pass filters. These are indicated in Figure 4.22 via colored lines that mark the bandwidth of the reconstructed signal for each of these three different cases.

Note that most of the concepts in this chapter have been illustrated in 1D, however they provide perfect understanding for 2D image phenomenon. Ponder over this deeply and carry this skill forward. DFTs can be extended to higher dimension and they are surprisingly widespread in their use. However, understanding 1D DFT goes a long way in understanding DFT in higher dimensions.

4.6 Duality

The most effective aspect of DFT is its *duality* which has been proved theoretically, but here we would only explore the concept without proof. Intuitively, you can see this by going back and examining the DFT synthesis and analysis equations (Equation 4.3 and 4.5) and appreciating the striking similarity between them. This duality gives rise to several interesting properties that become very

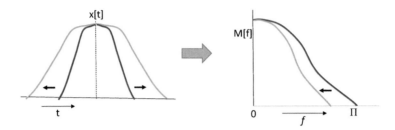

Figure 4.23. This shows the duality in expansion and compression of spatial domain signals that lead to the compression and expansion of their frequency domain counter parts.

handy tools in analyzing signals. With all the background that you have in DFT now, this should become a very intuitive now and let us explore this more in this section. However, note that when we are talking about duality we consider only amplitude and not the phase. As such, we saw earlier that phase information is not studied in frequency domain much. Also, since time and spatial parameters are common independent variables for 1D and 2D respectively, they are often referred to as time domain and spatial domain functions respectively. But the terms are used interchangeably since they both refer to the primal space. The dual space is given by the frequency domain representation.

First, let us consider a δ in the spatial domain. We know that it is the sharpest signal possible and therefore is created by assembling all the different cosine and sine waves – from the smoothest to the sharpest of them. Therefore, its frequency response is a constant. Now consider a signal which is constant in time domain. This means that it is of frequency zero and hence its amplitude frequency response is a spike at 0. This is called duality — the frequency response of a spike is a constant and that of a constant is a spike.

Next, let us consider a signal in spatial domain, as in Figure 4.23. Note that as the signal expands in the spatial domain, its frequency response compresses and vice versa. This can also be explained intuitively. When the signal compresses in the spatial domain, it gets sharper which indicates that the higher frequency increases. Therefore, the response in frequency domain expands. In fact, this is the basis of low pass filtering. Widening kernel indicates smaller range of frequency getting passed and therefore a more drastic low pass filtering.

Given this duality, let us define some Fourier pairs (Figure 4.24). These are specific functions and their DFT amplitude response in frequency domain. For example, the DFT of a Gaussian filter is a Gaussian. However, their widths are inversely related; i.e. if the width of the spatial domain Gaussian increases, the width of its frequency response decreases. This is a Fourier pair. Similarly, the DFT of the box filter is a Sinc function given by $\frac{Sin[f]}{f}$, and that of a sinc function

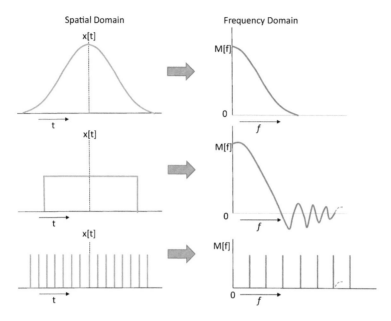

Figure 4.24. This figure shows three Fourier pairs from top to bottom. (a) A Gaussian is a Fourier pair of a Gaussian; (b) A box is a Fourier pair of a sinc and (c) a comb is a Fourier pair of a comb.

in the spatial domain is a box function, thus forming another Fourier pair. Note that the sinc is an infinite function that asymptotically approaches to zero with increasing f. Another dual function is a comb function. The frequency response of a comb function is also a comb function, only that their intervals are inversely related similar to the Gaussian function duality. In other words, if the spatial domain comb gets denser, its frequency response gets sparser and vice versa.

These Fourier pairs can give us a lot of insights. Recall from the low pass filtering discussion that the box filter is not the greatest low pass filter. Now the time has come to explain why. An ideal low pass filter will be a box filter in the frequency domain which would completely cut off frequencies above a certain threshold and retain the ones below the threshold. However, a box in frequency domain is a sinc in spatial domain. Sinc is an infinite function and no finite digital kernel can be developed for it. So, an ideal low pass filter is impossible to build. Further, the most commonly used low pass filter is a box filter (often achieved by averaging neighboring pixels in a square region of an image). However, note that the frequency response of a box is a sinc. Convolution by a box in spatial domain is a multiplication by sinc in the frequency domain. Since sinc is an infinite function, it means that by low pass filtering by a box, some

high frequencies will always remain in the filtered function due to multiplication
with an infinite function. Therefore, a box filter in reality is not ideal since it
leads to *leakage* of high frequency in the filtered signal. Practically, a Gaussian
filter offers us the best of both worlds since it has very little high frequency leaks.
However, the higher frequencies can get significantly attenuated. In fact, a box
multiplied by a Gaussian in spatial domain (sinc convolved with a Gaussian in
frequency domain) turns out to be one of the best low pass filters since it reduces
the attenuation of the higher frequencies without compromising the leakage of
high frequencies. More complex filters exist, offering different tradeoffs between
the amount of attenuation of the higher frequencies and the amount of leakage
of higher frequencies.

Fun Facts

Did you know that your ears do Fourier transform automatically! There are
little hairs (cilia) in you ears which vibrate at specific (and different) frequencies. When a wave enters your ear, the cilia will vibrate if the wavefunction
"contains" any component of the correponding frequency! Because of this,
you can distinguish sounds of various pitches!

In fact, Fourier transform is one of the most widely used mathematical
tools. It has been used to study the vibrations of submersible structures
interacting with fluids, to try to predict upcoming earthquakes, to identify
the ingredients of very distant galaxies, to search for new physics in the heat
remnants of the Big Bang, to uncover the structure of proteins from X-ray
diffraction patterns, to study the acoustics of musical instruments, to refine
models of the water cycle, to search for pulsars (spinning neutron stars), and
to understand the structure of molecules using nuclear magnetic resonance.
The Fourier transform has even been used to identify a counterfeit Jackson
Pollock painting by deciphering the chemicals in the paint. That is quite the
legacy for a little math trick.

4.7   Conclusion

Fourier analysis is a mathematical concept that can be applied to any function of any dimension. There are different flavors of Fourier transform based on
whether we are dealing with periodic signals or aperiodic signals, continuous or
digital signals. A large number of books has explored Fourier analysis from a
mathematical standpoint – [Tolstov 76, Spiegel 74, Morrison 94] and, in more
recent times, from the perspective of its application [Stein and Shakarchi 03, Folland 09, Kammler 08]. Other texts explore it from the context of signal pro-

cessing in the electrical engineering domain where it is mostly relevant for 1D signals [Smith 97, Proakis and Manolakis 06]. Few image processing books provide the insights of application of Fourier analysis to digital images. In this book, we have tried to provide you with this rare understanding of spectral analysis of images.

Bibliography

[Folland 09] Gerald B. Folland. *Fourier Analysis and Its Applications (Pure and Applied Undergraduate Texts)*. Pacific Grove, California and Wadsworth & Brooks/Cole Advanced Books and Software, 2009.

[Kammler 08] David W. Kammler. *A First Course in Fourier Analysis*. Cambridge University Press, 2008.

[Morrison 94] Norman Morrison. *Introduction to Fourier Analysis*. John Wiley and Sons, 1994.

[Proakis and Manolakis 06] John G. Proakis and Dimitris K Manolakis. *Digital Signal Processing*. Prentice Hall, 2006.

[Smith 97] Steven W. Smith. *The Scientist & Engineer's Guide to Digital Signal Processing*. California Technical Publishing, 1997.

[Spiegel 74] Murray Spiegel. *Schaum's Outline of Fourier Analysis with Applications to Boundary Value Problems*. Mcgraw Hill, 1974.

[Stein and Shakarchi 03] Elias M. Stein and Rami Shakarchi. *Fourier Analysis: An Introduction*. Princeton University Press, 2003.

[Tolstov 76] Georgio P. Tolstov. *Fourier Series (Dover Books on Mathematics)*. Prentice Hall, 1976.

Summary: Do you know these concepts?

- Discrete Fourier Transform

- Frequency Domain Response

- Spatial/Time Domain Response

- Aliasing

- Complex Conjugate

- Sampling and Reconstruction

- Periodicity of the Fourier Transform

Exercises

1. Consider an one dimensional signal x of length 16 where sample i is given by $x[i] = 2sin(\frac{\pi i}{4}) + 3cos(\frac{\pi i}{2}) + 4cos(\pi i) + 5$.

   (a) What is the length of each of the arrays x_c and x_s?

   (b) Write out the arrays x_c and x_s.

   (c) Convert the x_c and x_s representation to that of magnitude M and phase θ. Write out the arrays M and θ.

2. Consider the box filter in *spatial domain* for a low pass filter.

   (a) What is its frequency domain response?

   (b) Is the box filter an ideal low pass filter? Justify your answer.

   (c) Is a box filter in the *frequency domain* an ideal low pass filter? Justify your answer.

   (d) What is the frequency domain response of a Gaussian filter in the *spatial domain*?

   (e) How does it compare to a box filter in the *spatial domain* for low pass filtering? Justify your answer.

   (f) A multiplication of Gaussian and Sinc in the *spatial domain* is considered an ideal low pass filter. Express analytically the frequency domain response of this filter.

   (g) How does this filter compare with the Gaussian filter in the spatial domain? Justify your answer. (Hint: Use pictures of the frequency domain response to identify pros and cons)

3. (a) shows you a picture of Goofy. This was smoothed only in the horizontal direction to create the image (b). Consider the two amplitude responses in (c) and (d). One of them belongs to (a) and the other to (b). Match them and justify your answer.

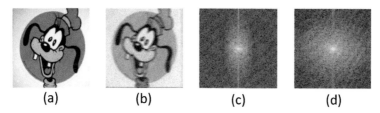

(a)   (b)   (c)   (d)

4. You want to digitize an analog signal of bandwidth 120Hz. The sampling frequency of your display is 100 Hz. The bandwidth of your reconstruction kernel is 80 Hz.

    (a) Why wont you be able to sample and reconstruct this signal without artifacts using this display?

    (b) How would you process the image to reconstruct it without any artifacts?

    (c) What kind of artifacts would the reconstruction kernel generate?

    (d) How would you change the reconstruction kernel to correct it?

5. (a) and (b) show two images. Consider the two amplitude responses in (c) and (d). One of them belongs to (a) and the other to (b). Match them and justify your answer.

     (a)   (b)   (c)   (d)

6. Match the images on the left with their amplitude responses on the right. Justify your answers.

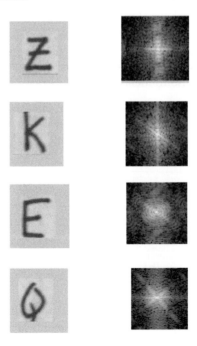

7. How would you get rid of the shadows of the bars in the left image to get the right image below?

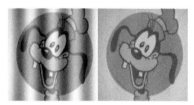

8. The amplitude response of the left image below is given by the right image which does not seem to suffer from the effects of periodicity. Justify this phenomenon.

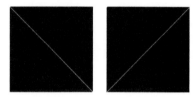

9. (a) and (b) show two images. Consider the two amplitude responses in (c) and (d). One of them belongs to (a) and the other to (b). Match them and justify your answer.

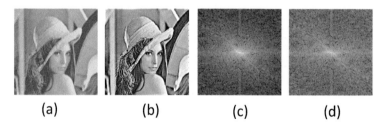

    (a)   (b)   (c)   (d)

10. Consider a filter in the *spatial domain* formed by the multiplication of a Gaussian and a box filter. What is the *frequency domain* response of this filter? Is this response sharper or smoother than a sinc? Justify your answer.

5

Feature Detection

Features of an image are defined as regions where some unusual action happens. For example, the brightness changes drastically to form an edge; or, gradients of edges change drastically to form corner. In this chapter, we will see how some of these features can be detected using convolution, which is a linear filtering process. Therefore, the properties of scale, shift invariance, and additivity are true for these filters. However, later in the chapter we will see that some features can only be detected via more complex non-linear filters.

Figure 5.1. This figure shows the salient features (right) detected by humans from an image of an object (left). We will see in this chapter that such clean and crisp detection of features is still not possible by computers.

5.1   Edge Detection

Edges have a special importance in our perception. Several cells in the human cortex have been found to be specialized for edge detection. Edges help us understand several aspects of an object including texture, lighting and shape. In this section, we will see how we can detect edges in an image using computers.

Figure 5.1 shows the kind of edge detection we would like to have. In the rest of this section we will explore algorithms to achieve this goal and see how close we can get to it. As humans it is very easy for us to perform this task. However, we will see in the following sections that the job is not that easy for computers and developing algorithms that can achieve similar performance is rather difficult.

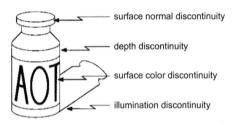

Edges can have multiple origins (Figure 5.2). They can be caused by discontinuity in depth or surface normals. They can be caused by a stark change in color or illumination. We will perform edge detection using the following steps. (a) Find all the pixels that are part of an edge. These are called *edgels*. We would take an image as input and create a binary im-

Figure 5.2. This shows the different kinds of edges and the causes of their origin for a simple figure of a bottle.

age as output where all the edgels will be marked as white and the non-edgels as black. We may also output a gray scale image where the color value will denote the strength of the edge. Black would indicate no edge at all. (b)Next, we will use methods to aggregate this edgels into longer edges, sometimes using a compact parametric representation.

5.1.1   Edgel Detectors

In this section, we will discuss different methods to detect edgels, their advantages and disadvantages. Finally, we will explore methods to detect edges at different levels of details.

Gradient Based Detectors Let us now take a close look at the edges in Figure 5.2. Despite having different origins they have one feature in common. Edges are formed when there is a drastic change in the brightness over one or more pixels as shown in Figure 5.3 creating a roof, ramp or step edge. Therefore, if we consider the image to be a function, a sharp change in its value causes an edgel. Change is measured by the first derivative. Therefore, a pixel is an edgel if the magnitude of the first derivative of the image at that pixel is large. Let us

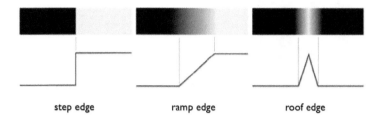

Figure 5.3. This shows edges in 1D functions. There can be three types of edges: step edge, ramp edge and roof edge.

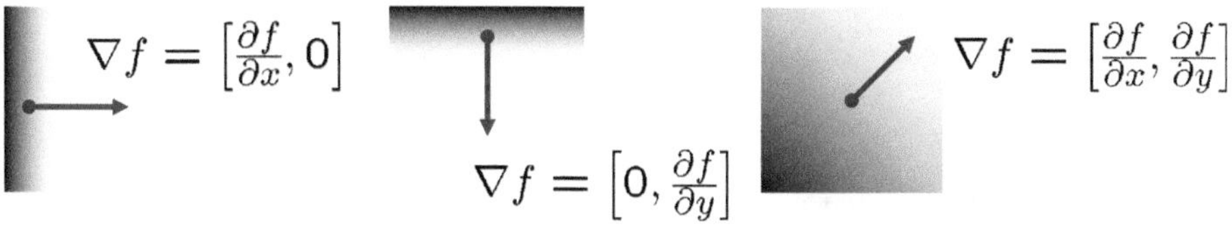

$$\nabla f = \left[\frac{\partial f}{\partial x}, 0\right]$$

$$\nabla f = \left[0, \frac{\partial f}{\partial y}\right]$$

$$\nabla f = \left[\frac{\partial f}{\partial x}, \frac{\partial f}{\partial y}\right]$$

Figure 5.4. This figure shows how the partial derivatives relate to the kind of edges we see in an image.

denote the image by a 2D function f. So, the gradient of f has a direction. We can compute gradient at a pixel in x-direction and y-direction separately and then combine them to get the gradient of the image at that point as follows.

$$\nabla f = \left(\frac{\partial f}{\partial x}, \frac{\partial f}{\partial y}\right) = (g_x, g_y) \tag{5.1}$$

The gradient points in the direction of most rapid change as shown in Figure 5.4. In fact, we can quantify the *direction* and *strength* of the edge respectively by

$$\theta = \tan^{-1}\frac{g_y}{g_x} \tag{5.2}$$

$$||\nabla f|| = \sqrt{g_x^2 + g_y^2} \tag{5.3}$$

Therefore, now that we know how to detect edges, the next question is how we evaluate the partial derivatives in the digital domain. For this, we use the method of *finite differences*. The difference between the pixel's function value and that of its right (or left) neighbor gives g_x at that pixel. Similarly, the difference between its value and that of its top (or bottom) neighbor gives g_y.

$$g_x = f(x+1, y) - f(x, y) \tag{5.4}$$
$$g_y = f(x, y+1) - f(x, y) \tag{5.5}$$

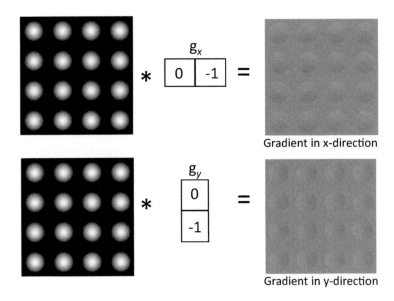

Gradient in x-direction

Gradient in y-direction

Figure 5.5. This shows the gradient function represented as kernels or filters which are then convolved with the image to give two images – the gradient in x-direction (top) and y-direction (bottom) at any pixel.

Figure 5.6. Left: The top row shows the same image with noise increased as we go from left to right. The bottom row shows the effect of applying a finite different x-gradient filter to the corresponding images on the top row. These images get grainier and grainier as the noise increases. Right:This shows the Sobel gradient operator for x and y direction.

In fact, we can express this as a kernel or filter as shown in Figure 5.5 and when an image is convolved with each of these filters, we get two gradient images - one each for x and y directions. Once the gradient is detected we identify a pixel as an edgel if the strength of the edge (Equation 5.3) is above a certain value. The value we choose is called a threshold and is a parameter to the edgel detection process. This process of generating a binary image by choosing a threshold is called *thresholding*. Following the thresholding, instead of generating a binary image, one can also generate an image in which the egdes are marked by different

Figure 5.7. From left to right: The original image, the gradient image obtained after applying the Sobel detector where the gray values denote the strength of the edges, the edgel binary image after thresholding with gray value of 64 and 96. Note fewer edges exist for larger thresholding value.

Figure 5.8. This figure shows the effect of noise on edge detection using the Sobel operator followed by thresholding with gray value of 150. From left to right: The original image, the edge detected image, the original image after noise is removed, the edge detected image.

gray levels based on the direction or strength of the edges. Therefore, the gray value would encode θ or $||\nabla f||$.

The above finite difference gradients are the simplest operator possible and do not take noise in the image into account. Noise has its origin in the physical device and can be modeled by a random value added to every pixel of an image. Noise shows up as graininess in the image and if sufficiently high, it can increase error of most image processing algorithms including edge detection. Figure 5.6 shows the effect of noise in finding the finite difference to generate the gradient images. The image becomes more and more grainy as the noise increases which indicates that the thresholding will start responding to noise rather than the content of the image. One way to make these filters more robust to noise would be to consider a bigger neighborhood. Several such filters exist and the one that is found to be very robust is the Sobel operator (Figure 5.6). The scale value of

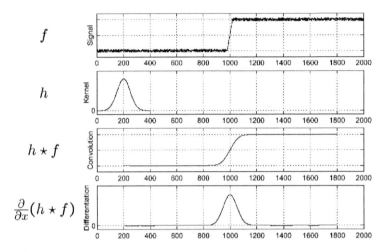

Figure 5.9. This shows the noisy function (f) first smoothed using a Gaussian filter (h) to give $h \star f$ which is then convolved with $g_x = \frac{\partial}{\partial x}$ giving $\frac{\partial}{\partial x} (h \star f)$.

$\frac{1}{8}$ does not make any difference in the edgel detection since the threshold can be adjusted to accommodate for it. It is there to normalize the gradient value. Figure 5.7 shows the results of the Sobel operator on an image where the gray values at the pixels indicate magnitude of the gradient at that pixel.

Nevertheless, noise poses a serious issue in edge detection and if it goes beyond a certain level, no operator can really achieve an accurate edgel detection (Figure 5.8). So, it is important to handle noise. The most common way to do it is to low pass filter the image first so that the noise is smoothed out. A Gaussian filter is usually used for this purpose due to the property of reducing high frequency leaks. The edge operator is then applied on the smoothed signal to achieve the edgel detection. This concept is explain pictorially using 1D signals in Figure 5.9.

However, the operation of smoothing and the gradient filter can be achieved in one filter. It can be showed that

$$\frac{\partial}{\partial x} (h \star f) = \frac{\partial h}{\partial x} \star f. \tag{5.6}$$

This means that the effect of applying a gradient filter ($\frac{\partial}{\partial x}$) to a function f convolved with a low pass filter h is the same as convolving f with a new filter formed by applying the gradient filter to the low pass filter ($\frac{\partial h}{\partial x}$). Therefore, we can take the derivative of the Gaussian low pass filter to create a single filter which can then be convolved with the image to provide our gradient images. This is explained in the Figure 5.10.

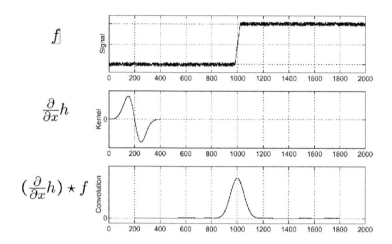

Figure 5.10. This shows the noisy function (f) convolved with the derivative of the Gaussian, $\frac{\partial h}{\partial x}$, providing $g_x = \frac{\partial h}{\partial x} \star f$.

Curvature Based Detectors The gradient based operators work well but they suffer from two problems. First, for step or ramp edges (brightness that smoothly changes from one level to another to create the edge rather than a sharp change from one level to another and back to the previous level), they have poor localization (i.e. precise location of the edge) after thresholding which becomes especially evident at thicker edges.

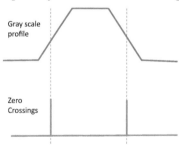

This is due to the fact that a step or ramp edge can trigger response on multiple adjacent pixels based on how smooth the step or ramp is. Second, the response of the gradient based operators is different for different direction edges. As a result, the gradient based operators can miss edges based on the thresholding value since the value favors some directions over others.

Curvature based edge detectors alleviate

Figure 5.11. This shows the zero crossing in the 2nd derivative for a feature edge in 1D.

these problems. The basic idea behind these operators lies in the realization that edges occur when the first derivative (or gradient) of an image reaches a maxima or a minima. This means at these points the second derivative or curvature should have a zero-crossing (Figure 5.11). The goal of the curvature based operators is to detect these zero crossings to find the precise location of the edge and mark these pixels. This edge detector is often called the

Marr-Hildreth edge detector based on the name of scientists who first proposed it. The advantage of this detector is that it responds similarly to all different direction edges and finds the correct positions of the edges.

The curvature of any 2D function is given by the sum of the directional curvatures in two orthogonal directions as

$$\nabla^2 f = \frac{\partial^2 f}{\partial x^2} + \frac{\partial^2 f}{\partial y^2} \tag{5.7}$$

0	-1	0
-1	4	-1
0	-1	0

-1	-1	-1
-1	8	-1
-1	-1	-1

Figure 5.12. The Laplacian operator considering only four perpendicular neighbors (left) and considering all eight neighbors (right).

The curvature at a pixel can be computed using finite differences for digital data. Curvature at a pixel (x, y) in x-direction is computed by the difference between the gradient g_x and g_{x+1} at (x, y) and $(x+1, y)$ respectively. Therefore, curvature in this direction is given by

$$g_x - g_{x-1} = f(x+1, y) - f(x) - (f(x-1, y) - f(x, y)) \tag{5.8}$$
$$= f(x+1, y) - 2f(x) + f(x-1, y) \tag{5.9}$$

If we consider the same finite differences gradient in the vertical direction and add it to the above formula, the curvature operator will involve the pixel itself and its 4-connected neighbors to the left, right, top and bottom. Similarly, if the diagonal neighbors are also considered, the finite difference formula for curvature involves all the eight neighbors around a pixel, referred to as the 8-connected neighbors. The filters thus formed as shown in Figure 5.12 are called the Laplacian operators. Once the image is convolved for Laplacian operator, pixels at which a *zero crossing* occurs are marked as edge pixels. In order to detect the zero crossings, first the convolved image is thresholded to retain values

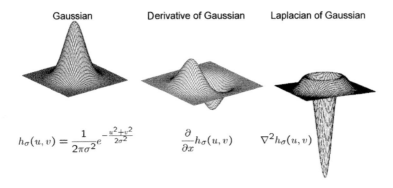

Gaussian   Derivative of Gaussian   Laplacian of Gaussian

$$h_\sigma(u, v) = \frac{1}{2\pi\sigma^2} e^{-\frac{u^2+v^2}{2\sigma^2}} \qquad \frac{\partial}{\partial x} h_\sigma(u, v) \qquad \nabla^2 h_\sigma(u, v)$$

Figure 5.13. This figure shows a 2D Gaussian filter, its derivative and a LoG filter along with their analog equations. Sampling these functions would generate the filters we have been discussing for edge detection.

Figure 5.14. The original image (left) and the edges detected by the curvature based detector (middle) and canny edge detector (right).

near the zero. Then the neighborhood of every marked pixel is examined to see if both positive and negative values exist to indicate a zero crossing.

However, in this case also, before the Laplacian operator is applied to the image, a convolution with a Gaussian is required to reduce the noise level. The effect of a convolution with a Gaussian followed by a convolution with a Laplacian operator on an image is equivalent to a single convolution of the image with a kernel that is a Gaussian function convolved with a Laplacian operator. This is called the *Laplacian of Gaussian* or LoG operator (Figure 5.13). However, this removes the option of two different sized operators for Gaussian and Laplacian.

Often the same effect of a LoG operator can be achieved by subtracting a delta function from a Gaussian filter (this is different than a high pass filter which is a Gaussian subtracted from delta). This is because the shape of this filter, as you can probably intuitively visualize, is very close to the LoG filter. Figure 5.14 shows the result of the curvature based detector on an image. However, the apparent advantage of a curvature based detector in finding the exact location of the edge becomes its undoing for feature edges (Figure 5.15). Feature edges are caused by the intensity moving from one level to another and then coming back close to the original (instead of a ramp edge where intensity changes just from one level to another). For feature edges, two zero crossings are detected for the same edge as shown in Figure 5.15. This phenomenon can often lead to spurious edges. Thresholding parameters can also allow some edges to get undetected.

These are called missed edges. You can see this easily in 5.14 where most edges have a ghost and there are several spurious edges.

Canny Edge Detector   The Canny edge detector tries to alleviate all the various problems of the gradient and curvature based detectors. Canny first formalized a set of properties an optimal edge detector should have and then worked towards a method that satisfies these properties. According to Canny, an optimal edge detector should have the properties of *good detection*, *good localization* and *minimal response*. Good detection says that the filter should respond to edges only and not noise. Therefore, edges should be found with the minimum of spurious edges. Good localization means that the detected edge is near the true edge. Finally, minimal response says that the edge's exact location is marked with a single point response. There is a tradeoff to be achieved between these different goals. In any real image, noise will play a role. Smoothing or low pass filtering will improve the detection at the cost of localization and minimal response. In fact, we see that both the methods we have discussed so far suffer from both inaccurate detection and localization.

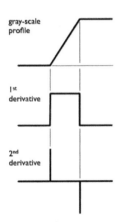

gray-scale profile

1st derivative

2nd derivative

Based on the above goals, Canny proposed a four step method: (a)Suppress noise using low pass filtering; (b) Compute gradient magnitude and direction images; (c) Apply non-maxima suppression to the gradient magnitude image; (d) Use hysteresis and connectivity analysis to detect edges.

In the first stage, the image is low pass filtered using a Gaussian filter. Next, the gradient strength and direction are calculated using the standard Sobel operator. At this point, we have achieved the property of good detection and localization, but the property of minimal response is still not ensured. In order to ensure minimal response, we apply the technique of non-maxima suppression. The exact location of the edge occurs wherever the gradient reaches a maxima.

Figure 5.15. This shows that the zero crossing may not happen at a single pixel location to provide a good edge localization.

This implies that if the strength of the gradient does not achieve a maxima at any of the marked pixels, it should be suppressed. Every marked pixel p can be part of an edge which can have one of four or eight orientations depending on 4 or 8 connectedness of the neighborhoods considered. The computed gradient p is binned to one of these values. Next the two neighbors with gradients in the same bin are considered. If the magnitude of their gradient is larger than that of p, the gradient at p is suppressed (made zero). Applying this operation to every marked pixel achieves the minimal response with each edge being detected by a single pixel.

Put a Face to the Name

David Courtney Marr (1945-1980) was a British neuroscientist and psychologist who was considered instrumental in the resurgence of interest in the discipline of computational neuroscience. He integrated results from psychology, artifical intelligence and neurophysiology to build new visual processing models. One of the great examples of his visionary contributions is the Marr-Heldrith edge detector which was designed with his student Ellen Heldrith. Marr and Heldrith modeled the edge detection operation in the human brain to be carried on by adjacently located minima and maxima detector cells whose response is then combined with a logical AND operation by closely located zero detector cells. The Marr-Heldrith operator that performs edge detection following this model was first proposed in 1980. Interestingly, scientists Hubel and Weisl found the existence of all these predicted cells in our visual pathways, though this happened after Marr's death. A process called lateral inhibition in the ganglion cells performs the convolution. Cells called simple cells in the cortex respond to maxima and minima of the signals sent from ganglion cells. Finally, cells called the complex cells have been found near these simple cells that act like zero detectors. Therefore, this edge detector is one that behaves closest to our human brain. Marr's life was ended prematurely at the age of 35 due to leukemia. His findings are collected in the book *Vision: A Computational Investigation into the Human Representation and Processing of Visual Information*, which was finished mainly on 1979 summer, was published in 1982 after his death and re-printed in 2010 by The MIT Press. The Marr Prize, one of the most prestigious awards in computer vision, is named in his honor. Ellen Hildreth is currently a professor in Wellesley College and continues to study computer modeling of human vision.

Finally, hysteresis is used to avoid streaking, a phenomenon of breaking of the edge contour when the output of the previous step fluctuates above and below a single threshold. In hysteresis, two thresholds are defined, L and H where $L < H$. The gradient at any pixel being higher than H is detected to be a strong edge pixel and hence marked. The gradient being lower than L assumes a non-edge pixel and is not marked. Any pixel with a gradient value between L and H is considered to be a weak edge pixel and its candidacy to the set of edgels is additionally evaluated using connectivity analysis. If a weak edge pixel is connected to at least one strong edge pixel, then it is considered to be continuation of a strong edge and hence marked. This removes spurious short edges under the assumption that edges are long lines. Figure 5.16 shows the

Figure 5.16. The figure shows the different steps of Canny edge detector. From left to right: The original image, the gradient magnitude image, the image after non-maximum suppression, the final image after applying hysteresis.

different steps of this process. Figure 5.14 compares the Canny edge detector with the Marr-Heldrith curvature based edge detector.

Put a Face to the Name

John F. Canny is an Australian scientist who currently is a professor in the EECS department of UC-Berkeley. He is known for designing the most effective edge detector for which he received the ACM Doctoral Dissertation and Machtey award in 1987. He also received the 2002 American Association for Artificial Intelligence Classic Paper award for the most influential paper from a 1983 national conference on artificial intelligence. He is known for his seminal contributions in robotics and human perception.

However, as a note of caution, every edge detector is dependent on several parameters starting from the size of the Gaussian filter used (determined often by the amount of noise), the thresholds chosen and the 4 or 8 connectivity of the neighborhood considered. Changing these parameters differently can lead to very different results. This is illustrated in Figure 5.18. That is the reason, it is important to take the results shown in this chapter with a pinch of salt. The results have been generated with efforts to get the parameters as comparable as possible and yet to extract the best out of each method. Therefore, we believe they do highlight the pros and cons of the methods fairly, but one can indeed tweak the parameters to achieve results close to what they specifically desire.

Figure 5.17. The figure shows the effect of multi-resolution edge detection on two different images. The images are low pass filtered using wider and wider Gaussian kernels (in a clockwise fashion for the wheel image and from left to right for the sculpture image) and the edge detection performed following that. As the kernel gets wider, finer edges disappear while the larger ones remain.

5.1.2 Multi-Resolution Edge Detection

Edges in an image can have different resolution, i.e. how sharp or smooth the edge is. Changes in intensity that occur over a larger space (i.e. smoother changes in the image intensity) form low-resolution edges, while intensity changes over a smaller space constitute low resolution edges. Perceptually, lower resolution edges are more important than the higher resolution ones in detecting objects, illumination and their interaction. However, a low resolution edge formed by a very gradual and slow ramp will be not be detected unless the image is low pass filtered to remove higher frequencies to enable resampling using a much smaller number of pixels where the same edge shows up as a step or a much sharper ramp and gets easily detected. This effect is shown in Figure 5.17. The image is low pass filtered with widening kernel indicating lower band of frequencies passed and the edge detection is applied on these low pass filtered images. This is what we call *multi-resolution* edge detection.

Incidentally, when discussing different edge detectors, we have discussed the role of low pass filtering for noise removal. However, this is not the only motivation behind using low pass filtering or smoothing. Its greater use lies in multi-resolution edge detection where a pyramid of edge images is created where each level of the pyramid provided edges of a particular resolution. As the levels increase, the low pass filter kernel is increased in size to create progressively low pass images (as in a Gaussian pyramid). As the images get more and more blurry higher up in the pyramid, higher resolution edges (finer scale edges) disappear.

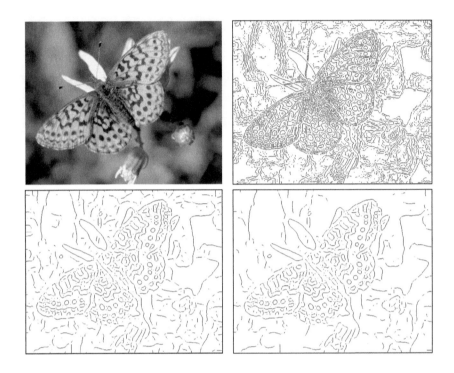

Figure 5.18. The figure shows the effect scale versus threshold. The images on the top right and bottom left are the results of applying edge detection applied on top left image at finer and coarser scale respectively. However, the two images on the bottom are at the same scale but the right one is the result of having a higher thresholding parameter.

The lower resolution (i.e. coarser scale) edges remain across all the different levels. These are the edges that are perceptually most salient and contributes more significantly to our perception. However, different kinds of edges can be detected by changing the parameters of resolution and thresholding as shown in Figure 5.18.

The next question is how easy is it to find the corresponding edges across the different levels of the multi-resolution edge pyramid? As it turns out, this is more complex than you think primarily due to the fact that the same edge can be localized at slightly different places in different levels of the pyramids. However, we humans achieve it with rather uncanny accuracy. You can convince yourself by staring at the images for a few seconds and you can easily detect this correspondence at least for the larger scale edges. A seminal work by Witkin

Figure 5.19. The figure shows the effect of local edge aggregation. Left: The original image; Right: The image with edgels detected by Canny edge detector which are then linked by local aggregation. Each linked edge is colored with a different color, not necessarily unique.

shows that the multi-resolution operation of edge detection occurs in our brain with continuous levels rather than discrete levels. The characteristics of edges across these continuous levels follow some very predictable patterns. The most important of these patterns are: (a) The edge position may shift as the scale becomes coarser but the shift will not be drastic or discontinuous; (b) two edges can merge with coarser scales but can never split. These patterns are exploited to find the correspondences by humans. Since generating close to continuous scales of edges is almost impossible in computers, this automatic detecting of corresponding edges across different levels still remains as a challenge.

5.1.3 Aggregating Edgels

Edge detectors produce edgels which lie on a edge. The next step is to collect this edgels together to create a set of longer edges. This may seem to be trivially achieved by just tracing the edges starting from a pixel. But this is only true in an ideal case. As you see in the examples shown in the previous section, edgels are not perfectly detected. Some parts of an edge may be missing or some small edges may appear to be present in a place where there are no edges in reality. Therefore aggregating edgels turn out to be more complex than naive. There are two types of aggregation method. The first applies local edge linkers to trace out longer edges while the latter uses global edge linkers to classify multiple edgels to belong to a single edge.

Path Tracing Via Local Aggregation Almost all edge detectors yield information about the magnitude and direction at an edgel during the process of detecting the edgel. Local edge linking methods usually start at some arbitrary edge point and consider and add those pixels from a local neighborhoods to the edge set whose edge direction and magnitude are similar to each other. The basic premise is that neighboring edgels with similar properties are likely to lie on the same edge. The neighbourhoods based around the recently added edgels are then considered in turn and so on. If the edgels do not satisfy the constraint then we conclude we are at the end of the edge, and so the process stops. A new starting edge point is found which does not belong to any edge set found so far, and the process is repeated. The algorithm terminates when all edgels have been linked to one edge or at least have been considered for linking once.

Thus the basic process used by local edge linkers is that of tracking and traversing a sequence of edgels. Branching edges are considered in a breadth or depth first fashion just as in tree traversal. An advantage of local aggregation methods is that they can readily be used to find arbitrary curves. Probabilistic methods can also be applied to achieve better estimates by global relaxation labeling techniques. An example of linked edges is shown in Figure 5.19 where each linked image is shown with a different, but not always unique, color.

Global Aggregation Via Hough Transform A complementary approach to edge linking is to identify parametric edges (e.g. lines and parametric curves like circles, parabolas) in an image so that we not only identify edges but also have a more compact representation of them which can be used for other purposes such as finding how an image was scaled or rotated to create another image). Such a representation is also called a vector representation of edges/images. The most popular way to compute this vector representation is using a voting based method called *Hough transform*.

To understand this, let us assume that we would like to find lines in the edgel image. Let us now consider the set of all the different lines that can be present in the image. Only small subset of these lines, some of which are shown in blue in Figure 5.20 can pass through the point (x, y). Also, a much larger subset of lines, some of which are shown in red in Figure 5.20 will not pass through (x, y). The first step of the Hough transform is to find the set of lines that would pass through (x, y). A line passing through (x, y) has the equation $y = mx + b$ where m is the slope and b is the y-intercept of the line. The set of all values of m and b that satisfy this equation for a given coordinate (x, y) defines the set of all lines that pass through (x, y).

Let us now consider an alternate 2D space spanned by m and b. Each line in the image space that passes through (x, y) will be defined by a specific slope and offset and therefore will be denoted by the point in the space spanned by (m, b). For example, the x-axis is a line with slope and offset both zero. Therefore, in the (m, b) space, it will be represented by the origin, $(0, 0)$. Now the line equation

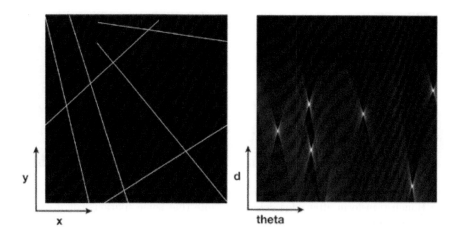

Figure 5.20. These plots explain the dual spaces. Left: An image space (x, y) with different lines detected as edgels. Right: The corresponding Hough space after voting. Note that the number of maximas are the same as the number of lines present in the image.

can be written as

$$b = y - mx \qquad (5.10)$$

Therefore, for a known (x, y) this will span a straight line in the (m, b) space defining the set of all lines that pass through the point (x, y). In other words, the (m, b) space, called the Hough space based on its inventor, is a dual of the (x, y) space because a point in the (x, y) space denotes a line in the (m, b) space and vice versa.

To identify a parametric edge we will do the following. For each detected edgel (x, y), we will vote on all the lines that would pass through that edgel (m, b) space $--$ this will be just a line given by $-b = mx - y$. If a line $y = mx + b$ is actually present in the edge image, all the edgels on it will vote for the same point (m, b) in the (m,b) space. Therefore, presence of an edge corresponds to a high number of votes for that (m, b) location. Once all the edgels have voted, we can detect the maximas in the (m, b) space to find the slope and offset and therefore the parametric equation of the detected lines. However the vertical lines have infinite m which makes it difficult to handle them in (m, b) space. Therefore, we use a polar notation for the lines where the lines are represented defined based on their distance from the origin (d) and the angle made with the x-axis (θ). Therefore, the Hough space is defined by (d, θ) rather than (m, b). The image space and its dual Hough space are shown for a simple edgel image in Figure 5.21. We visualize the Hough space using a gray scale image. There are five maximas (shown by the white bright spots) denoting the five lines in the

Figure 5.21. This figure shows the straight lines detected (right) in an image (left) using Hough transform after edge detection. Note that some spurious edges can also be detected (as shown in the left) due to inaccuracies resulting from thresholding.

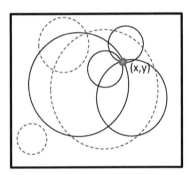

Figure 5.22. This shows an edgel (x, y). The blue lines denote some of the possible circles that can pass through (x, y). Some of the circles that do not pass through (x, y) are shown in red.

image. We also show the results of finding lines using Hough transform.

Similar techniques can be used to detect other parametric entities like circles and parabolas. Lets us consider briefly one such case. Let us consider the circle whose equation is given by $(x - c_x)^2 + (y - c_y)^2 = r^2$ where (c_x, c_y) is its center and r its radius. Therefore, the Hough space is a 3D space spanned by (c_x, c_y, r). Therefore, binning and counting of the votes cast and maximas have to be detected in this 3D space to find the circles in the image. Also $r = \sqrt{(x - c_x)^2 + (y - c_y)^2}$ is a conic. Therefore, a point in the image space corresponds to a conic in the Hough space.

5.2 Feature Detection

Feature, in general, refers to a pixel or a set of pixels that stand out from their surrounding. So far in this chapter, we have focused on the special feature of edges. Though features can often be formed by intersection of multiple lines that can be identified via edge detectors, in this section we will explore some non-linear operators for feature detection in general. The first of these we will discuss is called the Morovac operator and it measures the self-similarity of an image near a point. So, what does self similarity mean? If you consider a pixel (x, y), self-similarity defines how similar are the patches that are largely overlapping (x, y). Most of the pixels in an image have high self-similarity. Pixels at an edge are not similar in the direction perpendicular to the edge. And corners are not similar in any direction. In fact, a general feature is a point of interest where neighboring patches overlapping the pixel have a high degree of variance.

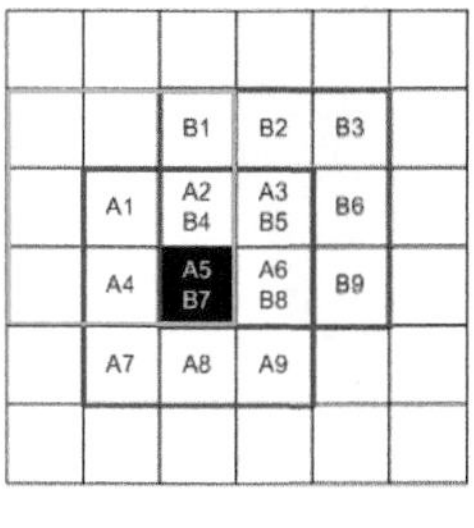

The next question is how do we compute self-similarity. Let us show an example in Figure 5.23. Here we are computing the self similarity of $A5$. Let us call the 3×3 neighborhood around the pixel A5, which includes pixels $A_1...A_9$, as patch A. Let us also consider other patches of size 3×3 that overlap with A. Let one such overlapping neighboring patch be B. Nine such neighboring patches exist, another one of which is shown in green. The similarity of A with B is defined as

$$S_{AB} = \sum_{i=1}^{n}(A_i - B_i)^2 \qquad (5.11)$$

Figure 5.23. This shows the self-similarity operator. A_5 belongs to a 3×3 patch shown in red. Two other neighboring patches are shown in blue and green. The corresponding pixels in A and B whose squared differences measure self-similarity is also indicated.

If we add up the similarity of all the nine neighboring patches of A and sum them together, it will provide an estimate of how similar A is to all its neighboring patches. If we likewise calculate the self-similarity for every pixel in an image, a maxima in this self-similarity image is a corner.

However, the Morovac operator has some limitations. If a one pixel noise is present, the Morovac filter will respond to that. It will also be triggered for an edge. Further, this filter is not *isotropic*. What this means is that the classification of pixels will change if the image is rotated. Therefore, this operator is not rotationally invariant as shown in Figure 5.24.

To alleviate this problem, the Harris and Stephens-Plessey corner detector was proposed. This starts with first generating the gradient images g_x and g_y. Next, for every pixel (u, v) the geometry of the surface near the pixel is defined by the matrix

$$A = \sum_u \sum_v w(u,v) \begin{pmatrix} g_x(u,v)^2 & g_x(u,v)g_y(u,v) \\ g_x(u,v)g_y(u,v) & g_y(u,v)^2 \end{pmatrix} \qquad (5.12)$$

where $w(u,v)$ is a weight that decreases with distance from (u,v). The two eigenvalues, λ_1 and λ_2, of this matrix are proportional to the principal curvatures at (u,v). If the magnitude of both are small, no feature exists at (u,v). If one of them has a large magnitude, an edge exists at (u,v). And only if both eigenvalues are large a corner is detected at (u,v). This is illustrated in Figure 5.26. Interestingly, it can be theoretically proved that if w is a Gaussian then this corner detector is isotropic - i.e. rotationally invariant. Figure 5.25 shows the results of this corner detector.

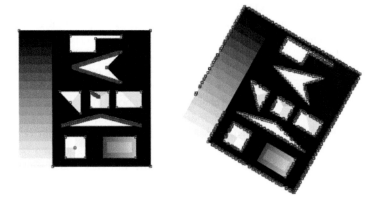

Figure 5.24. This shows the features detected by the Morovac filter (in red) as the image orientation is changed by rotating it by 30 degrees. Vastly different pixels are now detected as corners.

Figure 5.25. This shows the results of the Harris Stephens and Plessey corner detector.

5.3   Other Non-Linear Filters

Let us now discuss a few non-linear filters which are not used for feature detection. This will give you an idea of how such filters can be used for other domains as well.

For this we will first explore a filter called the *median filter*. This filter is very similar to the linear filter which is the *mean* or *box* filter. In a box filter, the mean of all the values in the neighborhood of a pixel is used to replace the value of the pixel. This effectively achieves a smoothing of the function and is often used for reducing noise. In a median filter, the pixel is replaced by the median (instead of mean) of all the values in its neighborhood. Using the median makes

Figure 5.27. This figure illustrates the effect of a median filter. From left to right: the original image; the original image with salt and pepper noise added to it; the noisy image processed using a box or mean filter; the noisy image processed using the median filter to provide an image almost identical to the original image.

the filter a non-linear filter.

A median filter is used to remove outliers, i.e. values that are drastically different from their neighborhood. They can be due to different device limitations in different applications. For example, a dead pixel in a camera can cause the value of that pixel to be a one or a zero at all times which will turn out to be an outlier. In the case of images, such outliers define a noise which is often called the *salt and pepper noise* – pixels turned either black or white due to system issues. Unlike a median filter, a mean filter reduces a Gaussian noise effectively, but does not work well for salt and pepper noise since the mean tends to spread the contribution from the very localized salt and pepper locations to their neighborhood. However, a median filter works much better since the median is usually unaffected by variation in the values in the neighborhood. Figure 5.27 shows an example.

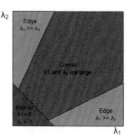

Figure 5.26. This shows how a corner is detected at a pixel in a Harris and Stephens-Plessey corner detector based on the magnitude of the two eigenvalues λ_1 and λ_2.

The median filter is in fact a specific type of a more general type of non-linear filters called order statistics filters. For example, instead of the median, we can replace the pixel with the value of the minimum or maximum value in its neighborhood. These are called minimum and maximum filter the respectively. For regular images (not having salt and pepper noise), these achieve the results of morphological operators of erosion and dilation. Erosion is the process of suppressing higher values thereby darkening the image, while dilation is the process of growing regions of higher values thereby brightening the image. These are illustrated in Figure 5.28. They form the building blocks of a set of image processing operations called morphological operators.

Figure 5.28. This figure illustrates the effect of the minimum (middle) and maximum (right) filter on an image (left) to achieve the effects of erosion and dilation respectively.

5.4   Conclusion

Feature detection is considered to be part of low level processing in human vision. In this chapter we have discussed the basic techniques to simulate human vision which have been combined to provide a much more sophisticated feature detector, relatively more popular of which is called the Scale Invariant Feature Transform (SIFT) [Lowe 04]. Such low level feature detection processes then become important for image segmentation and identifying objects often aided heavily by prior knowledge learned. To learn more about these advanced steps of computer vision, please refer to [Forsyth and Ponce 11, Prince 12].

Bibliography

[Forsyth and Ponce 11] David A. Forsyth and Jean Ponce. *Computer Vision: A Modern Approach.* Pearson, 2011.

[Lowe 04] David G. Lowe. "Distinctive Image Features from Scale-Invariant Keypoints." *International Journal Computer Vision* 60:2 (2004), 91–110.

[Prince 12] Simon J. D. Prince. *Computer Vision: Models, Learning, and Inference.* Cambridge University Press, 2012.

Summary: Do you know these concepts?

- Edge Detection
- Sobel Operator
- Laplacian Operator
- Canny Edge Detector
- Multi-resolution Edge Detection
- Morovac Operator
- Corner Detection
- Median Filter
- Erosion and Dilation

Exercises

1. You would like to detect edges in an image. You can use a curvature based method C or a gradient based method G.

   (a) Would using C require using a single or multiple convolution operations? Justify your answer.

   (b) Would using G require using a single or multiple convolution operations? Justify your answer.

   (c) Edge detector filters usually combine a low pass filter with a curvature or gradient filter. Why?

   (d) How does the width of this low pass filter affect the resolution of the edges you would detect?

2. In gradient-based edge detection algorithms, a gradient is approximated by a difference. Three such difference operations are shown below. This difference can be viewed as a convolution of $f(x, y)$ with some impulse response of a filter $h(x, y)$. Determine $h(x, y)$ for each of the following difference operators.

   (a)

$$f(x, y) - f(x - 1, y)$$

   (b)

$$f(x + 1, y) - f(x, y)$$

   (c)

$$f(x + 1, y + 1) - f(x - 1, y + 1) + 2[f(x + 1, y) - f(x - 1, y)] \\ + f(x + 1, y - 1) - f(x - 1, y - 1)$$

3. Consider a binary image created by an edge detection method that marks all the edge pixels. I would like to use a Hough transform to see if this image has any circles. The equation of a circle with center (a,b) and radius c is given by $(x - a)^2 + (y - b)^2 = c^2$.

   (a) What is the dimension of the Hough space?

   (b) Write the equation of the corresponding Hough space entity for each pixel (x, y) ?

   (c) Infer from this equation the shape in the Hough space that corresponds to each pixel (x, y)?

4. The Harris corner detector is invariant to which of the following transformations: Scaling, Translation and Rotation. Justify your answer.

5. Consider a parabola given by equation $y = ax^2 + bx + c$.

   (a) What is the dimension of the Hough space?

   (b) What is the entity in Hough space to which the parabola corresponds to?

   (c) What is the equation of the entity in Hough space?

6. Consider the Harris corner detector where $M(x, y)$ is the Hessian of at pixel (x, y).

   (a) Is a pixel at location (x, y) a corner when the largest eigenvalue of $M(x, y)$ is much larger than the smallest eigenvalue of $M(x, y)$? Justify your answer.

   (b) Are all eigenvalues of $M(x, y)$ positive? Does the criterion of selecting corners that you specied above work for a negative eigenvalue?

7. Explain from properties of convolution why

$$f \star \frac{\partial h}{\partial x} = \frac{\partial f}{\partial x} \star h.$$

Part III

Geometric Visual Computing

6

Geometric Transformations

Geometric transformation, in general, means transforming a geometric entity (e.g. point, line, object) to another. This can happen in any dimension. For example, a 2D image can be transformed to another by translating it or scaling it or applying a unique transformation to each of its pixels. Or, a 3D object like a cube can be transformed into a parallelepiped or sphere. All of these will be considered as geometric transformations. Often a 2D image transformation is also called an *image warp*.

6.1 Homogeneous Coordinates

Before we start with geometric transformations, we will first introduce a very important concept of homogeneous coordinates. Let us consider the very simple case of the 1D world (see the red line in Figure 6.1left). Let us consider a point P' on this line. The coordinate of this point will be one dimensional. Let it be (p). Now consider a higher dimensional 2D world in which this line is embedded at $y = 1$. Draw a ray from the origin of this world through the point P'. Consider

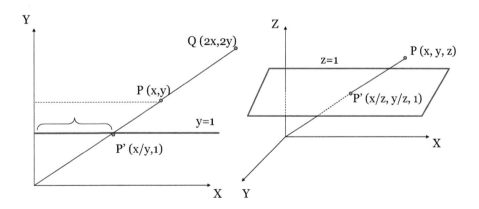

Figure 6.1. Homogeneous coordinates in 1D (left) and in 2D (right).

a point $P(x, y)$ on this line. Now, find out what would be the coordinate of P' in the 1D world of the red line when expressed using x and y. Using similar triangles you can see that $p = \frac{x}{y}$. Further, any point on this ray from the origin can be expressed as (kx, ky) where $k \neq 0$, and the value of p is invariant to the location of P on this ray.

Therefore, a point in the 1D world, such as P', when embedded in 2D, can be thought of as the projection of all points on a ray in the 2D world and the 2D coordinate of the projection is given by $(\frac{x}{y}, 1)$. Thus, any point in 1D can be considered as the projection of a ray in the 2D world on to a specific 1D world, in this case, the world of $y = 1$. This is called the $(n+1)$D homogeneous coordinate of a n dimensional point. Therefore, $(\frac{x}{y}, 1)$ is the 2D homogeneous coordinate of the 1D point P'. Also, $(k\frac{x}{y}, k)$ refers to the same ray but the projection is now on the $y = k$ plane. Since, they refer to the same ray, these two homogeneous coordinates are considered equivalent.

Let us now extend this concept to the next dimension using the right figure of Figure 6.1). Consider the 2D world of the red plane and consider the point P' on this plane. This point can be considered to be the projection of a ray from the origin through P' in the 3D world. Therefore, we will have a 3D homogeneous coordinate for the 2D point P'.

Extending this idea to the 3D world, it is evident that we will get $4D$ homogeneous coordinates for 3D points. In homogeneous coordinate representation, the last coordinate denotes the lower dimensional hyperplane in which the point resides and therefore need not be 1. However, when we deal with objects, it is important for us to consider them residing in the same hyperplane. The easiest way to achieve this is by *normalizing* the homogeneous coordinates i.e. when considering the 4D homogeneous coordinate (x, y, z, w) where $w \neq 1$, we convert it to $(\frac{x}{w}, \frac{y}{w}, \frac{z}{w}, 1)$.

Though at this point it may seem strange to look at points as rays in a higher dimension space, it is not as ad-hoc as it may seem. Intuitively, it stems from considerations in computer vision, which is the science of recovering 3D scene from 2D images. In our visual system, the 3D scene is projected as an image on the retina of the eye. Our brain identifies each point of the image in the retina as a ray into the 3D world. From only one eye and its retinal image, we cannot recover any information beyond the ray - i.e. we can only tell which ray contains the point but not where on the ray the point lies. In other words, we cannot decipher the depth of the point. But, when two eyes see the same point in 3D, we get two different rays from the two image projections on the retinas. Intersection of these two rays gives the exact position (depth) of the point in 3D. This is called stereo vision. Hopefully this will convince you of the importance of rays in computer vision and visual perception to motivate this ray representation.

There are several other practical advantages of this representation. In this chapter we will consider the 3D world with 4D homogeneous points for all our discussions. First, let us consider how you would represent points at infinity using 3D coordinates? The only option we have is (∞, ∞, ∞). Now, this representation is pretty useless since it is the same for all points in infinity, even if they are in different directions from the origin. However, using a 4D homogeneous coordinate, points at infinity can be represented as $(x, y, z, 0)$, where (x, y, z) is the direction of the point from the origin. When we normalize this to get the 3D point back, we get (∞, ∞, ∞) as expected. As a consequence, homogeneous coordinates provides a way to represent directions (vectors) and distinguish them from the representation of points. $w \neq 0$ signifies a point and $w = 0$ signifies a direction.

In the rest of the chapter we are going to represent points or vectors as 4×1 column vectors. Therefore a point $P = (x, y, z, 1)$ will be written as

$$P = \begin{pmatrix} x \\ y \\ z \\ 1 \end{pmatrix} \tag{6.1}$$

If P is a vector instead of a point, then the last coordinate will be 0.

6.2   Linear Transformations

Linear transformation is a special kind of transformation. Given two points P and Q, the transformation $\mathcal{L}$ is considered a linear transformation if

$$\mathcal{L}(aP + bQ) = a\mathcal{L}(P) + b\mathcal{L}(Q) \tag{6.2}$$

where a, b are scalars. In other words, the transformation of the linear combination of points is the linear combination of transformation of points. This holds true for multiple points and not just two.

The implications of linear transformations are quite important. $aP + bQ$ defines a plane and a line if $a + b = 1$. Linear transformation implies that to transform a line or a plane, we do not need to sample multiple points inside it, transform them and then connect them to get the transformed entity. Instead, it says that the same result will be achieved if the points are transformed and connected via a straight line or plane passing through it. Computationally, this has a huge impact since now, we save on computing the transformations of a bunch of points on the line and instead need to compute only two transformations. Second, linear transformation also implies that a line transforms to a line and a plane transforms to a plane. In fact, this can be generalized to higher orders of functions. If you consider a curve of degree n (e.g. a straight line is a function of degree 1, circle is of degree 2 and so on), a linear transformation

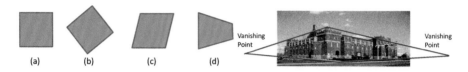

Figure 6.2. Left: This figure illustrates different kinds of linear transformation. Consider the square object in (a). (b) is a Euclidean transformation (angles and lengths preserved), (c) is an affine transformation (ratio of angles and lengths preserved) and (d) is a projective transformation (parallel lines became non-parallel). Right: This shows the projective transformation of a camera captured image and the relevant vanishing points.

will not change the degree of the curve. Finally, a linear transformation can be represented as a matrix multiplication where a $(n + 1) \times (n + 1)$ matrix representing the transformation converts a homogeneous coordinate represented as a $(n + 1) \times 1$ column vector to another. Therefore, linear transformations are represented by 3×3 matrices in 2D and 4×4 matrices in 3D.

Next we will discuss three type of linear transformations: *Euclidean, affine and projective*. Euclidean transformations preserve lengths and angles. For example, a square will not be changed to a rectangle by an Euclidean transformation. Translation and rotation are Euclidean transformations. Affine transformation preserves the ratios of lengths and angles. Therefore, a square can be converted to rectangle or a rhombus by an affine transformation, but cannot be transformed to a general quadrilateral. Examples of affine transformations are shear and scaling which will still retain the parallel sides of a rectangle and it will still remain a parallelogram. Both Euclidean and affine transformations cannot transform points within finite range to points at infinity and vice versa. This can only be achieved by projective transformations. What does this mean? This means that parallel lines will remain parallel and intersecting lines will remain intersecting with Euclidean or affine transformations. However, with projective transformation, parallel lines can become intersecting and vice versa. This is the kind of transformation we see in a camera image where parallel lines bounding rectangular sides of buildings become non-parallel and tend to meet somewhere within or outside the image called the *vanishing point*. This is illustrated in Figure 6.2.

Put a Face to the Name

Euclid is known as Father of Geometry. He was a Greek mathematician from Alexandria, Egypt who lived in 300 B.C. (yes! more than 2000 years back). He is best known for his work, Elements, where he collected the work of many mathematicians who preceded him. The whole new stream of geometry established by him is known as Euclidean Geometry. Basically the modern 2D geometry is actually adopted from Euclidean Geometry. Elements, a set of 13 books, is one of the most influential and successful textbooks ever written. Euclid proved that it is impossible to find the "largest prime number," because if you take the largest known prime number, add 1 to the product of all the primes up to and including it, you will get another prime number. Euclid's proof for this theorem is generally accepted as one of the "classic" proofs because of its conciseness and clarity. Millions of prime numbers are known to exist, and more are being added by mathematicians and computer scientists even today.

6.3 Euclidean and Affine Transformations

In this section we will explore the different Euclidean and affine transformations in detail. For each of these, we will start with the simpler case of 2D transformations and then extend them to 3D.

6.3.1 Translation

Translation is as simple as it sounds. Translate a point from one location to another. Let us consider the 2D point $P = (x, y)$ transformed to $P' = (x', y')$, such that

$$x' = x + t_x \tag{6.3}$$
$$y' = y + t_y \tag{6.4}$$

The matrix form of this transformation of a 2D point P represented as 3×1 homogeneous coordinates is given by

$$P' = \begin{pmatrix} x' \\ y' \\ 1 \end{pmatrix} = \begin{pmatrix} x + t_x \\ y + t_y \\ 1 \end{pmatrix} = \begin{pmatrix} 1 & 0 & t_x \\ 0 & 1 & t_y \\ 0 & 0 & 1 \end{pmatrix} \begin{pmatrix} x \\ y \\ 1 \end{pmatrix} = \mathcal{T}(t_x, t_y)P \tag{6.5}$$

Note that since the last element of P' is 1, the last row of the matrix should be $(0, 0, 1)$. We denote this translation matrix by $\mathcal{T}$. Any translation matrix will have the same format where the last column will have the translation parameters. Therefore, we denote this with $\mathcal{T}(t_x, t_y)$. Anytime we use this notation we will refer to a matrix where the top left sub-matrix is identity and the translation parameters go to the last column.

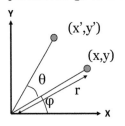

Figure 6.3. This shows a point (x, y) being rotated by an angle θ to be transformed to the point (x', y').

Every transformation has an inverse. This is defined as the transformation that takes the transformed point P' back to P. It is intuitive that the inverse of a translation would be another translation whose parameters are negated. Or, in other words,

$$\mathcal{T}^{-1}(t_x, t_y) = \mathcal{T}(-t_x, -t_y) \tag{6.6}$$

This is consistent with the math, since $x = x' - t_x$ and $y = y - t_y$. We can extend this to 3D as

$$\mathcal{T}(t_x, t_y, t_z) = \begin{pmatrix} 1 & 0 & 0 & t_x \\ 0 & 1 & 0 & t_y \\ 0 & 0 & 1 & t_z \\ 0 & 0 & 0 & 1 \end{pmatrix} \tag{6.7}$$

and

$$T^{-1} = T(-t_x, t_y, -t_z) \tag{6.8}$$

You can verify this by finding the inverse of $\mathcal{T}$ using standard matrix algebra.

6.3.2 Rotation

Next, we will consider the case of another Euclidean transformation, the rotation. Here also we will first consider the easier case of 2D rotation. For this, please take a look at Figure 6.3. This shows a point $P = (x, y)$ being rotated by an angle θ to be transformed to the point $P' = (x', y')$.

Consider the point P in polar coordinates given by the length r and angle ϕ. P is then expressed as

$$x = r cos(\phi) \tag{6.9}$$
$$y = r sin(\phi) \tag{6.10}$$

The rotation can be expressed using the equations

$$x' = r cos(\theta + \phi) \tag{6.11}$$
$$= r cos(\theta) cos(\phi) - r sin(\theta) sin(\phi) \tag{6.12}$$
$$= x cos(\theta) - y sin(\theta) \tag{6.13}$$
$$y' = r sin(\theta + \phi) \tag{6.14}$$
$$= r sin(\theta) cos(\phi) + r cos(\theta) sin(\phi) \tag{6.15}$$
$$= x sin(\theta) + y cos(\theta) \tag{6.16}$$

Therefore, using the same technique as we used before, we can find the rotation matrix $\mathcal{R}$ as

$$P' = \begin{pmatrix} x' \\ y' \\ 1 \end{pmatrix} = \begin{pmatrix} cos(\theta) & -sin(\theta) & 0 \\ sin(\theta) & cos(\theta) & 0 \\ 0 & 0 & 1 \end{pmatrix} \begin{pmatrix} x \\ y \\ 1 \end{pmatrix} = \mathcal{R}(\theta)P \qquad (6.17)$$

Clearly, the inverse transformation of $\mathcal{R}$ is a rotation by $-\theta$. Therefore,

$$\mathcal{R}(\theta)^{-1} = \mathcal{R}(-\theta) \qquad (6.18)$$

Now plug in $-\theta$ in $\mathcal{R}$ and you will see that the matrix is a transpose of $\mathcal{R}$, i.e.

$$\mathcal{R}(\theta)^{-1} = \mathcal{R}(-\theta) = \mathcal{R}(\theta)^T \qquad (6.19)$$

This property of inverse of rotation matrix being its transpose is a very special and useful one and is true for all rotation matrices, even in higher dimensions!

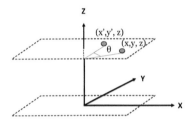

Figure 6.4. This shows the 3D rotation of a point (x, y, z) about the z-axis resulting in the point (x', y', z).

Now, let us extend this concept to 3D. While a rotation about a plane occurs around a point, a 3D rotation occurs about an axis. Figure 6.4 shows a rotation about the z-axis. For a rotation about the z-axis, the z-coordinates of the points remain unchanged. The rotation still affects the x and y coordinates just the same way as it would in a 2D on the xy plane. Therefore the 3D rotation can be represented by the following equations.

$$x' = xcos(\theta) - ysin(\theta) \qquad (6.20)$$
$$y' = xsin(\theta) + ycos(\theta) \qquad (6.21)$$
$$z' = z \qquad (6.22)$$

In 3D we distinguish the rotations with their axes. Therefore, the 3D rotation about z-axis, $\mathcal{R}_z$ is given by

$$\mathcal{R}_z(\theta) = \begin{pmatrix} cos(\theta) & -sin(\theta) & 0 & 0 \\ sin(\theta) & cos(\theta) & 0 & 0 \\ 0 & 0 & 1 & 0 \\ 0 & 0 & 0 & 1 \end{pmatrix} \qquad (6.23)$$

Also, in this case of 3D rotation,

$$\mathcal{R}_z(\theta)^{-1} = \mathcal{R}_z(-\theta) = \mathcal{R}_z(\theta)^T \qquad (6.24)$$

Similarly, a 3D rotation about Y axis keeps the y-coordinate unchanged while the rotation happens in the xz plane giving us the matrix

$$\mathcal{R}_y(\theta) = \begin{pmatrix} cos(\theta) & 0 & -sin(\theta) & 0 \\ 0 & 1 & 0 & 0 \\ sin(\theta) & 0 & cos(\theta) & 0 \\ 0 & 0 & 0 & 1 \end{pmatrix} \tag{6.25}$$

Try to write out the matrix for rotation about X-axis.

6.3.3 Scaling

Scaling is the transformation by which a point is scaled along one of the axes directions. Figure 6.5 shows an example. In this case, we will go directly to 3D scaling. The equations defining a scaling of s_x, s_y and s_z along the X, Y and Z axes respectively to transform P to P' are given by

$$x' = s_x x \tag{6.26}$$
$$y' = s_y y \tag{6.27}$$
$$z' = s_z z \tag{6.28}$$

The matrix for this is given by

$$P' = \begin{pmatrix} x' \\ y' \\ z' \\ 1 \end{pmatrix} = \begin{pmatrix} s_x & 0 & 0 & 0 \\ 0 & s_y & 0 & 0 \\ 0 & 0 & s_z & 0 \\ 0 & 0 & 0 & 1 \end{pmatrix} \begin{pmatrix} x \\ y \\ z \\ 1 \end{pmatrix} = \mathcal{S}(s_x, s_y, s_z)P \tag{6.29}$$

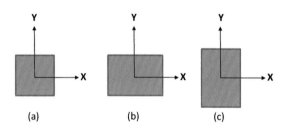

(a)   (b)   (c)

Figure 6.5. This shows an example of scaling. A square (a) is scaled along X and Y axes to create the rectangles in (b) and (c) respectively.

Clearly, the scale factors form the parameters of the scaling matrix. If $s_x = s_y = s_z$, then we call this a *uniform* scaling, otherwise *non-uniform*. Also, intuitively, the inverse matrix of scaling would be a scaling with reciprocal of the scale factors. You can verify that

$$\mathcal{S}(s_x, s_y, s_z)^{-1} = \mathcal{S}(\frac{1}{s_x}, \frac{1}{s_y}, \frac{1}{s_z}) \tag{6.30}$$

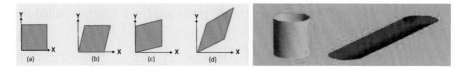

Figure 6.6. Left: This shows an example of 2D shear. A square (a) undergoes a Y-shear to create the rhombus in (b). In Y-shear the y-coordinate remains unchanged while the x-coordinate gets translated proportional to the value of the y-coordinate. Therefore, all the points on the X-axis whose y-coordinate is 0 remain unchanged. But as the y-coordinate increases, the x-coordinate moves, in this example, towards the right, creating the shear. Similarly, (c) shows an X-shear. (d) shows the result of first applying a X shear followed by a Y shear. Therefore, the only unchanged point is the origin where both x and y coordinates are 0. Note that if the proportionality constant for the shear is negative, then the Y-shear will move the square to the left instead of right. Right: This shows an example of Z-shear in 3D where Z is the axis of the cylinder.

6.3.4 Shear

Shear is a transformation where one coordinate gets translated by an amount that is proportional to the other coordinate. Figure 6.6 shows an example of 2D shear. The shear is identified by the coordinate that remains unchanged due to the shear. So, a Y-shear keeps the y-coordinate unchanged and translates the x-coordinate proportional to the y-coordinate.

The equations that describe the transformation of point P to P' due to a Y-shear is given by

$$x' = x + ay \tag{6.31}$$
$$y' = y \tag{6.32}$$

where a is the parameter of the shear. Therefore, the shear matrix is given by

$$P' = \begin{pmatrix} x' \\ y' \\ 1 \end{pmatrix} = \begin{pmatrix} 1 & a & 0 \\ 0 & 1 & 0 \\ 0 & 0 & 1 \end{pmatrix} \begin{pmatrix} x \\ y \\ 1 \end{pmatrix} = \mathcal{H}_y(a)P \tag{6.33}$$

When extending this to 3D, two coordinates should be translated proportional to the third one. So, for Z shear, the z-coordinate remains unchanged while the x and y coordinates are translated proportional to z. However, the constant of proportionality can be different and therefore the shear matrix would have two parameters. The matrix for a 3D Z-shear is given by

$$P' = \begin{pmatrix} x' \\ y' \\ z' \\ 1 \end{pmatrix} = \begin{pmatrix} 1 & 0 & a & 0 \\ 0 & 1 & b & 0 \\ 0 & 0 & 1 & 0 \\ 0 & 0 & 0 & 1 \end{pmatrix} \begin{pmatrix} x \\ y \\ z \\ 1 \end{pmatrix} = \mathcal{H}_z(a, b)P \tag{6.34}$$

where a and b are the two parameters of the shear matrix. It can be verified that the inverse of a shear matrix is

$$\mathcal{H}_z(a,b)^{-1} = \mathcal{H}_z(-a,-b). \tag{6.35}$$

6.3.5   Some Observations

This brings us to the end of the discussion on basic Euclidean and affine transformations. Here are a few observations from this discussion. First, since Euclidean transformations preserve lengths and angles, they will automatically preserve the ratio of lengths and angles. Therefore, *Euclidean transformations are a subset of affine transformations*. Euclidean transformations are often called *rigid body transformation* since the shape of the object cannot be changed by these transformations.

Next, all affine transformations of 3D space we have discussed have the last row predefined to be $(0,0,0,1)$. This is not a coincidence. Affine transformation in 3D can be represented as a linear transformation in 4D, and is a subspace of all 4D linear transformations. In other words, in affine transformation, we have the liberty to change only 12 parameters of the 4×4 matrix and still be in this subspace. This is often described as the degrees of freedom of a class of transformations. In other words, *affine transformations in 3D have 12 degrees of freedom*. However, it is a coincidence that for affine transformations the number of degrees of freedom is the same as the number of entries that can be changed in the matrix. We will have an in-depth discussion on the degrees of freedom at the end of this chapter.

Next, some points or lines are fixed under certain transformations i.e. they do not change position with the transformation. For example, the origin is fixed under scaling and shear in 3D. The axis of rotation is fixed under 3D rotation and the origin is fixed under 2D rotation. These are called *fixed points of mappings*. It can be seen that there are fixed points under translation.

Finally, the translation matrix cannot be expressed as a 3×3 matrix while scaling or rotation or shear can be. Homogeneous coordinates are essential to express translation in 3D as a linear transformation in 4D. This is another practical importance of having homogeneous coordinates.

6.4   Concatenation of Transformations

You now know all the basic affine transformations. The next step is how to use this basic knowledge to find the matrices for more complex transformations like scaling or rotation about an arbitrary axes. To achieve this, we need to learn how to concatenate transformations.

Let us consider a case where a point P is first translated and then rotated. Now let us see how to find the final point. Let the translated point be P' and it is given by

$$P' = \mathcal{T}P \tag{6.36}$$

P' is then rotated to produce $P"$ given by

$$P" = \mathcal{R}P' = \mathcal{R}\mathcal{T}P \tag{6.37}$$

Therefore, to concatenate the effect of a translation followed by rotation, we have to premultiply the respective matrices based on the order of transformations. Of course, the order of this multiplication is critical since we know matrix multiplication is not commutative, i.e.

$$\mathcal{R}\mathcal{T}P \neq \mathcal{T}\mathcal{R}P \tag{6.38}$$

Therefore, you will end up with grossly inaccurate transformations if you do not pay special attention to the order. Further, the inverse of the transformation to get to P back from $P"$ is given by multiplying the inverse matrices in the reverse order due.

$$P = \mathcal{T}^{-1}\mathcal{R}^{-1}P" \tag{6.39}$$

Let us now try to find the matrices for more complex transformations. The algorithm to achieve this is as follows.

1. **Step 1:** Apply one or more transformations to get to a case where you can apply basic known affine transformations. Let this set of transformations be denoted by $\mathcal{F}$.

2. **Step 2:** Apply the basic affine transformation $\mathcal{B}$.

3. **Step 3:** Apply the inverse of F to undo the effect of $\mathcal{F}^{-1}$.

4. **Step 4:** Since they are applied in order, the matrix that needs to be premultiplied with the point is given by $\mathcal{F}^{-1}\mathcal{B}\mathcal{F}$.

Let us illustrate the use of concatenation of transformations to design more complex transformations.

6.4.1 Scaling About the Center

Let us consider the 2D transformation where we want to scale a square of size 2 units with its bottom left corner coincident with the origin by a factor 2 about its center $(1, 1)$ as shown in Figure 6.7. It is pretty intuitive to figure out the final transformed square. But we will learn here how to find the matrix that would achieve this transformation.

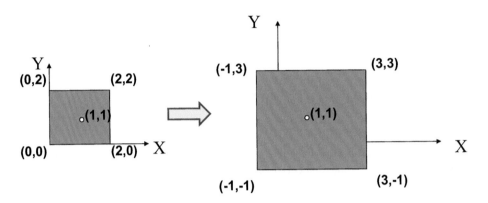

Figure 6.7. Left: This shows a square of size 2 units with its bottom left corner aligned with the origin. Right: This shows the same square after transformation by a factor of 2 about its origin.

1. **Step 1:** We know that scaling keeps the origin fixed. Therefore, if we want to keep the center of the square fixed, the first transformation we need to apply should bring this point at the origin. Now, the center of the square is $(1,1)$. So, the transformation to achieve this would be $\mathcal{T}(-1,-1)$. So, our F is $\mathcal{T}(-1,-1)$.

2. **Step 2:** Now with the center at origin, you can apply the basic transformation asked of you, i.e. scaling with factor 2 along both X and Y directions. Therefore $\mathcal{B} = \mathcal{S}(2,2)$.

3. **Step 3:** Now we need to apply $\mathcal{F}^{-1} = \mathcal{T}(-1,-1)^{-1} = \mathcal{T}(1,1)$ to undo the effect of $\mathcal{F}$.

4. **Step 4:** Therefore the final concatenated transformation is given by $\mathcal{T}(1,1)\mathcal{S}(2,2)\mathcal{T}(-1,-1)$. If you write these out completely, the 3×3 matrix for this transformation is given by

$$\begin{pmatrix} 1 & 0 & 1 \\ 0 & 1 & 1 \\ 0 & 0 & 1 \end{pmatrix} \begin{pmatrix} 2 & 0 & 0 \\ 0 & 2 & 0 \\ 0 & 0 & 1 \end{pmatrix} \begin{pmatrix} 1 & 0 & -1 \\ 0 & 1 & -1 \\ 0 & 0 & 1 \end{pmatrix} \tag{6.40}$$

All these steps are illustrated in Figure 6.8.

6.4.2 Rotation About an Arbitrary Axis

Now, let us consider a more complex case of rotation by an angle θ about an arbitrary axis instead of one of the three coordinate axes. Let us consider an axis

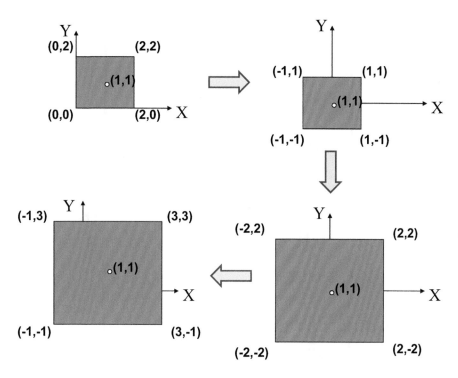

Figure 6.8. This shows the different steps of using the concatenation of transformation to achieve the scaling about the center. First is the original square which is then translated by $(-1, -1)$ to get the center coincident with the origin. Next it is scaled by 2 and then translated back by $(1, 1)$ to undo the effect of the earlier translation.

rooted at the point (x, y, z) with direction specified by the unit vector (a, b, c). It is important to normalize the axis to be a unit vector. Otherwise a scale factor equivalent to the magnitude of the vector will creep into the transformation. When we derived the matrix for 3D rotation earlier, we did assume unit vectors as axes.

The arbitrary axes are illustrated in Figure 6.9. The goal here would be to first take this arbitrary axis to a position where the desired transformation can be related to the basic transformations we know. Since we know the matrices for transformation along one of the coordinate axes, the first step would be to design $\mathcal{F}$ such that the arbitrary axis is aligned with one of the coordinate axes. Without loss of generality, we will try to align it with the Z axis. Once this is achieved, we will apply the rotation about the Z axis by an angle θ. Therefore, $\mathcal{B} = \mathcal{R}_z(\theta)$. Next, we will apply $\mathcal{F}^{-1}$ to undo the effect of $\mathcal{F}$.

Now, let us focus on finding $\mathcal{F}$. Let us see what we need to do to get the

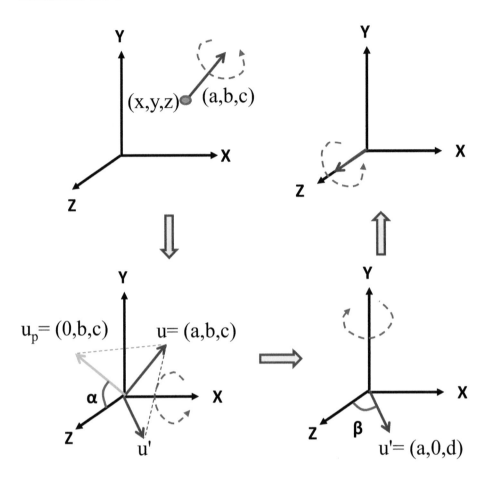

Figure 6.9. This shows an arbitrary axis of rotation rooted at (x, y, z) and directed towards the unit vector (a, b, c) and the transformations it undergoes to get aligned with the Z axis. First, the vector is translated to be rooted at the origin. Next it is rotated by α about X axis to lie on the XZ plane. Next it is rotated by β about Y axis to be coincident with the Z axis.

arbitrary axis aligned with Z-axis, as illustrated in Figure 6.9. First we translate the scene by $(-x, -y, -z)$ so that the base point of the vector moves to origin. This transformation is therefore $\mathcal{T}(-x, -y, -z)$. The arbitrary axis now becomes the unit vector (a, b, c). We will align this vector with the Z axis in two steps – first we will rotate the vector about X axis so as to get it coincident with the XZ plane. Next we will rotate this vector on the XZ plane about the Y axis to get it coincident with the Z axis. Let us now compute the angles we have to rotate – α about X and β about Y – to achieve this. Once we figure this out, the former rotation will be given by $\mathcal{R}_x(\alpha)$ and the latter by $\mathcal{R}_y(\beta)$. Thus,

$$\mathcal{F} = \mathcal{R}_y(\beta)\mathcal{R}_x(\alpha)\mathcal{T}(-x, -y, -z) \tag{6.41}$$

From this we can find $\mathcal{F}^{-1}$ as

$$\mathcal{F}^{-1} = \mathcal{T}(x, y, z)\mathcal{R}_x(-\alpha)\mathcal{R}_y(-\beta) \tag{6.42}$$

Therefore the complete transformation, $\mathcal{F}^{-1}\mathcal{B}\mathcal{F}$, will be given by

$$\mathcal{T}(x, y, z)\mathcal{R}_x(-\alpha)\mathcal{R}_y(-\beta)\mathcal{R}_z(\theta)\mathcal{R}_y(\beta)\mathcal{R}_x(\alpha)\mathcal{T}(-x, -y, -z) \tag{6.43}$$

$$= \mathcal{T}(x, y, z)\mathcal{R}_x(\alpha)^T\mathcal{R}_y(\beta)^T\mathcal{R}_z(\theta)\mathcal{R}_y(\beta)\mathcal{R}_x(\alpha)\mathcal{T}(-x, -y, -z) \tag{6.44}$$

Now that we have deciphered the complete transformations, let us find out the matrices $\mathcal{R}_x(\alpha)$ and $\mathcal{R}_y(\beta)$. Please refer to Figure 6.9. We first consider the projection of u on the YZ plane, u_p, denoted by the blue vector. This is computed by setting the x coordinate of u as 0. Therefore, $u_p = (0, b, c)$. The angle α that u has to rotate about X to get to u' is the same that u_p has to rotate to be coincident with the XZ plane. Therefore, if we consider $\sqrt{c^2 + b^2} = d$, then $sin(\alpha) = \frac{b}{d}$ and $cos(\alpha) = \frac{c}{d}$. Therefore,

$$\mathcal{R}_x(\alpha) = \begin{pmatrix} 1 & 0 & 0 & 0 \\ 0 & \frac{b}{d} & -\frac{c}{d} & 0 \\ 0 & \frac{c}{d} & \frac{b}{d} & 0 \\ 0 & 0 & 0 & 1 \end{pmatrix} \tag{6.45}$$

Premultiplying u with $\mathcal{R}_x(\alpha)$ gives us $u' = (a, 0, d)$. Next, we need to find the matrix $\mathcal{R}_x(\beta)$. This is pretty straightforward since u' is already in the XZ plane. So, $sin(\beta) = \frac{a}{\sqrt{a^2+d^2}}$ and $cos(\beta) = \frac{d}{\sqrt{a^2+d^2}}$ where $\sqrt{a^2 + d^2} = \sqrt{a^2 + b^2 + c^2} = 1$ since u is an unit vector. Therefore,

$$\mathcal{R}_y(\beta) = \begin{pmatrix} d & 0 & -a & 0 \\ 0 & 1 & 0 & 0 \\ a & 0 & d & 0 \\ 0 & 0 & 0 & 1 \end{pmatrix} \tag{6.46}$$

Using the values of the transformations from Equations 6.45 and 6.46 in Equation 6.44 we can get the complete transformation.

6.5   Coordinate Systems

Throughout all the discussions in this chapter, we have assumed that we have a reference – an orthogonal coordinate system. For n dimensional world, this is made of n orthogonal unit vectors, $u_1, u_2, \ldots, u_n$ and an origin R. When considering the 3D world, we would have three vectors u_1, u_2 and u_3. Let each vector u_i in homogeneous coordinates be given by $(u_{ix}, u_{iy}, u_{iz}, 0)$ and the origin R be given by $(R_x, R_y, R_z, 1)$. The coordinates of a point P in the standard coordinate system $X = (1, 0, 0, 0)$, $Y = (0, 1, 0, 0)$, and $Z = (0, 0, 1, 0)$ is expressed by its coordinates (a_1, a_2, a_3) in the u_1, u_2, u_3 coordinates system as linear combination of its axes and the origin as

$$P = a_1 u_1 + a_2 u_2 + a_3 u_3 + R \qquad (6.47)$$

This can further be expressed as a matrix as

$$P = \begin{pmatrix} u_1 & u_2 & u_3 & R \end{pmatrix} \begin{pmatrix} a_1 \\ a_2 \\ a_3 \\ 1 \end{pmatrix} = \begin{pmatrix} u_{1x} & u_{2x} & u_{3x} & R_x \\ u_{1y} & u_{2y} & u_{3y} & R_y \\ u_{1z} & u_{2z} & u_{3z} & R_z \\ 0 & 0 & 0 & 1 \end{pmatrix} \begin{pmatrix} a_1 \\ a_2 \\ a_3 \\ 1 \end{pmatrix} = M_u C_u.$$

$$(6.48)$$

where M_u denotes the matrix that defines the coordinate system and C_u defines the coordinates of P in the coordinate system denoted by M_u. This is a very important relationship since now you can see that even coordinate systems can be defined using matrices. For the X, Y, and Z axes respectively defined by vectors $(1, 0, 0, 0)$, $(0, 1, 0, 0)$ and $(0, 0, 1, 0)$ and origin is $R = (0, 0, 0, 1)$, and the matrix representing this coordinate system is essentially an identity matrix, i.e. $M_u = I$.

6.5.1   Change of Coordinate Systems

Coordinate systems are reference frames. Think of them as reference points that you use when you tell someone your home address. You may say that from the University High School take a left and then immediate right to get to our house. However, you may want to use a completely different reference point, say Trader Joes, and say from Trader Joes take an immediate right and then the second left. As your reference changes, the coordinates of your house with respect to that reference also changes. This does not mean that your house has moved – it is still in the same location — just the way your address the house differs due to the change in reference.

Something similar happens when you work with multiple coordinate systems. A point P will have different coordinates in different coordinate systems though its actual location remains the same. Let us now consider a second coordinate system made of vectors v_1, v_2, v_3 and origin Q. Let the coordinate of the same point P in this coordinate system be (b_1, b_2, b_3) which means

$$P = \begin{pmatrix} v_1 & v_2 & v_3 & Q \end{pmatrix} \begin{pmatrix} b_1 \\ b_2 \\ b_3 \\ 1 \end{pmatrix} = \begin{pmatrix} v_{1x} & v_{1y} & v_{1z} & Q_x \\ v_{2x} & v_{2y} & v_{2z} & Q_y \\ v_{3x} & v_{3y} & v_{3z} & Q_z \\ 0 & 0 & 0 & 1 \end{pmatrix} \begin{pmatrix} b_1 \\ b_2 \\ b_3 \\ 1 \end{pmatrix} = M_v C_v.$$

$$(6.49)$$

From Equations 6.48 and 6.49, we see that $M_u C_u = M_v C_v$. Therefore, the co-ordinate C_v of the same point P in a new coordinate system when the coordinate C_u in the first coordinate system is known, is given by

$$C_v = M_v^{-1} M_u C_u \qquad (6.50)$$

Now let us ponder a little bit on the matrix $M = M_v^{-1} M_u$ that achieves the change of coordinates. What kind of transformation does this matrix represent? Let us say a point P is at the origin in one coordinate system and the same point has the coordinate $(5, 0, 0)$ in another coordinate system. Going from one 'version' of P to another, i.e. to get a point from $(0, 0, 0)$ to $(5, 0, 0)$, we need to translate it by 5 units. From the perspective of the point P (if it did not move), but the same change in coordinates can be achieved by moving one coordinate system by -5 units along the X direction. This can be viewed as the transformation that takes one coordinate system and transforms it in such a way so as to coincide with the second coordinate system.

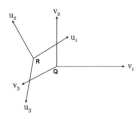

Using the same concept, the matrix M that transforms the coordinates from one system to another can also be viewed as a transformation that transforms one coordinate system to another. So, now let us explore, what kind of transformation would it take to align one coordinate system to another. Please refer to the illustration in Figure 6.10. This shows the two coordinate systems represented by M_u and M_v. The transformation needed to align the two orthogonal coordinate systems is given by a translation to align the origin, followed by a rotation to align the axes. Let us consider a rotation matrix $\mathcal{R}_c$ and a translation matrix $\mathcal{T}_c$ such that

Figure 6.10. This shows example of two different coordinate systems that can represent M_u and M_v.

$$\mathcal{R}_c = \begin{pmatrix} r_{11} & r_{12} & r_{13} & 0 \\ r_{21} & r_{22} & r_{23} & 0 \\ r_{31} & r_{32} & r_{33} & 0 \\ 0 & 0 & 0 & 1 \end{pmatrix} \qquad \mathcal{T}_c = \begin{pmatrix} 1 & 0 & 0 & t_1 \\ 0 & 1 & 0 & t_2 \\ 0 & 0 & 1 & t_3 \\ 0 & 0 & 0 & 1 \end{pmatrix} \qquad (6.51)$$

then the coordinate transformation matrix $M = M_v^{-1} M_u$ in Equation 6.50 is given from the above two equations as

$$M = \mathcal{R}_c \mathcal{T}_c. \qquad (6.52)$$

Let us take a closer look at the matrices $\mathcal{R}_c$ and $\mathcal{T}_c$ and compute their multiplication. At this juncture, we would like to introduce a different kind of matrix notation. Let us consider $\mathbf{R}$ to be the top left 3×3 matrix of $\mathcal{R}_c$, $\mathbf{T}$ to be the translation vector given by the 3×1 column vector $(t_x, t_y, t_z)^T$, $\mathbf{I}$ be a 3×3 identity matrix, and $\mathbf{O}$ to be the 1×3 row vector given by $(0,0,0)$. Therefore $\mathcal{R}_c$ and $\mathcal{T}_c$ can now be expressed as

$$\mathcal{R}_c = \left(\begin{array}{c|c} \mathbf{R} & \mathbf{O^T} \\ \hline \mathbf{O} & 1 \end{array} \right) \quad \mathcal{T}_c = \left(\begin{array}{c|c} \mathbf{I} & \mathbf{T} \\ \hline \mathbf{O} & 1 \end{array} \right). \tag{6.53}$$

This is a notation using sub-matrices to write a matrix. The sizes of the sub-matrices should be consistent to yield the correct dimension of the matrix. For example, the size of $\mathbf{R}$ is 3×3, $\mathbf{O^T}$ is 3×1, size of $\mathbf{O}$ is 1×3 and 1 is just a scaler of dimension 1×1 leading to the dimension of $\mathcal{R}_c$ to be 4×4 as is consistent for a 3D rotation matrix.

Now let us consider the multiplication of these two matrices which can also be represented in terms of sub-matrices as

$$M = \mathcal{R}_c \mathcal{T}_c = \left(\begin{array}{c|c} \mathbf{RI} + \mathbf{O^T O} & \mathbf{RT} + \mathbf{O^T} \\ \hline \mathbf{OI} + \mathbf{O} & \mathbf{OT} + 1 \end{array} \right) = \left(\begin{array}{c|c} \mathbf{R} & \mathbf{RT} \\ \hline \mathbf{O} & 1 \end{array} \right) \tag{6.54}$$

Verify that all the sub-matrix multiplications are consistent in their dimension. Coming back to the composition of M, you can see that it is created by a rotation and a translation matrix. Later in this book, we will make use of this sub-matrix representation to learn about decomposition of this matrix M during camera calibration.

Next, let us look at one more issue. How can we create an orthogonal coordinate system from as minimal information as possible? Let us consider the origin to be at $(0,0,0)$. Suppose we are given one vector unit u_1. Is there are a simple way to find two other orthogonal unit vectors u_2 and u_3 which can form a coordinate system together with u_1. As it turns out, it is pretty simple. First, find u_2 via a cross product of u_1 and any of the X, Y or Z axes. Therefore $u_2 = u_1 \times u_x$ where u_x is the unit vector in the direction of X axis. By design, u_1 and u_2 are orthogonal. Next, find the third vector $u_3 = u_1 \times u_2$. Also, by design, u_3 is orthogonal to both u_1 and u_2 and therefore they form a coordinate system. This is illustrated in Figure 6.11.

Finally, consider the matrix $M = M_v^{-1} M_u$ one more time. Let us consider these two coordinate systems (shown in Figure 6.10 to have the same origin. In that case, M_v and M_u are each a rotation matrix and M_v^{-1} is M_v^T. Now consider v_1, v_2 and v_3 to be the standard X, Y, Z coordinate axes. Therefore $M_v = M_v^T = I$. Now consider this situation of having one coordinate system which is our standard XYZ coordinate system and we have another coordinate system rooted at the same origin defined by u_1, u_2 and u_3. Therefore, the transformation to make this coordinate system coincident with XYZ coordinates is given by

$M = IM_u = M_u$. Interestingly, if u_1, u_2 and u_3 are known, this rotation matrix is simply given by plugging in these vectors as

$$M_u = \begin{pmatrix} & u_1 & & \\ & u_2 & & \\ & u_3 & & \\ 0 & 0 & 0 & 1 \end{pmatrix} \tag{6.55}$$

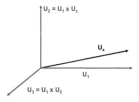

Figure 6.11. This shows how to create a 3D coordinate system using a single vector u_1.

Let us relate this back to another situation we face when finding the matrix of rotation about an arbitrary axes. After we rooted the arbitrary axes to the origin, we could have used an alternate way to find the matrix that would align u with one of the coordinate axes. We could have considered $u = u_3$ and created a coordinate system using $u_2 = u_3 \times (1, 0, 0)$ and $u_1 = u_3 \times u_1$. Then we could have put these vectors in the Equation 6.55 to generate the rotation matrix to align u with Z-axis. This matrix would be equivalent to what you achieved by $\mathcal{R}_y(\beta)\mathcal{R}_x(\alpha)$ in Equation 6.41 and its inverse given by M_u^T will be exactly what you would achieve by $\mathcal{R}_x(\alpha)^{-1}\mathcal{R}_y(\beta)^{-1}$ in Equation 6.42. In other words, we can just define a coordinate system using the arbitrary axis and the vectors of this coordinate system will define the rotation matrix as one of the orthogonal axes computing α and β.

6.6   Properties of Concatenation

Now that we have learned about both coordinate systems and concatenation of transformations, we will now explore some relationships between them. We have already seen before that since concatenation of transformation is represented by matrix multiplication and since matrix multiplication is not commutative, the order in which we concatenate transformations is critical to arrive at the desired transformation.

However, though matrix multiplication is not commutative, it is associative. To understand the implication of this associative law, let us consider two different transformations – $\mathcal{T}_1$, $\mathcal{T}_2$, and a point P. These transformations can be any linear transformation and so can be represented by a matrix. Now, due to the associative law, we can say that

$$\mathcal{T}_1\mathcal{T}_2 P = (\mathcal{T}_1(\mathcal{T}_2 P)) = ((\mathcal{T}_1\mathcal{T}_2)P) \tag{6.56}$$

The above equation says that it really does not matter if the multiplications are performed from the left to right or right to left i.e. $\mathcal{T}_1$ and $\mathcal{T}_2$ can be multiplied first and then the result post-multiplied by P or $\mathcal{T}_2$ and P can be multiplied first and then the result pre-multiplied by $\mathcal{T}_1$ to get the same answer. Therefore, whether the multiplications are performed as a post or pre-multiplication does not really matter as long as their order is preserved. Though this may seem to be of trivial consideration, it has a rather deep geometric interpretation.

6.6.1   Global vs Local Coordinate System

The transformation $\mathcal{T}_1\mathcal{T}_2 P$ transforms the point P and whether it is done using pre-multiplication or post-multiplication, the result will be the same. However, the geometric interpretation of the intermediate steps are dependent on the pre or post multiplication.

So far, in this chapter, we have been doing concatenation as a pre-multiplication, i.e. pre-multiply $\mathcal{T}_2$ with P first and then pre-multiply the result with $\mathcal{T}_1$. When we perform each of these steps we consider the coordinate system to be constant. Hence, the coordinate systems remain *global* across the different transformations. This is usually easy to understand since we typically work with a standard frame of reference.

However, the post multiplication also has a interpretation. It means that the *coordinate system itself is being transformed*. Therefore when you post multiply $\mathcal{T}_1$ with $\mathcal{T}_2$, it means that you have first applied the transformation $\mathcal{T}_1$ to the coordinate system followed by $\mathcal{T}_2$ to the transformed coordinate system and then placed P in this transformed coordinate system. Here, the coordinate system remains *local* to each transformation and changes from one transformation to another.

However, the result of implementing the transformation both in global or local coordinates achieves the same result. To illustrate this, please see Figure 6.12. We consider the transformation $\mathcal{R}\mathcal{T}P$ in 2D for this object and perform it in both global and local coordinate systems to achieve the same result.

6.7   Projective Transformation

This brings us to the end of affine transformation. Now, we will explore projective transformation. Projective transformations are most general linear transformations which take points $P = (x, y, z, w)$ to points $P' = (x', y', z', w')$. Projective transform $\mathcal{P}$ is expressed as

$$\begin{pmatrix} x' \\ y' \\ z' \\ w' \end{pmatrix} = \begin{pmatrix} p_{11} & p_{12} & p_{13} & p_{14} \\ p_{21} & p_{22} & p_{23} & p_{24} \\ p_{31} & p_{32} & p_{33} & p_{34} \\ p_{41} & p_{42} & p_{43} & p_{44} \end{pmatrix} \begin{pmatrix} x \\ y \\ z \\ w \end{pmatrix} \tag{6.57}$$

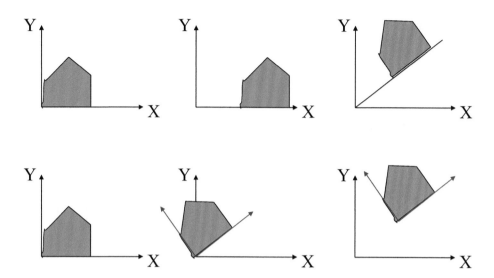

Figure 6.12. We consider the transformation $\mathcal{RTP}$ in 2D for this object. Top: This shows the transformation performed in a global coordinate system, i.e. using pre-multiplication. Therefore, the object is first translated and then rotated. Bottom: This shows the transformation performed in a local coordinate system i.e. using post-multiplication. Therefore, the coordinate system is first rotated relative to its own co-ordinate system. The object will also change position due to this since its coordinates with respect to the local coordinate system have not changed. Next the coordinate system is translated relative to itself. The changing coordinates are shown in red. Note the final location of the object is the same which is due to the associative nature of matrix multiplication.

The most important difference of projective transformation is that it can take finite points to points at infinity. The implication of this is that non-parallel lines can become parallel and vice versa. However, it still does not change the degree of a curve. So, a line cannot become a curve. A circle can become an ellipse (none of its points go to infinity) or even a parabola (where some of its points go to infinity) but it cannot become a degree-3 polynomial. Please see exercise problems to check it for yourselves.

The most common projective transformation we face is when we deal with a camera. A camera projects the 3D objects in the world on a 2D image plane to create an image. The most basic camera model is called the *pin-hole camera* where the camera is considered to be a simple *pinhole*. Think of a box with a hole in one of its face and the opposite face acting as an imaging plane and voila! There we have a pin-hole camera as illustrated in Figure 6.13. Here O denotes the pinhole. Rays of light from 3D points A, B and C come through O and

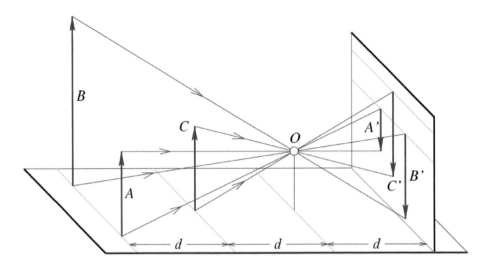

Figure 6.13. This shows a pinhole camera where O is the pinhole. Rays of light from 3D points A, B and C come through O and intersect the image plane behind it form their 2D image at A', B' and C' respectively.

intersect the image plane behind it to form their 2D image at A', B' and C' respectively under a projective transformation. It is important to note that such a projective transformation changes the size of the image based on the distance from the pinhole. For example, objects B and C appear to be of the same height in the image B' and C'. However, B is double the size of C in 3D, but it is double distance away. Further, multiple points on the same ray will have the same image on the image plane therefore losing their depth information. We will explore the camera projective transformation in more details in the next chapter.

6.8   Degrees of Freedom

Degrees of freedom defines the number of parameters that can be changed during a transformation. Let us consider for example, a 2D rigid body transform. This will be represented by a 3×3 matrix when considering homogeneous coordinates. This is often referred to as the 3×3 homogeneous transformation. As you know, for rigid body transformation, the object can undergo only translation (2 parameters) or rotation (1 parameter). Therefore, this matrix has three degrees of freedom. This matrix will have the translation parameters on the last column and the rotation parameters will be used to fill the top left 2×2 submatrix. Therefore, though six entries of this matrix can be changed, they are not

Figure 6.14. Left two images show non-linear lens distortion that changes straight lines to curves. Right two images show the non-linear transformation (right) of a cubical 3D color gamut (left) during color management.

completely independent. Therefore, the degree of freedom of this matrix is three although the number of matrix entries that can be changed is six.

However, the degrees of freedom cannot be greater than the number of matrix elements that get affected by the transformation. Let us consider a 2D affine transformation represented using a 3 × 3 matrix. Since affine transformation allows scaling and shear, it may seem that we will have an additional four parameters we can control (2 each for scaling and shear) in addition to the three parameters for rigid-body transformation. Therefore, this transformation has 7 degrees of freedom. However, since only six entries of the matrix have been affected, it has 6 degrees of freedom. On deeper analysis, you can see that the rotation can be expressed as a combination of scaling and shear. For example, x coordinate will be transformed as $ax + by$ where a is considered the scale factor and b, the shear factor. But, they will be similar to cosine and sine of the angle of rotation. Therefore, the rotational degree of freedom in absorbed by the scaling and shear parameters, thereby providing the transformation with only six degrees of freedom.

From a matrix computations perspective, any constraint imposed on a matrix reduces its degrees of freedom. So, for example, if you had a matrix of degree 7 with the special constraint that the matrix is rank deficient (i.e. determinant of the matrix is 0), each deficiency in rank would be translated to a reduction of degree of freedom by one. As we go into the next chapters on geometric visual computing, we will be discussing degrees of freedom in several occasions to provide a more comprehensive understanding.

6.9 Non-Linear Transformations

The discussion in this chapter is incomplete without discussing non-linear transformations. Any transformation that changes the degree of a curve (e.g. a line to a curve) is called a non-linear transformation. Distortion due to the lens of a camera is a good example of a non-linear distortion. This is the distortion introduced by the camera lens following the projective transformation from 3D to 2D. Such a distortion is shown in Figure 6.14 for a checkerboard pattern and of an architectural site.

Non-linear transformations cannot be achieved as simple as a matrix multiplication. Typically, points on the objects should be sampled; each of them should be transformed and then another surface should be found via surface fitting to find the transformed object. Such transformations are common is applications like modeling, surface design, color management and simulation. In this book, we will focus mostly on linear transformations.

Fun Facts

The word geometry comes from the Greek words geo, meaning earth, and metria, meaning measure. Geometry was one of the two fields of pre-modern mathematics, the other being the study of numbers (arithmetic). The earliest recorded beginnings of geometry can be traced to early peoples, who discovered obtuse triangles in the ancient Indus Valley Civilization (now in India and Pakistan) and ancient Babylonia (now in Iran) from around 3000 BC. Ancient Egyptians used geometric principles as far back as 3000 BC, using equations to approximate the area of circles among other formulas. The Babylonians may have known the general rules for measuring areas and volumes. They measured the circumference of a circle as three times the diameter and the area as one-twelfth the square of the circumference, which would be correct if π is estimated as 3. Greek philosopher and mathematician Pythagoras lived around the year 500 BC and is known for his Pythagorean theorem relating to the three sides of a right angle triangle: $a^2 + b^2 = c^2$. Archimedes of Syracuse lived around the year 250 BC and played a large role in the history of geometry including a method for determining the volume of objects with irregular shapes.

When Europe began to emerge from its Dark Ages, the Hellenistic and Islamic texts on geometry found in Islamic libraries were translated from Arabic into Latin. The rigorous deductive methods of geometry found in Euclids *Elements of Geometry* were relearned, and further development of geometry in the styles of both Euclid (Euclidean geometry) and Khayyam (algebraic geometry) was continued by Rene Descartes (1596 - 1650) and Pierre de Fermat (1601 - 1665) in analytical geometry and by Girard Desargues (1591 - 1661) in projective geometry.

6.10 Conclusion

In this chapter we covered geometric transformation that forms the fundamental of computer vision and graphics. Matrices provide us a formal framework to work with difficult geometric problems in these domains. Advanced concepts in these directions can be explored in computer vision books like [Faugeras 93] or computer graphics books like [Hughes et al. 13, Shirley and Marschner 09].

Bibliography

[Faugeras 93] Olivier Faugeras. *Three-dimensional Computer Vision: A Geometric Viewpoint*. MIT Press, 1993.

[Hughes et al. 13] John F. Hughes, Andries van Dam, Morgan McGuire, David F. Sklar, James D. Foley, Steven K. Feiner, and Kurt Akeley. *Computer Graphics: Principles and Practice (3rd ed.)*. Addison-Wesley Professional, 2013.

[Shirley and Marschner 09] Peter Shirley and Steve Marschner. *Fundamentals of Computer Graphics*, Third edition. A. K. Peters, Ltd., 2009.

Summary: Do you know these concepts?

- Homogeneous Coordinates
- Linear Transformations
- Euclidean Transformations
- Rigid Body Transformations
- Affine Transformations
- Projective Transformations
- Degrees of Freedom
- Concatenation of Transformations
- Coordinate Systems
- Changing Coordinate Systems
- Creating Coordinate Systems
- Global vs Local Coordinate Systems
- Non-Linear Transformation

Exercises

1. Consider the following matrix [Note: $1/2 = 0.707$].

$$\begin{pmatrix} 0.707 & 0 & 0.707 & 0 \\ 0 & 2 & 0 & 0 \\ -0.707 & 0 & 0.707 & 0 \\ 0 & 0 & 0 & 1 \end{pmatrix} \tag{6.58}$$

   What transformation does the matrix achieve? What is the order of this transformation in the local coordinate system?

2. Consider the figure below. Give a matrix, or a product of matrices, which will transform the square ABCD to the square ABCD. Show what happens if the same transformation is applied to the square ABCD.

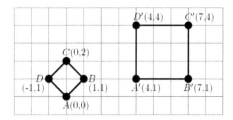

3. Consider a 2D rectangle $ABCD$ where $A = (0,0)$, $B = (2,0)$, $C = (2,1)$ and $D = (0,1)$. We want to apply a 2D transformation to this rectangle which makes it a parallelopiped $ABEF$ where $E = (4,1)$ and $F = (2,1)$. a. What kind of transformation is this? What is the 3x3 matrix M achieving this transformation? c. What additional transformation N would we need to apply to $ABEF$ to get the parallelopiped $A'B'E'F'$ where $A' = (1,2)$, $B' = (3,2)$, $E' = (5,3)$, and $F' = (3,3)$? d. What is the final concatenated matrix in terms of M and N that will transform $ABCD$ to $A'B'E'F'$?

4. Derive the scaling matrix for scaling an object by a scale factor 3 along an arbitrary direction given by vector $u = (1,2,1)$ rooted at $(5,5,5)$.

5. Explain what transformation is produced by each of the following matrices when applied on a 4x1 homogeneous coordinate.

$$\begin{pmatrix} 1 & 0 & 0 & 0 \\ 0 & 1 & 0 & 0 \\ 0 & 0 & 1 & 0 \\ 0 & 0 & 1 & 0 \end{pmatrix} \quad \begin{pmatrix} 1 & 0 & p & -p(1+r) \\ 0 & 1 & q & -q(1+r) \\ 0 & 0 & 1+r & -r(1+r) \\ 0 & 0 & 1 & -r \end{pmatrix} \tag{6.59}$$

6. Consider the 3×3 transformation

$$T = \begin{pmatrix} 2a & a & a \\ a & a & 0 \\ 2a & a & a \end{pmatrix} \qquad (6.60)$$

Is this a euclidean, affine or projective transformation? Prove or justify your answer.

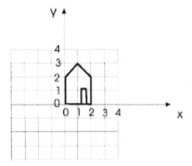

7. Consider the house given above. Assume the Z axis to be coming out of the page assuming the page to be the XY plane. Draw the house after the following transformations performed in the local coordinate system: $\mathcal{T}(1,0,0)$, $\mathcal{R}_z(90)$, $\mathcal{T}(0,2,0)$.

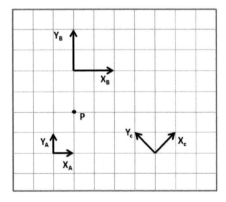

8. In the 2D space above, consider the point P and the three different coordinate axes, A, B and C. What is the coordinate of P in coordinate system

A, B and C? What is the 3×3 matrix to convert from homogeneous coordinates in coordinate system A to those in coordinate system B and from coordinate system C to those in coordinate system A?

9. How many degrees of freedom does a 3D homogeneous affine transformation have? Justify your answer. What do you think would be the degrees of freedom of a 3D projective transformation?

10. Consider the following matrix

$$
\begin{pmatrix}
a & 0 & p & x \\
0 & b & q & y \\
0 & 0 & 1 & z \\
0 & 0 & 0 & 1
\end{pmatrix}
\tag{6.61}
$$

What are the different fundamental transformations involved in this matrix and their parameters?

11. Consider a 3×3 projective transformation M that transforms a point (x, y, w) to (x', y', w'). M^{-1} is given by

$$
\begin{pmatrix}
2 & 1 & 0 \\
1 & 1 & 0 \\
2 & 1 & 1
\end{pmatrix}
\tag{6.62}
$$

(a) Show that the circle $x^2 + y^2 = 1$ is transformed to a parabola by this projective transformation. Find the equation of this transformed parabola.

(b) Consider two parallel lines given by $4x + y = 5$ and $4x + y = 3$. Show that these two lines are transformed to intersecting lines by the projective transformation. Find the equation of these two intersecting lines.

Hint: Note that we are concerned about $\frac{x'}{w'}$ and $\frac{y'}{w'}$ and $w = 1$.

12. In 3D, show that $R_z(\theta_1)R_z(\theta_2) = R_z(\theta_2)R_z(\theta_1)$. What does this tell about the properties of rotation around coordinate axes? Show that $R_z(\theta_1+\theta_2) = R_z(\theta_1)R_z(\theta_2)$. Using this property show that rotation about any arbitrary axis denoted by R_a also follows the property, $R_a(\theta_1+\theta_2) = R_a(\theta_1)R_z(\theta_2) = R_a(\theta_2)R_z(\theta_1)$.

7

The Pinhole Camera

The pinhole camera model, introduced in the last chapter, is by far the most popular model for a camera. The pin-hole camera is modeled as a closed box with a tiny hole punched with a pin on one of its faces. Light rays from any point in a scene enters the box only through this pinhole forming an inverted image on the opposite face of the box which is therefore termed the image plane. This image is formed by the intersection of the light rays passing through the pinhole with the image plane, as illustrated in Figure 6.13. The advantage of the pinhole camera is that every point in the scene, irrespective of its distance from the pinhole, will form a crisp or focused point image on the image plane. Depth of field of a camera is defined as the range of depth of the scene points which the camera can image in a focused manner (without blurring it). Therefore, a pinhole camera has an infinite depth of field. However, a pinhole camera is very light inefficient – very little light can enter through a pin-hole. Therefore, lenses are used to make the camera more light efficient. The result of this is a camera that no longer has an infinite depth of field, but still acts like a pinhole camera for all the points within its depth of field – i.e within the range of depth that the camera can image in a focused manner. In this chapter, we will first develop a mathematical model for pinhole camera. Then we will discuss how the deviation from the pinhole model affects image capture or acquisition from practical cameras.

7.1 The Model

Figure 7.1 shows a schematic for the pinhole camera model. The image plane is brought to the side of the scene to avoid inversion. In a practical camera, this is confirmed via the use of a complex lens system. O is the center of projection of the camera and the principal axis is parallel to the Z axis. The image plane is at a distance f from O and perpendicular to the principal axis (parallel to the XY plane). f is referred to as the focal length of the camera. To find the image of a 3D point $P = (X, Y, Z)$ on the camera's image plane, a straight line is drawn from P to O and its point of intersection with the image plane defines the image

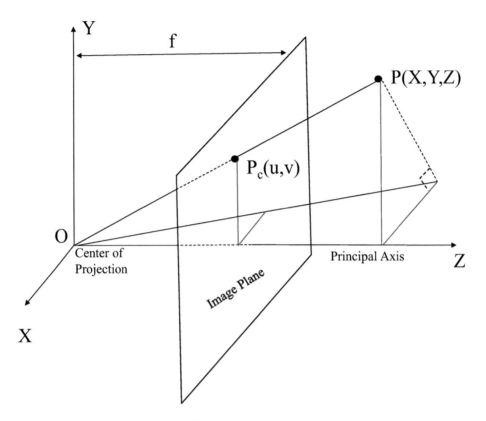

Figure 7.1. The Pinhole Camera

of the point P, denoted by the image plane coordinate $P_c = (u, v)$.

First, we will derive this function that maps the 3D point P to its 2D projection P_c. Considering the origin $(0, 0)$ of the image plane to be at the point on the image plane where the principal axis intersects it, we find using similar triangles

$$\frac{f}{Z} = \frac{u}{X} = \frac{v}{Y} \tag{7.1}$$

which gives us

$$u = \frac{fX}{Z} \tag{7.2}$$

$$v = \frac{fY}{Z} \tag{7.3}$$

Using homogeneous coordinates for P_c, we can write this as

$$\begin{pmatrix} u \\ v \\ w \end{pmatrix} = \begin{pmatrix} f & 0 & 0 \\ 0 & f & 0 \\ 0 & 0 & 1 \end{pmatrix} \begin{pmatrix} X \\ Y \\ Z \end{pmatrix} \tag{7.4}$$

You can verify that the above equation indeed generates the point $P_c = (u, v, w) = (\frac{fX}{Z}, \frac{fY}{Z}, 1)$. However, note that P is not expressed in homogeneous coordinates.

Fun Facts

Figure 7.2. Left: An artist in 18th century using camera obscura; Right: The Brownie of 1900.

The early cameras were called *camera obscura* which was essentially a pinhole camera that was used extensively by artists to create paintings by tracing out the image formed by the camera obscura. An arab physicist, Ibn al-Haytham, is credited with inventing the first camera obscura. He published the first Book of Optics in 1021 AD. Before the invention of the photographic film there was no way to preserve the image formed other than tracing it out. Several people worked hard on developing the photographic process including Nicphore Nipce in 1810s, Louis Daguerre and Henry Fox Talbot in 1830s and Richard Leach Maddox in 1870s. The use of photographic film was finally pioneered by George Eastman in 1889. His first camera, called "Kodak", was first offered for sale in 1889. It came preloaded with film to capture barely 100 pictures. In 1900, Eastman took mass-market photography one step further with the Brownie, a simple and very inexpensive box camera that introduced the concept of the 'snapshot'. The Brownie was extremely popular and various of its models remained on sale until the 1960s. Oskar Barnack, who was in charge of research and development at Leitz, commercialized the first 35mm camera, the Leica, in 1925. This was the early form of consumer film cameras that was in use even in the late 1990s.

This defines an ideal situation where the camera image plane is parallel to the XY plane and its origin is at the intersection of the principal axes with the image plane. Next, we will deviate from this ideal situation to add new parameters to the model. Let the origin of the image plane not coincide with the point where the Z axis intersects the image plane. In that case, we need to translate P_c to the desired origin. Let this translation be defined by (t_u, t_v). Hence, now (u, v) is given by

$$u = \frac{fX}{Z} + t_u \tag{7.5}$$

$$v = \frac{fY}{Z} + t_v \tag{7.6}$$

This can be expressed in a similar form as Equation 7.4 as

$$\begin{pmatrix} u \\ v \\ w \end{pmatrix} = \begin{pmatrix} f & 0 & t_u \\ 0 & f & t_v \\ 0 & 0 & 1 \end{pmatrix} \begin{pmatrix} X \\ Y \\ Z \end{pmatrix} \tag{7.7}$$

In the above equation, P_c is expressed in inches. Since this is a camera image, we need to express it in pixels. For this we will need to know the *resolution* or density of pixels in the camera (pixels/inch). If the pixels are square the resolution will be identical in both u and v directions. However, for a more general model, we assume rectangle (and not square) pixels with resolution m_u and m_v pixels/inch in u and v directions respectively. Therefore, to measure P_c in pixels, its u and v coordinates should be multiplied by m_u and m_v respectively. Thus

$$u = m_u \frac{fX}{Z} + m_u t_u \tag{7.8}$$

$$v = m_v \frac{fY}{Z} + m_v t_v \tag{7.9}$$

which are then expressed as

$$\begin{pmatrix} u \\ v \\ w \end{pmatrix} = \begin{pmatrix} m_u f & 0 & m_u t_u \\ 0 & m_v f & m_v t_v \\ 0 & 0 & 1 \end{pmatrix} \begin{pmatrix} X \\ Y \\ Z \end{pmatrix} = \begin{pmatrix} \alpha_x & 0 & u_o \\ 0 & \alpha_y & v_o \\ 0 & 0 & 1 \end{pmatrix} P = KP \tag{7.10}$$

K in the above equation only depends on the internal camera parameters like its focal length, principal axis, pixel size and resolution. These are called the intrinsic parameters of the camera. If the image plane is not a perfect rectangle, i.e. if the image plane axes are not orthogonal to each other, then K also includes a skew parameter s as

$$K = \begin{pmatrix} \alpha_x & s & u_o \\ 0 & \alpha_y & v_o \\ 0 & 0 & 1 \end{pmatrix} \tag{7.11}$$

Note that K is an upper triangular 3×3 matrix and is usually called the *intrinsic parameter* matrix for the camera.

Now, consider the situation where the camera's center of projection is not at $(0,0,0)$, the principal axis is not coincident with the Z-axis, and the image plane – though still orthogonal to the principal axis – is not parallel to the XY plane. In this case, we have to first use a matrix to coincide the camera's center of projection with $(0,0,0)$, its principal axis with the Z-axis and u-axis of the image plane align with the X-axis (or v-axis of the image plane align with the Y-axis) to make its image plane parallel to the XY plane. This transformation is achieved by a translation that moves the center of projection to the origin followed by a rotation to align the principal axis and the image plane. Let this translation be $T(T_x, T_y, T_z)$. Let the the rotation applied to coincide the principal axis with the Z axis be given by a 3×3 rotation matrix R. Then the matrix formed by first applying the translation followed by the rotation expressed using multiplication of sub-matrices is given by the 3×4 matrix

$$E = (R \mid RT). \tag{7.12}$$

E is called the *extrinsic parameter* matrix. Note that since translation is used now, we have to move to homogeneous coordinate for P as well. So, the complete transformation of P to P_c is now given by

$$P_c = K(R \mid RT)P = (KR \mid KRT)P = KR(I \mid T)P = CP \tag{7.13}$$

where the 3×4 matrix C is usually called the *camera calibration* matrix. Here, P is in 4D homogeneous coordinates $(X, Y, Z, 1)$ and P_c derived by CP is in 3D homogeneous coordinates (u, v, w). Therefore, the exact 2D location of the projection on the camera image plane will be obtained by normalizing the 3D homogeneous coordinates $(\frac{u}{w}, \frac{v}{w}, 1)$. The intrinsic parameter matrix has five degrees of freedom (2 for the location of the principal center, two for the size of pixels in two directions and one skew factor) while the extrinsic matrix has six degrees of freedom (3 each for translation and rotation). Therefore, C has 11 degrees of freedom. It can be shown that this implies that the the bottom right element of C will always be 1.

7.1.1 Camera Calibration

In this section, we will see how to find C (i.e. the 11 entries of C) for a particular camera and decompose it to get the intrinsic and extrinsic parameters. This process is called *camera calibration*. The first step of camera calibration is to find what is termed as *correspondences*. Correspondences are defined by the 3D points and their corresponding 2D projections on the camera image plane. If we know a $3D$ point P_1 is corresponding to P_{c_1} on the camera image coordinate, then

$$P_{c_1} = CP_1 \tag{7.14}$$

Or,

$$\begin{pmatrix} u_1 \\ v_1 \\ w_1 \end{pmatrix} = C \begin{pmatrix} X_1 \\ Y_1 \\ Z_1 \\ 1 \end{pmatrix} \qquad (7.15)$$

The normalized 2D camera image coordinates $(\frac{u_1}{w_1}, \frac{v_1}{w_1})$ are given by (u'_1, v'_1). This normalization is critical to assure that all the correspondences lie on the same 2D plane.

In order to find C we have to solve for its 11 unknowns. Let the rows of C be given by r_i, $i = 1, 2, 3$. Thus,

$$C = \begin{pmatrix} r_1 \\ r_2 \\ r_3 \end{pmatrix}. \qquad (7.16)$$

Since we know the correspondence P_1 and P_{c_1}, we know

$$u'_1 = \frac{u_1}{w_1} = \frac{r_1.P_1}{r_3.P_1} \qquad (7.17)$$

$$v'_1 = \frac{v_1}{w_1} = \frac{r_2.P_1}{r_3.P_1}. \qquad (7.18)$$

This gives us two linear equations

$$u'_1(r_3.P_1) - r_1.P_1 = 0 \qquad (7.19)$$

$$v'_1(r_3.P_1) - r_2.P_1 = 0 \qquad (7.20)$$

In the above equations, the unknowns are the elements of r_1, r_2 and r_3. Each 3D to 2D correspondence thus generates two linear equations. To solve for 11 unknowns, we will need at least six such correspondences. Usually for better accuracy, many more than six correspondences are used and the over-constrained system of linear equations thus formed is solved using linear regression methods for 11 entries of C. The correspondences can be determined using *fiducials* or *markers*. Markers are placed in known 3D locations in the 3D scene. Their coordinates in the image are determined either manually or automatically via image processing techniques to find the corresponding 2D locations.

Once C is recovered, the next step is to break it up into its intrinsic and extrinsic component. Since

$$C = (\ KR \mid KRT\ ) = (M \mid MT), \qquad (7.21)$$

where $KR = M$, we can find M as the left 3×3 sub matrix of C. Next, we use RQ decomposition to break M into two 3×3 matrices $M = AB$, where A is upper triangular and B is an orthogonal matrix (i.e. $B^T B = I$). This upper triangular A corresponds to K and B corresponds to the rotation R. Let c_4 denote the last column of C. From the previous equation, we can then find T from

$$MT = c_4 \tag{7.22}$$

$$T = M^{-1}c_4 \tag{7.23}$$

Thus, we recover the intrinsic and extrinsic parameters of the camera.

7.1.2   3D Depth Estimation

In the previous section we saw how given 3D to 2D correspondences, we can calibrate a camera. In this section, we will see how we can recover the 3D position (depth) of a scene seen by more than one calibrated cameras. In other words, given P_c and C of each camera, i.e. using 2D images of the 3D world formed by *calibrated cameras*, we will estimate the exact location of points in 3D. Let us assume an unknown position of a $3D$ point P, defined by homogeneous coordinates,(X, Y, Z, W) and its known image on the image plane of a camera defined by the matrix C_1 given by homogeneous coordinates $P_{c_1} = (u_1, v_1, w_1)$. Note that w_1 may not be 1.

Fun Facts

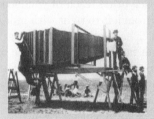

In 1900, George R. Lawrence built a mammoth 900 lb. camera, then the worlds largest, for $5,000 (enough to purchase a large house at that time!) It took 15 men to move and operate the gigantic camera. A photographer was commissioned by the Chicago & Alton Railway to make the largest photograph (the plate was 8′ × 4.5′ in size!) of its train for the companys pamphlet The Largest Photograph in the World of the Handsomest Train in the World.

Therefore, we know

$$P_{c_1} = \begin{pmatrix} u_1 \\ v_1 \\ w_1 \end{pmatrix} = C_1 \begin{pmatrix} X \\ Y \\ Z \\ W \end{pmatrix} \tag{7.24}$$

The corresponding 2D image points detected in the camera image coordinates is given by $(\frac{u_1}{w_1}, \frac{v_1}{w_1}) = (u_1', v_1')$. Representing the rows of the calibration matrix C_1 as $r_i^{C_1}$, $i = 1, 2, 3$, from Equation 7.24, we get two linear equations as follows.

$$u_1'(r_3^{C_1}.P) - r_1^{C_1}.P = 0 \tag{7.25}$$

$$v_1'(r_3^{C_1}.P) - r_2^{C_1}.P = 0 \qquad (7.26)$$

Therefore, from each camera we can generate two linear equations for P. We have 4 unknowns to be solved for P given by X, Y, Z, W. Therefore, we need at least two cameras with different calibration matrices (i.e. two cameras at two different positions) to find the 3D location of P. This provides what we call binocular cues or disparity. Further, also note that to recover P, we do need to find the point on the second camera's image that corresponds to the same $3D$ point P. Finding the image of the same $3D$ point on the images of two or more cameras is often termed as the *correspondence problem* and is considered a hard problem due to the large search space provided by an image. If there is no prior knowledge, every pixel in the image of the second camera is a candidate for being the 2D image of the P.

You may feel that sometimes humans can perceive depth even with a single eye. How is it possible if we say that at least two cameras (eyes in the case of humans) are needed for this purpose? It is not entirely true or entirely wrong that humans do have depth perception even with a single eye. Actually, humans have *some* depth perception with single eye due to several oculomotor (cues due to movement of the muscles holding the cornea) and monocular cues (cues of the eyeball moving inwards or outwards). These are not present for a camera and hence depth estimation is not possible with a single camera. However, try the following experiment with your friend to realize that we do not get an accurate depth perception without both eyes. Sit in front of each other, each of you close one of your eyes and both of you bring your right arm from left to right with index finger pointing to the left, and attempt to exactly touch each other's index finger tip. Attempt the same with both eyes open. You will notice the importance of depth perception in correctly judging the exact position of your friend's finger tip. In the absence of monocular or oculomotor cues, often more than two cameras are used (called stereo rigs) for greater accuracy and singular value decomposition is used to solve the over-constrained system of linear equations that result.

7.1.3   Homography

Homography is a mathematical relationship between the position and orientation of two cameras in a constrained situation where two cameras see the same points on a plane. This relationship can be easily recovered without going through an explicit camera calibration. Figure 7.3 illustrates the situation. Let us assume a point P_π on the plane π. Let the normal to the plane be appropriately defined as $N = (a, b, c)$ such that the plane equation can be written

$$(\ N\ \ 1\ ).P = 0 \qquad (7.27)$$

where P is any point on the plane. Let the two cameras be defined by calibration matrices C_1 and C_2. Without the loss of generality, we can assume that the origin

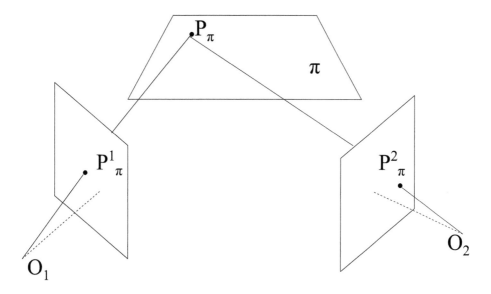

Figure 7.3. Homography between two cameras through a plane.

of the global coordinate system in which P_π is defined coincides with O_1, the center of projection of C_1. Let the image of P_π on camera C_1 and C_2 be P_π^1 and P_π^2 respectively. Therefore,

$$P_\pi^1 = \begin{pmatrix} u_1 \\ v_1 \\ w_1 \end{pmatrix} = C_1 . P_\pi, \qquad (7.28)$$

implying that the point P_π lies on the ray $(u_1, v_1, w_1, 0)^T$ in 3D. Let this point be at a distance τ on this ray. This implies

$$P_\pi = \begin{pmatrix} u_1 \\ v_1 \\ w_1 \\ \tau \end{pmatrix} = \begin{pmatrix} P_\pi^1 \\ \tau \end{pmatrix} \qquad (7.29)$$

Since P_π satisfies the plane equation, we get τ from Equation 7.27 as

$$\tau = -N . P_\pi^1. \qquad (7.30)$$

Therefore

$$P_\pi = \begin{pmatrix} u_1 \\ v_1 \\ w_1 \\ \tau \end{pmatrix} = \begin{pmatrix} I \\ -N \end{pmatrix} P_\pi^1 \qquad (7.31)$$

Note that I is a 3×3 matrix and N is a 1×3 matrix. Hence, $(I - N)^T$ is a 4×3 matrix.

Let $C_2 = (\ A_2 \quad a_2\ )$, where A_2 is the 3×3 matrix and a_2 is a 3×1 vector. Then,

$$P_\pi^2 = C_2.P_\pi \qquad\qquad (7.32)$$

$$= (\ A_2 \quad a_2\ ) \begin{pmatrix} I \\ -N \end{pmatrix} P_\pi^1. \qquad\qquad (7.33)$$

Using multiplication of sub-matrics we get a 3×3 matrix, what we call the homography H, as follows.

$$P_\pi^2 = (A_2 - a_2 N)P_\pi^1 = HP_\pi^1. \qquad\qquad (7.34)$$

a_2 is a 3×1 matrix and N is a 1×3 matrix. Thus, $a_2 N$ would generate a 3×3 matrix that can be subtracted from 3×3 matrix A_2 to generate H. Therefore, H is a 3×3 matrix that relates one camera image with another called the homography. Using this matrix, the image from one camera can be warped to produce the image from another camera. Therefore, for the special case when the scene observed by the two cameras is planar, instead of going through a full camera calibration, we can relate the image in one camera to another.

Homography is a 2D projective transformation and therefore has eight degrees of freedom that is equivalent to having the bottom right element as 1. Therefore, when computing H the number of unknowns is 8. From each correspondence, using Equation 7.33, we can generate two linear equations. To find the 8 unknowns in H, we need just 4 correspondences. However, it is always advisable to use more than four correspondences to create an over-constrained system which would yield a more robust estimate of H.

Now let us consider an alternate scenario where the location of the two cameras are the same (i.e. they have the same center of projection) but they have different orientations. In this case, the extrinsic parameters of these two cameras will differ only by a rotation, represented by a 3×3 matrix. The camera calibration matrix of these two cameras will be related by an invertible 3×3 matrix, and therefore a homography. This is the situation in the common application of panoramic image generation. A camera is usually mounted on a tripod or held in hand and rotated about a fixed center of projection to capture multiple images. Therefore, each camera position can be related to another via a homography. Though each image covers a narrow field of view, the multiple images can be stitched together to achieve an image with much larger field of view, more commonly called a *panorama*. In this application usually adjacent images have a considerable overlap. Common features in these overlaps are matched (manually or using automatic methods) and then used to recover the homography between adjacent camera locations. This homography is then used to transform the images to the reference coordinate system of one of the cameras to achieve

Figure 7.4. Three images (together) stitched together using homographic transformations to create a panorama (bottom). The red boundaries show the original images and the blue boundary shows a rectangular section cut off from the non-rectangular panorama.

a stitched panorama. This is illustrated in Figure 7.4. The overlap regions are blended together (using methods discussed at length in Chapter 11) to achieve a smooth color transition.

Put a Face to the Name

George Eastman (July 12, 1854 March 14, 1932) was an American innovator and entrepreneur who founded the Eastman Kodak Company and popularized the use of roll film making photography mainstream. Roll film was also the basis for the invention of motion picture film in 1888. Eastman was born in Waterville, New York as the youngest child of George Washington Eastman and Maria Eastman at the 10-acre farm which his parents bought in 1849. He was largely self-educated, although he attended a private school in Rochester after the age of eight. In the early 1840s his father had started a business school, the Eastman Commercial College in Rochester, New York, described as one of the first "boomtowns" in the United States, based on rapid industrialization. As his father's health started deteriorating, the family gave up the farm and moved to Rochester in 1860 where his father died of a brain disorder in May 1862. To survive and afford George's schooling, his mother took in boarders. The second of his two older sisters contracted polio when young and died in late 1870s when George was 16 years old. The young George left school early and started working to help support the family. As Eastman began to experience success with his photography business, he vowed to repay his mother for the hardships she had endured in raising him. He was a major philanthropist contributing to the establishments of many institutions, the most notable of them being the Eastman School of Music, schools of dentistry and medicine at the University of Rochester, the Rochester Institute of Technology (RIT), some buildings in MIT's second campus on the Charles River and historically black institutions of the South Tuskegee and Hampton universities. In his final two years, Eastman was in intense pain caused by a disorder affecting his spine. On March 14, 1932, Eastman shot himself in the heart, leaving a note which read, "To my friends: my work is done. Why wait?"[2]. The George Eastman House, now operated as the International Museum of Photography and Film, has been designated a National Historic Landmark.

7.2   Considerations in the Practical Camera

A pinhole camera is extremely light inefficient since very little light enters through the pinhole. Therefore, the design of the practical camera needs to deviate from this ideal pinhole camera model as shown in Figure 7.5. It consists of

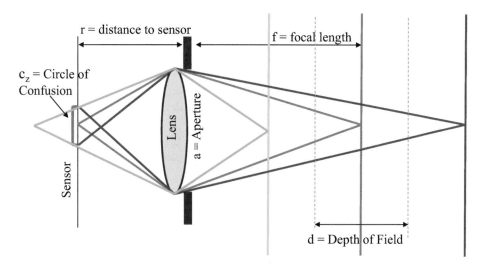

Figure 7.5. This shows a practical camera with an opening called an aperture to let in light, a lens to focus the light

a circular hole called the aperture to let the light in. This is usually made of a diaphragm to allow changing of its size, denoted by radius a, thereby allowing control of the amount of light to be let in (Figure 7.7). This is followed by a lens which allows the focusing of light on the sensor behind it. Let us denote the focal length of the lens with f. Let us consider a point in the 3D scene at a distance f away from the lens. The rays of light from this point are collected by the lens and focused on the sensor to form a sharp image (shown by the blue rays). For a focused image, this camera behaves just like a pinhole camera and its model that we developed in the previous section is valid. However, we need to consider the issues related to the parts of the scene which are not in focus which will be discussed in the rest of this section.

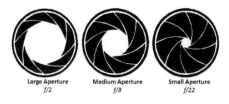

Figure 7.6. This shows how the aperture opening is changed using a diaphragm.

Consider Figure 7.6. Let the distance between the sensor and the lens be r. Now, let us consider a point at a depth z that is farther away or closer than f, as shown by the red and green rays. Notice that these rays focus before or behind the sensor. Therefore, instead of having a sharp focused image they create a blurry circle on the image plane called the *circle of confusion*. Let the radius of the circle of confusion for a point at depth z be denoted

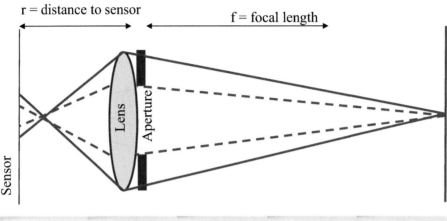

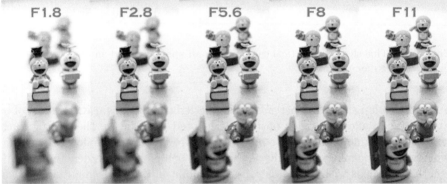

Figure 7.7. Top: This shows the effect of reducing the aperture size on the circle of confusion. The bold line shows the original aperture in Figure 7.5 and the dotted line shows the effect of reducing the aperture size that reduces the circle of confusion. Bottom: This shows some images taken from a camera using varying apertures decreasing from left to right.

by c_z. Using thin lens equation, it can be shown that

$$c_z = ar \left(\frac{1}{f} - \frac{1}{z} \right), \tag{7.35}$$

where a is the aperture of the lens. If c_z is less than the size of the pixel p, the image will look focused. The range of depth for which $c_z < p$ can be shown to be from $(f - d)$ to $(f + d)$ where $d = \frac{pf^2}{pf - ar}$. This range of depth from $f - d$ to $f + d$ is called the depth of field of the camera.

Next, let us see how these different parameters like aperture and focus have an effect on the picture captured by a camera. First let us check what happens when

Figure 7.8. The left three images show the effect of the focal length on the depth of field. Note that as the focal length increases the depth of field also increases. The right three images show the same effect for a smaller aperture. Note that for the same focal length, the smaller aperture have larger depth of field.

the aperture of the camera is reduced in size. Since c_z is directly proportional to aperture, the size of the circle of confusion goes down with reduced aperture. The implication of this is that if the pixel size remains the same, points at greater distance from f can now produce circle of confusion within p and therefore the depth of field of the camera will increase. This is also consistent with the pinhole camera model since as the aperture goes towards 0 as is the case in pinhole camera, the depth of field goes towards infinity. Usually aperture is expressed as a fraction of the focal length. An f2 aperture means an aperture size of $f/2$. These are usually specified as f-numbers. Typical f-numbers are f2, f4, f8, f16, f2.8, f5.6, f11 and so on. Figure 7.6 shows this effect.

Now, lets see what happens when the focal length of the lens is changed. With decrease in focal length, the $\frac{1}{f}$ term in Equation 7.35 increases thereby increasing c_z. Therefore, the depth of field of the camera reduces. This effect is shown in Figure 7.8. It also illustrates the combined effect of focal length and aperture on the depth of field.

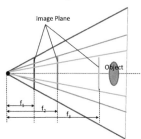

Figure 7.9. This figure illustrates the effect of focal length on the field of view if the sensor size remains the same.

The change of the focal length also has an effect on the field of view captured by the camera. The longer the focal length, the smaller the field of view. To understand this, let us go back to the pinhole camera as shown in Figure 7.9. Let the image plane be moved at different focal lengths, f_1, f_2 and f_3 such that $f_1 < f_2 < f_3$. Therefore, if the sensor size remains constant, as the focal length increases the field of view - the angle between the lines passing from the center of projection through the extremities of the sensor – decreases. This effect makes the relative size of the flowers in 7.8 become bigger as the focal length increases.

Figure 7.10. This figure illustrates motion blur. The color wheel in the left is static and its motion increases in the following figures from left to right, the rightmost one being the fastest. The picture is taken with the same shutter speed creating more blur for faster motion.

Finally, we will discuss one more parameter of a practical camera, the *shutter speed*. The sensor in the camera needs to be exposed for a limited time to capture the image. This exposure time is controlled by the shutter. When you hear a camera 'click', the shutter opens and remains so for sometime exposing the sensor to the light and then closes.

The time the shutter is open has a linear effect on the amount of light that is let in. Usually the shutter is open for a fraction of a second (e.g. $\frac{1}{30}$, $\frac{1}{60}$). If any object in the scene moves during the time the shutter is open, the image of the object is captured at multiple locations creating an effect called the motion blur as shown in Figure 7.10.

7.3   Conclusion

In this chapter we covered the fundamental model of a pinhole camera and its application in 3D depth reconstruction and homography based modeling. A more mathematical treatise of this model is available in [Faugeras 93], the classical book on 3D computer vision. More about stereo reconstruction and camera calibration is available at [Szeliski 10]. Details about the practical camera can be explored further by taking a course on computational photography – [Lukac 10] offers a in depth treatise in this direction.

Bibliography

[Faugeras 93] Olivier Faugeras. *Three-dimensional Computer Vision: A Geometric Viewpoint*. MIT Press, 1993.

[Lukac 10] Rastislav Lukac. *Computational Photography: Methods and Applications*, First edition. CRC Press, Inc., 2010.

[Szeliski 10] Richard Szeliski. *Computer Vision: Algorithms and Applications*. Springer-Verlag New York, Inc., 2010.

Summary: Do you know these concepts?

- Pinhole Camera

- Intrinsic and Extrinsic Camera Parameters

- Camera Calibration

- Depth Estimation or Reconstruction

- Stereo Camera Pair

- Homography

- Focal Length

- Aperture

- Depth of Field

- Shutter Speed

- Motion Blur

- Field of View

Exercises

1. Consider the following 3×4 camera matrix

$$C = \begin{pmatrix} 10 & 2 & 11 & 19 \\ 10 & 5 & 10 & 50 \\ 5 & 14 & 2 & 17 \end{pmatrix} \qquad (7.36)$$

   Consider the 3D point in homogeneous coordinates $X = (0, 2, 2, 1)^T$.

   (a) What are the Cartesian coordinates of the point X in 3D?

   (b) What are the Cartesian image coordinates of the projection of X?

2. Consider an ideal pinhole camera with focal length of $5mm$. Each pixel is $0.02mm \times 0.02mm$ and the image principal point is at pixel $(500, 500)$. Pixel coordinates start at $(0, 0)$ in the upper-left corner of the image.

   (a) What is the 3×3 camera calibration matrix, K, for this camera configuration?

   (b) Assuming the world coordinate frame is aligned with the camera coordinate frame (i.e., their origins are the same and their axes are aligned), and the origins are at the cameras pinhole, what is the 3×4 matrix that represents the extrinsic, rigid body transformation between the camera coordinate system and the world coordinate system?

   (c) Combining your results from the previous two questions, compute the projection of scene point $(100, 150, 800)$ into image coordinates.

3. A camera is rigidly mounted so that it views a planar table top. A projector is also rigidly mounted above the table and projects a narrow beam of light onto the table, which is visible as a point in the image of the table top. The height of the table top is precisely controllable but otherwise the positions of the camera, projector, and table are unknown. For table top heights of $50mm$ and $100mm$, the point of light on the table is detected at image pixel coordinates $(100, 250)$ and $(140, 340)$ respectively.

   (a) Using a projective camera model specialized for this particular scenario, write a general formula that describes the relationship between world coordinates (x), specifying the height of the table top, and image coordinates (u, v), specifying the pixel coordinates where the point of light is detected. Give your answer using homogeneous coordinates and a projection matrix containing variables.

   (b) For the first table top position given above and using your answer in the previous question, write out the explicit equations that are generated by this one observation. How many degrees of freedom does this transformation have?

(c) How many table top positions and associated images are required to solve for all of the unknown parameters in the projective camera model?

(d) Once the camera is calibrated, given a new unknown height of the table and an associated image, can the height of the table be uniquely solved for? If so, give the equation(s) that is/are used. If not, describe briefly why not.

(e) If in each image we only measured the u pixel coordinate of the point of light, could the camera still be calibrated? If so, how many table top positions are required? If not, describe briefly why not.

4. Assume a camera with camera matrix $C = K[r_1r_2r_3t]$, where K is the intrinsic parameter matrix and r_1, r_2, and r_3 are the columns of the rotation matrix. Let π be the XY plane at $Z = 0$. We know that any point P in this plane can be related to the camera image point P by a homography H, i.e. $P = HP$. Show that $H = K[r_1r_2t]$.

5. Consider a panoramic image generation application where the camera is placed on a tripod and rotated to capture multiple images for panoramic image generation. Can two adjacent images in this sequence be related by a homography? If so, under what conditions is this possible?

6. Four projectors are tiled in a 2×2 array to create a tiled display on a flat wall. The projectors have some overlap between each other. What is the minimum dimension of the matrix that relates pixel (x, y) in one projector to a pixel (x', y') in another. Justify your answer.

7. What are the two parameters in a practical camera that allow you to control the amount of light reaching the sensor? How does the elements or events of a scene guide the choice of which parameter you would use to control the amount of light?

8. Freezing motion is a technique to choose the correct shutter speed for capturing a moving object so that they appear to be static or frozen in the image. You are asked to freeze motion for a moving car, a person jogging in the park, a person taking a stroll on the beach, and a fast moving train. You are allowed to choose between four shutter speeds of $\frac{1}{125}$, $\frac{1}{250}$, $\frac{1}{500}$, and $\frac{1}{1000}$. Which speed will you choose for which object?

9. Consider a parametric line $P_0 + \alpha(P_1 - P_0)$ in the 3D scene. Consider a point P moving on this line as α goes from 0 to 1. Show that its projection under the camera calibration matrix will converge to a vanishing point.

10. Are the intrinsic and extrinsic parameter matrices affine, Euclidian or projection? Justify your answer. We know that the camera calibration matrix is a projective transformation matrix? Which of the intrinsic and the extrinsic parameter matrices contributes to it being a projective transformation? Justify your answer.

11. The camera calibration matrix is a 3×4 matrix whose inverse is not defined. What is the geometric interpretation of this in the context of reconstructing 3D geometry from the 2D image of a single camera?

12. Explain why a portraits eyes appear to "follow you around the room". Give your answer in terms of a homography relationship between the viewer and the picture.

Epipolar Geometry

In the previous chapter we learnt how we can use two or more cameras to re-construct the geometry of a scene. This is often considered as one of the most important goals of computer vision. Scene reconstruction is a fundamental step towards automated scene understanding. Only when the basic scene geometry is reconstructed, we can delve deeper in other aspects like understanding objects, their movements and interactions with other elements of the scene – all of which are related to much higher levels of cognition in humans as well.

Epipolar geometry defines geometric constraints across multiple cameras capturing the same scene. This enables simplification of common problems (like finding correspondences) when dealing with important vision tasks like motion estimation or 3D depth reconstruction. It is fascinating to see how even relatively simple constraints can make such hard problems tractable. In this chapter we will cover the fundamental concepts of epipolar geometry. We start this chapter by defining the notations we will be using.

8.1 Background

Let us consider a line defined by two 2D points, $A(x, y, t)$ and $B(u, v, w)$, in homogeneous coordinates, as shown in Figure 8.1. Therefore, the normalized homogeneous coordinates that provide the projection of these points in the 2D plane defined by $Z = 1$ is given by $A' = (\frac{x}{t}, \frac{y}{t})$ and $B' = (\frac{u}{w}, \frac{v}{w})$. Let the line between A' and B' be M_l. M_l is shown in red in Figure 8.1. The normal to the plane OAB is given by

$$B \times A = \begin{pmatrix} yw - tv \\ tu - xw \\ xv - yu \end{pmatrix}. \tag{8.1}$$

Any point lying on M_l should be the projection of a point $P = (p, q, r)$ that lies on the plane OAB. Therefore, P will satisfy the plane equation defined by the above normal as

$$p(yw - tv) + q(tu - xw) + r(xv - yu) = 0 \tag{8.2}$$

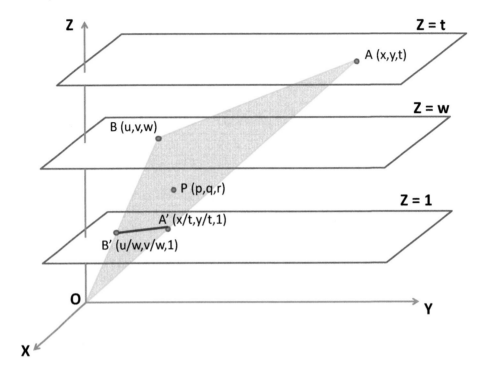

Figure 8.1. This figure shows two 2D points, A and B, in homogeneous coordinates and how a point P lying in the plane OAB relates to A and B.

Therefore, $B \times A$ provides the coefficients of the equation of the plane OAB. In other words, P would satisfy the following equation.

$$P^T \begin{pmatrix} yw - tv \\ tu - xw \\ xv - yu \end{pmatrix} = P^T(B \times A) = 0. \tag{8.3}$$

Now, consider the line M_l on the plane $Z = 1$ formed by the normalized homogeneous coordinates $A' = (\frac{x}{t}, \frac{y}{t}, 1)$ and $B' = (\frac{u}{w}, \frac{v}{w}, 1)$. Therefore, the slope m and offset c of this line M_l is given by

$$m = \frac{tv - yw}{tu - xw} \tag{8.4}$$

$$c = \frac{yu - xv}{tu - xw}. \tag{8.5}$$

Therefore, the equation of M_l is given by

$$(tv - yw)x_l + (xw - tu)y_l + (yu - zw) = 0 \tag{8.6}$$

where (x_l, y_l) is the 2D point on M_l denoted by the normalized homogeneous coordinates $(x_l, y_l, 1)$. Alternatively,

$$(x_l \; y_l \; 1) \begin{pmatrix} yw - tv \\ tu - xw \\ xv - yu \end{pmatrix} = (x_l \; y_l \; 1)(B \times A) = 0. \tag{8.7}$$

Therefore, $B \times A$ defines the coefficients for both the equation of the line M_l or equation of the plane OAB based on whether we are considering normalized or un-normalized homogeneous coordinates. We will use this fact effectively in many places when working with epipolar geometric constraints. Also, *we will be using the coefficient matrix of a line to describe the line itself.* Therefore, where we define a line l as

$$l = \begin{pmatrix} a \\ b \\ c \end{pmatrix}, \tag{8.8}$$

we refer to a line l with slope $\frac{-a}{b}$ and offset $\frac{-c}{b}$, as derived from the above equations. This notation will be used frequently in the rest of this chapter.

Now note that

$$B \times A = \begin{pmatrix} yw - tv \\ tu - xw \\ xv - yu \end{pmatrix} = \begin{pmatrix} 0 & w & -v \\ -w & 0 & u \\ v & -u & 0 \end{pmatrix} \begin{pmatrix} x \\ y \\ t \end{pmatrix} = [B]_X A \tag{8.9}$$

The left matrix is a special matrix with only the coordinates of B as its entries and hence is called $[B]_X$. Note that $[B]_X$ is a symmetric matrix, i.e. $[B]_X = [B]_X^T$. Now, since P satisfies Equation 8.3, the following equation will hold.

$$P^T([B]_X A) = 0 \tag{8.10}$$

In fact, it can also be shown that

$$P^T([B]_X A) = (A^T [B]_X^T) P = 0 \tag{8.11}$$

Take a special note of the dimensions of the matrices and you will find the result to be a 1×1 scalar. Also, the determinant of $[B]_X$ is 0 and all the 2×2 sub-matrices have non-zero determinant. Therefore, $[B]_X$ is a rank 2 matrix.

8.2 Correspondences in Multi-View Geometry

Consider two cameras, C_1 and C_2 for stereo depth reconstruction (Figure 8.2). Let their center of projection be O_1 and O_2 and image planes be I_1 and I_2 respectively. The line segment $O_1 O_2$ is called the *baseline*. The baseline should be of non-zero length in order to perform stereo reconstruction i.e. $O_1 \neq O_2$. Let us consider a 3D point P and let its image on C_1 and C_2 be p_1 and p_2 respectively. Let us now make some observations about this geometric setup.

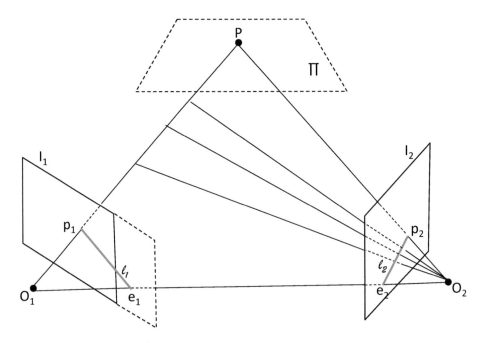

Figure 8.2. This figure illustrates the setting for finding the epipolar constraints in a two camera system. These two cameras are defined by their center of projection O_1 and O_2 and their image planes I_1 and I_2 respectively. Both of them are seeing the 3D point P. The image of the point P is given by p_1 and p_2 on the two cameras. e_1 and e_2 define the epipoles of the two cameras and l_1 and l_2 provide the epipolar lines for searching for the correspondences for the 3D point P.

1. PO_1O_2 forms a plane. As the location of the 3D point P changes, this plane changes but it rotates about the baseline O_1O_2.

2. The image of points on the ray O_2P falls on the line l_1 in C_1. Similarly, the image of any point on O_1P falls on the line l_2 in C_2.

3. The line joining O_1O_2 intersects the image plane I_1 and I_2 at points e_1 and e_2 respectively. These are called the *epipoles*. Note that the epipole need not be located on the physical image plane of the camera, but can be on the extension of its plane, as is the case for C_1.

4. The line l_1 and l_2 are given by e_1p_1 and e_2p_2 respectively and are called the *epipolar lines*. Note that as the plane PO_1O_2 change with a change in the position of P, the epipoles do not change since the baseline O_1O_2 does not change. Therefore, all epipolar lines pass through the epipole of the image.

Figure 8.3. On the top we see the marked features in the left image which lie on the epipolar lines in the right image and these lines can then be searched to find the corresponding feature. In the bottom, we show two images captured from a stereo camera pair and the epipolar lines on each of them. Note that the epipole in both the images lie outside the physical image.

The question is, why is this important? The importance of the above constraints is that they reduce the search space for correspondence when using a *calibrated* stereo camera pair. Since we consider calibrated cameras, the position of each of the camera can be projected on the image plane of the other thus giving us the epipoles e_1 and e_2. Next, if we detect the feature p_1 (image of P) in camera C_1, then its correspondence is bound to lie on the line $e_2 p_2 = l_2$. Therefore, instead of searching the entire image, we can now search on the line l_2 for the correspondence. Therefore, our search space for finding correspondence has reduced from 2D to 1D which leads to significant computational savings when finding the depth of points in the scene. This is illustrated in Figure 8.3. In the following sections, we will learn the mathematical foundations for reducing the search space for correspondences from 2D to 1D. Towards that, we need to first learn about fundamental matrix.

8.3   Fundamental Matrix

Fundamental matrix, F, is a 3×3 matrix that helps us to find the line l_2 on which a correspondence p_2 of p_1 in camera C_1 lies in C_2. We will show that l_2 is given by Fp_1.

In order to define the concept of *fundamental matrix* of a camera, we will use the same geometric setup in Figure 8.2. Let $p_1 = (x_1, y_1, t_1)$ and $e_1 = (u, v, w)$. From derivations in Section 8.1 we know that the line l_1 defined by its endpoints e_1 and p_1 is given by

$$l_1 = [e_1]_X p_1 = \begin{pmatrix} 0 & w & -v \\ -w & 0 & u \\ v & -u & 0 \end{pmatrix} \begin{pmatrix} x_1 \\ y_1 \\ t_1 \end{pmatrix} = Lp_1, \qquad (8.12)$$

where $L = [e_1]_X$. Now we know from the concepts presented in the previous chapter that since l_1 and l_2 are coplanar, there is a 3×3 homography or a 2D affine transformation A that maps l_1 to l_2. Therefore,

$$l_2 = Al_1 \qquad (8.13)$$
$$= ALp_1 \qquad (8.14)$$
$$= Fp_1 \qquad (8.15)$$

Now let $p_2 = (x_2, y_2, t_2)$. Since p_2 lies on the line l_2, it will satisfy the line equation

$$p_2^T l_1 = p_2^T Fp_1 = 0 \qquad (8.16)$$

Here, F related the two correspondences p_1 and p_2 and is called the fundamental matrix. Since A and L are both 3×3 matrices, the fundamental matrix F is also a 3×3 matrix. Also L is a rank 2 matrix and A is a rank 3 matrix. Therefore, F is a rank 2 matrix. A being a homography has eight degrees of freedom. When multiplied by rank 2 matrix L, the resultant F has an additional constraint of $det(F) = 0$. This reduces the degrees of freedom of F by 1 resulting in a matrix with seven degrees of freedom.

Fun Facts

Epipolar geometry seems to have been first uncovered by von Guido Hauck in 1883. He wrote several papers on the trilinear relationships of points and lines seen in three images. In his work, Hauck did not deeply analyze these trilinear relationships theoretically, that was done later via trifocal tensors in the the 1990s. He rather concentrated on the application of these relationships to generate a third image from two given ones (often called trifocal transfer in computer vision). This concept is the mainstay of the field of image based rendering in computer graphics developed in 1990s. Epipolar geometry was also explored by Hesse, in a limited manner, in 1863 in response to a challenging problem posed by French mathematician, Michel Chasles, where he challenged mathematicians to determine two pencils of 2D lines in homographical relationship given seven pairs of matching points such that matching lines are incident with matching points.

8.3.1 Properties

We can summarize the main properties of the fundamental matrix F as follows.

1. F is a rank 2 matrix with seven degrees of freedom.

2. The epipolar lines l_1 and l_2 on which p_1 and p_2 respectively lie are given by

$$l_2 = Fp_1 \tag{8.17}$$
$$l_1 = F^T p_2 \tag{8.18}$$

3. Therefore, two corresponding points p_1 and p_2 are related by

$$p_2^T F p_1 = 0 \tag{8.19}$$

or

$$p_1^T F^T p_2 = 0. \tag{8.20}$$

Also, depending on which of the above two equations is being used, F^T can also be considered a fundamental matrix.

4. The epipoles are related to F by

$$F e_1 = F^T e_2 = 0 \tag{8.21}$$

At this point, you may be wondering that both the homography and fundamental matrix are 3×3 matrices that define constraints across two stereo cameras. So, what is the difference between these two. Note that homography helps you to find the corresponding *point* in another camera given a point in the first one. Therefore, homography maps a point to another point. And this constraint is imposed due to a more restrictive scene composition realized by either C_1 and C_2 having a common center of projection or all 3D points lying on a plane. The fundamental matrix, on the other hand, only defines a *line* on which the correspondence will lie in the second camera. Therefore, fundamental matrix maps a point to a line, and not another point as in homography.

8.3.2   Estimating Fundamental Matrix

The next obvious question is, how do we estimate the fundamental matrix? In this section, we will explore estimation of fundamental matrix for different camera pair setups.

Calibrated Camera If we have a *calibrated pair* of stereo cameras, finding the fundamental matrix is relatively easy. Let the 3×4 calibration matrix of camera C_1 and C_2 be given by A_1 and A_2 respectively. Therefore,

$$p_1 = A_1 P \tag{8.22}$$

Since A_1 is not a square matrix, it is not invertible. But we can find its pseudo-inverse A_1^+ which is a 4×3 matrix such that $A_1 A_1^+ = I$ where I is the 3×3 identity matrix. It can be shown that $A_1^+ = (A_1^T A_1)^{-1} A_1^T$. Using this pseudo inverse A_1^+, we can write Equation 8.22 as

$$P = A_1^+ p_1 \tag{8.23}$$

Now, the image p_2 of P in the second camera can be expressed as

$$p_2 = A_2 P \tag{8.24}$$
$$= A_2 A_1^+ p_1 \tag{8.25}$$

The line l_2 can be defined by its endpoints e_2 and p_2 and also by Equation 8.18 giving

$$l_2 = e_2 \times p_2 = [e_2]_X A_2 A_1^+ p_1 = F p_1 \tag{8.26}$$

Therefore, F can be derived from the above equation as

$$F = [e_2]_X A_2 A_1^+. \tag{8.27}$$

In fact, it can be shown that $A_2 A_1^+$ is a full rank 3×3 matrix and is the homography A defined in Equation 8.14.

First Camera Aligned with World Coordinate Next, we will simplify the camera setup even more. Let us consider C_1 to be located at the origin aligned with the coordinate axes. Let the intrinsic matrix for C_1 be K_1. Therefore, A_1 is given by

$$A_1 = K_1 (I|O) \tag{8.28}$$

where I is the 3×3 identity matrix and O is $(0,0,0)^T$. Further A_1^+ is given by

$$A_1^+ = \begin{pmatrix} K_1^{-1} \\ O \end{pmatrix}. \tag{8.29}$$

Let the translation, rotation and intrinsic matrix of C_2 be given by T, R and K_2 respectively. Therefore, the calibration matrix of C_2, A_2, is given by

$$A_2 = K_2 (R|T) = (K_2R|K_2T). \tag{8.30}$$

From the above equations we can find the homography A as

$$A = A_2 A_1^+ = K_2 R K_1^{-1}. \tag{8.31}$$

Let us now consider e_2, the image of O_1 on I_2. e_2 is given by

$$e_2 = A_2 \begin{pmatrix} 0 \\ 0 \\ 0 \\ 1 \end{pmatrix} = (K_2R|K_2T) \begin{pmatrix} 0 \\ 0 \\ 0 \\ 1 \end{pmatrix} = K_2T. \tag{8.32}$$

Therefore, from Equation 8.27, we can now derive the fundamental matrix F as

$$F = [e_2]_X A_2 A_1^+ \tag{8.33}$$

$$= [K_2T]_X K_2 R K_1^{-1} \tag{8.34}$$

This shows that when using a simplified setup of calibrated cameras where one of them is aligned with the world coordinate axes, finding fundamental matrix is even easier.

8.3.3 Camera Setup Akin to Two Frontal Eyes

Next, we will explore a very specific type of camera setup like the two frontal eyes in animals like humans (and not lateral eyes in animals like rabbits). In this case, the two cameras can be assumed to have the same intrinsic matrix, i.e. $K_1 = K_2 = K$. This assumption is not as unlikely as it may seem. Even in consumer devices, most cameras of the same make often have the same intrinsic matrix. When considering frontal eyes, the relative orientation of the second camera can be defined by just a translation with respect to the first camera and no rotation. Therefore, both the cameras are in exactly the same orientation coincident with the coordinate axes, but while one is at the origin the other is translated to another location. Under these assumptions, Equation 8.34 becomes

$$F = [e_2]_X K K^{-1} = [e_2]_X. \tag{8.35}$$

Now let us simplify the setup further by assuming that the translation is parallel to the X axis – exactly the way the human eye is. In this scenario, the epipole e_2 will be on the X axis but at infinity. Therefore, $e_2 = (1, 0, 0)^T$ and

$$F = [e_2]_X = \begin{pmatrix} 0 & 0 & 0 \\ 0 & 0 & -1 \\ 0 & 1 & 0 \end{pmatrix}. \tag{8.36}$$

Let us now consider two corresponding pixels (x_1, y_1) and (x_2, y_2) as camera C_1 and C_2 respectively. Plugging the above fundamental matrix in Equation 8.19 gives

$$\begin{pmatrix} x_2 & y_2 & 1 \end{pmatrix} \begin{pmatrix} 0 & 0 & 0 \\ 0 & 0 & -1 \\ 0 & 1 & 0 \end{pmatrix} \begin{pmatrix} x_1 \\ y_1 \\ 1 \end{pmatrix} = 0 \qquad (8.37)$$

$$Or, \qquad \begin{pmatrix} 0 & 1 & -y_2 \end{pmatrix} \begin{pmatrix} x_1 \\ y_1 \\ 1 \end{pmatrix} = 0 \qquad (8.38)$$

$$Or, \qquad y_1 - y_2 = 0 \qquad (8.39)$$

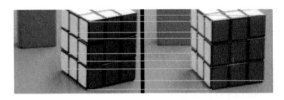

Figure 8.4. This figure shows this case of the frontal eye and how the correspondences lie on the same rater lines (shown by the green lines).

Therefore, for this setup, the epipolar lines are raster-lines (lines parallel to the X axis) and the epipoles are at infinity. Therefore, correspondences lie on the same raster lines on the two images and hence finding them is very easy as illustrated in Figure 8.4.

Uncalibrated Camera More often than not, we face a situation where the cameras are not calibrated. The question is, how do we find the fundamental matrix if the camera calibration matrices are unknown?

For that, let us consider two points in the images of C_1 and C_2 respectively, given by (x_1, y_1) and (x_2, y_2) to be corresponding features detected manually or any software assisted process. These two points will satisfy Equation 8.19 and therefore

$$\begin{pmatrix} x_2 & y_2 & 1 \end{pmatrix} \begin{pmatrix} f_1 & f_2 & f_3 \\ f_4 & f_5 & f_6 \\ f_7 & f_8 & f_9 \end{pmatrix} \begin{pmatrix} x_1 \\ y_1 \\ 1 \end{pmatrix} = \begin{pmatrix} x_2 & y_2 & 1 \end{pmatrix} F \begin{pmatrix} x_1 \\ y_1 \\ 1 \end{pmatrix} = 0 \tag{8.40}$$

where $f_1, \ldots f_9$ denote the entries of the fundamental matrix F. From the above equation we can generate the following linear equation

$$x_1 x_2 f_1 + x_1 y_2 f_2 + x_1 f_3 + y_1 x_2 f_4 + y_1 y_2 f_5 + y_1 f_6 + x_2 f_7 + y_2 f_8 + f_9 = 0 \tag{8.41}$$

Thus, every pair of correspondences detected creates a linear equation. Therefore, with adequate correspondences we can estimate F. Though F has seven degrees of freedom, it can be shown that it has eight parameters in the matrix that can be affected by the seven degrees of freedom. Therefore, we need at least eight correspondences to estimate F using this method. In the subsequent sections we will see how F is used for different purposes in different situations.

8.4 Essential Matrix

Essential matrix E is defined as the fundamental matrix of normalized cameras. A normalized camera is achieved when the coordinates of the camera are normalized and hence the name. Therefore, the intrinsic matrix K of a normalized camera is identity, i.e. $K = I$. Therefore, when considering two normalized correspondences $\hat{p}_1$ and $\hat{p}_2$, all the properties of fundamental matrix described in Section 8.3.1 are now applicable to essential matrix, the most useful of them being

$$l_2 = E\hat{p}_1 \tag{8.42}$$

$$l_1 = E^T\hat{p}_2 \tag{8.43}$$

$$\hat{p}_2{}^T E\hat{p}_1 = 0 \tag{8.44}$$

$$\hat{p}_1{}^T E^T \hat{p}_2 = 0. \tag{8.45}$$

The normalization removes two scale factors that denote the size of the pixel. Therefore, two degrees of freedom are reduced from F to yield E. Hence, E has five degrees of freedom.

Since any pair of camera can be reduced to the situation where C_1 is aligned with the world coordinate axes and C_2 translated by T and rotated by R with respect to C_1, we derive the essential matrix E by replacing $K_1 = K_2 = I$ in Equation 8.34 giving

$$E = [T]_X R \tag{8.46}$$

Since R and $[T]_X$ are both symmetric matrices, we can find E^T as

$$E^T = ([T]_X R)^T = R^T [T]_X^T = R[T]_X \tag{8.47}$$

The five degrees of freedom of E can also be seen from the above equations. Since E depends on R and T that have two and three degrees of freedom respectively, E has five degrees of freedom.

8.5 Rectification

Rectification is a process by which we take the images from a pair of stereo cameras and apply appropriate transformations such that they simulate the case of the frontal eyes and therefore the correspondences lie on the raster lines. In this section we will learn how to rectify images from two normalized uncalibrated cameras.

We can assume without loss of generality that one of these normalized cameras is aligned with the world coordinates while the other is translated by T and rotated by R with respect to it. Therefore, correspondences in these two normalized cameras will be related by essential matrix as explained by equations in the previous section. However, since these are uncalibrated cameras, we do not know R and T. But we can use a few normalized correspondences to estimate the essential matrix using the method to estimate fundamental matrix for uncalibrated cameras explained in Section 8.3.3.

Now, since we know that this E is related to R and T by Equations 8.46 and 8.47, if we can recover R and T from the computed E, we can apply the appropriate transformation to the second normalized camera image plane to convert the configuration of the camera pair to that of the frontal eye where the two cameras only differ by a translation along the X axis. This process is called *rectification*. The correspondences now lie on raster lines of the two images and are therefore significantly easier to locate.

Therefore, the next question is *how do we find the rotation and translation for the second camera from the estimated essential matrix?*. For this, we first use SVD decomposition to decompose E into

$$E = U\Sigma V^T \tag{8.48}$$

Note that U and V are orthogonal matrices and therefore their transpose is equal to their inverse. Let us also define a matrix W as

$$W = \begin{pmatrix} 0 & -1 & 0 \\ 1 & 0 & 0 \\ 0 & 0 & 1 \end{pmatrix} \tag{8.49}$$

such that $W^{-1} = W^T$.

It can be shown that the U, V, W and Σ thus defined above can be combined to create four different sets of solutions for R and T that will satisfy the Equations 8.46 and 8.47. These solutions are enumerated below.

$$\begin{aligned} &\textit{Solution 1}: \quad R = UW^{-1}V^T \quad [T]_X = VW\Sigma V^T & (8.50)\\ &\textit{Solution 2}: \quad R = UWV^T \quad [T]_X = VW^{-1}\Sigma V^T & (8.51)\\ &\textit{Solution 3}: \quad R = UWV \quad [T]_X = V^TW^{-1}\Sigma V^T & (8.52)\\ &\textit{Solution 4}: \quad R = UW^{-1}V \quad [T]_X = V^TW\Sigma V^T & (8.53) \end{aligned}$$

Let us verify one of these by plugging in the values of R and T given by the

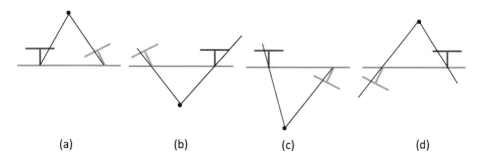

(a)   (b)   (c)   (d)

Figure 8.5. This shows that four solutions provided for the location of the second (green) camera when considering the rectification to transform it so that the red and green camera together yield the frontal eye configuration. Each camera is depicted by a "T" where the bottom of the "T" is the center of projection and the line of the top is the image plane.

first solution to Equation 8.47.

$$R[T]_X = UW^{-1}V^T VW\Sigma V^T \tag{8.54}$$
$$= UW^{-1}V^{-1}VW\Sigma V^T \tag{8.55}$$
$$= U\Sigma V^T \tag{8.56}$$
$$= E^T \tag{8.57}$$

which is the essential matrix itself. Similarly, any of the four solutions enumerated above will satisfy the equations 8.47. We use the equation for E^T in this case since it is the matrix that relates the correspondence from the second camera (which we plan to rectify) to the line on the first camera.

The normalized second camera depicted by these four solutions is shown in green with respect to the first camera in red (that is aligned with the world coordinate system) in Figure 8.5. Interestingly, though all four of these form valid theoretical solutions, only one of these is practically possible. This is the one illustrated in Figure 8.5(a) where the imaged point in black is in front of both the red and green camera. In (b), the point is behind both the cameras. In (c) and (d), the imaged point is behind one of the cameras. In fact, in (b) and (c), the imaged point is behind the baseline which is often referred to as the *baseline reversal*. Once rectified using the solution thus generated, all the epipolar lines in the two images are horizontal as illustrated in Figure 8.6.

Figure 8.6. This image shows the rectification. The left two images are unrectified and hence the epipolar lines corresponding to the features in the left image are not horizontal. The middle image is rectified to create the right image and the lines are now horizontal. The left and the right image together is called an rectified pair of images.

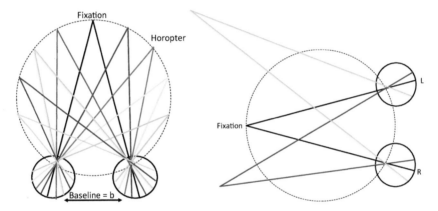

Figure 8.7. Left: This shows the horopter – the circle points of which get imaged at corresponding points in the left and right eye. Right: Points not on the horopter are imaged at non-corresponding points in the left and right eye.

8.6 Applying Epipolar Geometry

In this section we will see some applications of epipolar geometry. In particular, when dealing with uncalibrated cameras, epipolar geometry provides some constraints which can be used to derive different geometric scene parameters like depth.

8.6.1 Depth from Disparity

Reconstructing depth from disparity is one of the most significant application of epipolar geometry. In this section we are going to derive formally the equations we need to reconstruct depth from disparity.

Let us consider the two frontal human eyes shown by the two solid circles in Figure 8.6.1. First, we are going to introduce the concept of *corresponding points* that is different than correspondences that we have been discussing so far in this chapter. Corresponding points are defined as the points that coincide when two eyes are slipped on top of each other to overlap completely with each other. Interestingly, 3D points on the scene that lie on a specific circle at a particular radius from the eye are imaged at corresponding points in the two eyes. This is illustrated in the left figure of Figure 8.6.1. This circle is called the *horopter*.Points which are not on the horopter are imaged at non-corresponding points on the left and right eye as shown in the right figure of Figure 8.6.1. Imaging of such points are shown by green and red in this figure. The depth of these points can be deciphered by the difference of their distance from the image of the point of fixation given by black. This difference is called *disparity*.

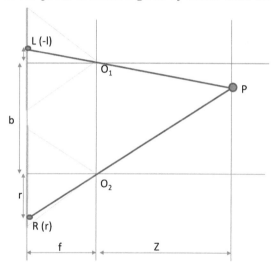

Let us now take this concept of disparity to rectified images and see how we can reconstruct 3D depth of the objects seen in an image using disparity using Figure 8.8. This figure shows two normalized rectified camera whose image planes and the field-of-view are shown by gray solid and dotted lines respectively. Since these are normalized cameras, their focal lengths are identical, given by f. The center-of-projections of these two cameras are given by O_1 and O_2 respectively. The image of the 3D point P is formed at L and R respectively in the two cameras, at coordinates $-l$ and r in the respective image planes, considering the principle center at the center of the image plane. b is the baseline, i.e. the distance between O_1 and O_2. The triangle PLR and PO_1O_2 and similar. Therefore, we find that

Figure 8.8. This shows two rectified stereo camera pair imaging the 3D point P.

$$\frac{b}{Z} = \frac{b+r-l}{Z+f}.$$
(8.58)

Disparity is formally defined as $(r - l)$. Therefore, from the above equation we can derive that

$$Z = b\frac{f}{d}.$$
(8.59)

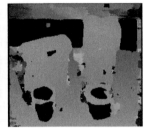

Figure 8.9. This image shows two rectified images (on left) and the depth reconstructed from them (on right).

Since b and f are constants, we see that the depths of the different points are inversely proportional to disparity and can therefore be deciphered from the image. Note that even when the baseline and focal length are not known, we can decipher the depth up to a scale factor (or constant of proportionaility). Therefore, we can recover the relative depth between objects in the scene, as illustrated in Figure 8.9.

8.6.2   Depth from Optical Flow

The next application we will focus on is depth from optical flow. Consider a single camera whose principal axis is aligned with the Z axis and which is moving along Z direction. Consider two different locations of this camera on the Z axes, the first one being the origin and the second one at distance t from the origin. The calibration matrices of the camera for these two locations, A_1 and A_2 respectively, are given by

$$A_1 = K[I|O] \tag{8.60}$$
$$A_2 = K[I|T_z] \tag{8.61}$$

where $T_z = (0, 0, t)^T$. Since the intrinsic parameters of the camera do not change with movement, the intrinsic matrix K remains the same. Let the image of the same point in these two camera positions be $p_1 = (x_1, y_1, 1)^T$ and $p_2 = (x_2, y_2, 1)^T$. Let us consider K as

$$K = \begin{pmatrix} f & 0 & 0 \\ 0 & f & 0 \\ 0 & 0 & 1 \end{pmatrix}. \tag{8.62}$$

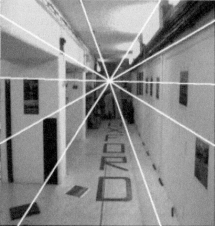

Figure 8.10. This image shows optical flow lines as the camera moves into the corridor from the left to the right image. Note that the frames on the wall closer to the camera and the letters on the floor closer to the camera undergo more displacement than more distant ones.

Therefore, for a 3D point (X, Y, Z)

$$(x_1, y_1) = K[I|O] \begin{pmatrix} X \\ Y \\ Z \\ 1 \end{pmatrix} \tag{8.63}$$

Similarly

$$(x_2, y_2) = K[I|T_z] \begin{pmatrix} X \\ Y \\ Z \\ 1 \end{pmatrix} \tag{8.64}$$

If we plug in the value of K and expand these equations we will get

$$x_1 - x_2 = \frac{t}{Z} x_2 \tag{8.65}$$

$$y_1 - y_2 = \frac{t}{Z} y_2 \tag{8.66}$$

The above equations define the displacement of the image of the same point from one camera position to another termed as the *optical flow*, as illustrated in Figure 8.10. From the above equations, we will make some nice observations as follows.

1. First, when $z = \infty$, then $p_1 = p_2$. Therefore, the image of a 3D point which is very far away will remain unchanged in the two images. This point is called *focus of expansion*.

2. As Z increases, the displacement of the image of the same point from one camera position to another decreases. That means, images of distant points undergo less displacement than images of closer points.

3. As t increases the displacement increases. Therefore, greater movement of the camera causes larger optical flow.

Now, if we are given images from these two locations and can find the correspondences using epipolar geometry, then we can decipher the optical flow. Now, if we know the amount of camera movement t, we can decipher the depth of the points Z from this optical flow information. This technique of deciphering depth by using a moving camera is called *structure from motion*.

8.7   Conclusion

Epipolar geometry explores the fundamental geometric constraints applicable for disparity based geometry reconstruction. In most of the treatise in this chapter we assume some known parameters, for e.g. the displacement t for structure from motion. It is possible to use all these techniques from uncalibrated situations where such parameters are often not known. However, those involve much complex optimization which is beyond the scope of this book. Advanced concepts of depth reconstruction with more unknowns (e.g. unknown camera parameters) are discussed in details in [Szeliski 10].

Epipolar geometry has deep mathematical implications, even in higher dimensions. In this book, we have strived to keep the treatise much more practical and therefore simpler by considering 2D cameras and practical calibrated scenarios. For readers who would like to study epipolar geometry in much more depth, a deeply mathematical treatise is available at [Hartley and Zisserman 03]. Epipolar geometry is the cornerstone of the domain of image based rendering in computer graphics first explored in depth by Leonard Mcmillan in his seminal paper [Mcmillan and Bishop 95] followed by a multitude of works in the last two decades which are summarized in [Shum et al. 07]. The mathematical finesse to implement the two applications discussed in this chapter (depth from disparity and optical flow) have been greatly simplified to get the fundamentals across. For a better treatise on the details of these methods, please consult [Hartley and Zisserman 03].

Bibliography

[Hartley and Zisserman 03] Richard Hartley and Andrew Zisserman. *Multiple View Geometry in Computer Vision.* Cambridge University Press, 2003.

[Mcmillan and Bishop 95] Leonard Mcmillan and Gary Bishop. "Plenoptic Modeling: An Image-based Rendering System." In *Proceedings of the 22nd Annual Conference on Computer Graphics and Interactive Techniques (SIG-GRAPH)*, pp. 39–46, 1995.

[Shum et al. 07] Heung Yeung Shum, Shing Chow Chan, and Sing Bing Kang. *Image Based Rendering.* Springer, 2007.

[Szeliski 10] Richard Szeliski. *Computer Vision: Algorithms and Applications.* Springer-Verlag New York, Inc., 2010.

Summary: Do you know these concepts?

- Epipoles and Epipolar Lines
- Fundamental Matrix
- Essential Matrix
- Normalized Cameras
- Rectification
- Disparity in Binocular Vision
- Depth from Disparity
- Optical Flow
- Structure from Motion

Exercises

1. Why is the epipolar constraint useful for stereo matching (i.e. finding corresponding points between the first and the second images)? What happens to the epipoles and the epipolar lines in the rectified images after applying image rectification? What are the advantages of applying image rectification before we do stereo matching?

2. Given a point $p = (x, y)$ in image 1, and the fundamental matrix

$$F = \begin{pmatrix} 0 & 1 & 0 \\ 1 & 0 & -1 \\ 0 & 1 & 0 \end{pmatrix} \tag{8.67}$$

   (a) Derive the equation of the corresponding epipolar line in image 2. Use your result to compute the equations of the epipolar lines corresponding to the points $(2, 1)$ and $(-1, -1)$.

   (b) Compute the epipole of image 2 using the equation of the two lines you derived for the previous question. In general, how can you determine the epipole from any fundamental matrix F? (Hint: the answer involves a term from linear algebra.)

   (c) The relationship between points in the second image and their corresponding epipolar lines in the first image is described by the transpose of the fundamental matrix F^T. Use this fact to compute the epipole in image 1 for the matrix F.

   (d) Which of the following points is the epipole of the first camera? (i) $(0.5, 0.5)$; (ii) $(1, 1)$; (iii) $(1, 0)$ and (iv) $(-1, 0)$.

   (e) Which of the following points is the eipole of the second camera? (i) $(0.5, 0.5)$; (ii) $(1, 1)$; (iii) $(1, 0)$ and (iv) $(-1, 0)$.

   (f) Consider the point $(0.5, 0.5)$ in the first camera. Find the slope and offset of the epipolar line in the second camera on which its correspondence will lie.

   (g) Let the two cameras have the focal lengths of 1 and 2 respectively with square pixels, principal center at the center of the image plane and no skew factor. Find the essential matrix that relates these two cameras.

3. Consider a pair of camera stereo rig whose rotation and translation with respect to global coordinate system is R_1, R_2, T_1 and T_2 respectively. Their intrinsic matrix is identity. Give an expression of the essential matrix and the length of the baseline in terms of these matrices.

4. Consider a camera attached to the front of the car as it is traversing on a straight bridge, the end of which is marked with a conspicuous building and the sides containing landmark paintings hanging from the bridge at equal intervals which is known. Consider two pictures taken from this camera. Which 3D scene location would appear at the focus of expansion? Assuming that you can detect the amount each landmark painting has moved from one picture to another, how can you find the speed of the car?

5. Consider a scene which is constant being observed by a pair of frontal stereo cameras. If the baseline of the camera is increased, how would the disparity of the 3D points change in the images captured? Justify your answer.

6. Suppose we would like to determine the size of a cube from a set of k calibrated cameras whose extrinsics are unknown (but whose intrinsics are known). Suppose each of the cameras can see the same m corners of the cube, and suppose there is no correspondence problem. How many cameras and how many corners do we need to determine the size of the cube? (Notice: a cube has only 8 corners, hence $m \leq 8$). If mulitple solutions exist, give them all. If no solution exists, explain why.

Part IV

Radiometric Visual Computing

9

Light

In this chapter we will discuss the science of light, more commonly termed as radiometry. We will discuss different radiometric quantities and how they are used in the domain of visual computing. Then we will see how radiometry leads to photometry which is the science of light in the context of the human visual system and hence human perception.

9.1 Radiometry

In radiometry light is considered to be a traveling form of energy and therefore the unit used to describe it is the SI unit of energy, *joule (J)*. This energy is associated with a source of origin, which is usually defined by its position, a direction of propagation and a wavelength λ. This conforms to the particle theory of light where the smallest unit of light is considered to be a photon or a quantum of energy. This also conforms to the wave theory of light, which assumes light to be a waveform traveling in a particular direction. λ is expressed by units of nanometers (nm). Light travels with a speed c_n in a medium of refractive index n. An invariant in this context is the frequency f of light is given by

$$f = \frac{c_n}{\lambda},\tag{9.1}$$

which does not change unlike c_n and λ. Another invariant is the energy carried by a photon,

$$q = \frac{hc}{\lambda}\tag{9.2}$$

where $h = 6.63 \times 10^{-34} J$ is the Planck's constant.

The *spectral energy* is a density function that gives the density of the quantum energy at an infinitesimal interval of wavelengths around λ with a width of $\Delta\lambda$. Note that due to the particle nature of light, spectral energy at a wavelength is a quantum value (either 0 or non-zero), but its density in an interval of wavelengths can be defined in a non-quantum fashion. This is similar to population where population at any point in space is either existent or non-existent, but the density of population over an area is always a non-quantum quantity. Just like population density, it is better to view the spectral energy which is a continuum and does not become granular even when the area is small. Therefore, spectral energy ΔQ is defined as

$$\Delta Q = \frac{\Delta q}{\Delta \lambda} \tag{9.3}$$

and has the unit of $J(nm)^{-1}$.

However, we are more interested in *spectral power* that is defined as the spectral energy over infinitesimal time Δt. It is given by $\frac{\Delta Q}{\Delta t}$ and has the unit of $W(nm)^{-1}$. Imagine a camera that leaves its shutter open for Δt time and has a filter of $\Delta \lambda$. Such a sensor would measure the spectral power.

Irradiance, H, is defined as the spectral power per unit area and is given by

$$H = \frac{\Delta q}{\Delta A \Delta t \Delta \lambda} \tag{9.4}$$

where ΔA can be considered as the finite area of the sensor measuring the spectral power, assuming that the sensor is parallel to the surface being measured. The unit of irradiance is therefore $Wm^{-2}(nm)^{-1}$ or $Js^{-1}m^{-2}(nm)^{-1}$. Irradiance is usually used to define the amount of spectral power *incident* or *hitting* a unit area. When the same quantity is used in the context of amount of spectral power *leaving* or *reflected* off a unit area, it is often called *radiant exitance, E*.

Irradiance only tells us about how much light hits a point, but it does not say much about the direction the light is coming from. Therefore, irrandiance can be considered to be the quantity measured by the sensor when a conical light limiter is placed on the sensor to limit the direction of the light it is measuring to $\Delta \sigma$. Therefore, *radiance* is defined as irradiance per unit direction as

$$R = \frac{\Delta H}{\Delta \sigma} = \frac{\Delta q}{\Delta A \Delta t \Delta \lambda \Delta \sigma} \tag{9.5}$$

and is measured by the unit $Wm^{-2}(nm)^{-1}(sr)^{-1}$ or $Js^{-1}m^{-2}(nm)^{-1}(sr)^{-1}$ where sr stands for steridian, the SI unit for a solid angle. It is analogous to radians used to define planar angles. A useful property of radiance is that it does not vary along a line in space. Consider the sensor with a conical light limiter of angle σ measuring the light hitting a surface from a distance d. Let the circular area subtended by this cone be ΔA. If we increase the distance by a factor of k to kd, the area being measured by the detector will increase by k^2 but the light reaching the detector will be attenuated by the same factor k^2 (due to distance attenuation of light being inversely proportional to the distance) thereby keeping the radiance constant. Here we consider the sensor to be parallel to the surface whose radiance is being considered. In other words, the normal to the surface is perpendicular to the sensor. However, the more general situation is when the sensor is tilted by an angle θ. In this case, the area sampled by the detector will no longer be a circle but an ellipse with a larger area of $\Delta A cos(\theta)$. Therefore, radiance will be defined by

$$R = \frac{\Delta H}{\Delta \sigma} = \frac{\Delta q}{\Delta A cos\theta \Delta t \Delta \lambda \Delta \sigma}. \tag{9.6}$$

As with irradiance, it is important to distinguish between the radiance incident from a point on the surface and radiance exiting from a point of the surface. The former is called *field radiance*, L_f, and the latter is called *surface radiance*, L_s. Therefore,

$$L_s = \frac{\Delta E}{\Delta \sigma cos\theta} \tag{9.7}$$

$$L_f = \frac{\Delta H}{\Delta \sigma cos\theta} \tag{9.8}$$

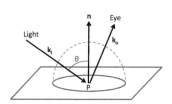

The reason radiance is considered to be the most fundamental radiometric quantity is that if we know R_f of a surface, we can derive all the other radiometric quantities from there. For example, irradiance can be expressed from the field radiance as

$$H = \int_{\forall \mathbf{k}} L_f(\mathbf{k_i}) cos\theta d\sigma \tag{9.9}$$

Figure 9.1. This figure shows the surface point P with normal $\mathbf{n}$ and the incident light direction $\mathbf{k_i}$ and viewing direction $\mathbf{k_o}$ for the BRDF $\rho(\mathbf{k_i}, \mathbf{k_o})$ at P.

where $\mathbf{k_i}$ is an incident direction that can be expressed as a (θ, ϕ) pair in the spherical coordinate system with respect to the normal at that point on the surface and is associated with a differential solid angle $d\sigma$. For example, if L_f is constant across all directions, we can compute the irradiance by replacing $d\sigma = sin\theta \, d\theta \, d\phi$ as

$$H = \int_{\phi=0}^{2\pi} \int_{\theta=0}^{\frac{\pi}{2}} L_f \, cos\theta \, sin\theta \, d\theta \, d\phi \tag{9.10}$$

$$= \pi L_f \tag{9.11}$$

Note the constant π will appear in many radiometric calculations and is the artifact of how we measure solid angles. We consider the area of the unit sphere to be a multiple of π rather than multiple of one. Similarly we can compute the spectral power incident on a surface by finding $\int_{\forall \mathbf{x}} H(\mathbf{x})dA$ where $\mathbf{x}$ is a point of the surface associated with a differential area of dA.

9.1.1 Bidirectional Reflectance Distribution Function

Bidirectional Reflectance Distribution Function or BRDF is a formal way to describe what we humans face everyday – objects look different when viewed from different directions as they are illuminated from different directions. Painters and photographers have for centuries explored the appearance of trees and urban areas under a variety of conditions, accumulating knowledge about "how things look" – which is nothing but BRDF related knowledge.

So, let us now define BRDF at a surface point P formally. Let us consider a point on a surface P with a normal $\mathbf{n}$, illuminated from a direction $\mathbf{k_i}$. This is achieved by placing a light source in this direction. Let the irradiance incident at P be H. Let the radiance going out towards a viewing direction $\mathbf{k_o}$ be L_s. This setup is illustrated in Figure 9.1. The BRDF ρ is defined as the ratio of L_s to H, i.e.

$$\rho(k_i, k_o) = \frac{L_s}{H}. \tag{9.12}$$

Therefore, BRDF gives us the fraction of light exiting towards the viewing direction $\mathbf{k_o}$ when illuminated from the incident direction $\mathbf{k_i}$. Note that both $\mathbf{k_o}$ and $\mathbf{k_i}$ are directions in 3D and can be represented with two angles in spherical coordinate. Let $\mathbf{k_i}$ be given by (θ_i, ϕ_i) and $\mathbf{k_o}$ be given by (θ_o, ϕ_o). Therefore, BRDF is a four dimensional function $\rho(\theta_i, \phi_i, \theta_o, \phi_o)$. Also, note the ρ is a ratio of radiance by irradiance. Therefore its unit is $(sr)^{-1}$.

Directional Hemispherical Reflectance Let us consider a simple question, "What fraction of the incident light is reflected?". It is evident that this number should be between 0 to 1 purely from the standpoint of conservation of energy. Let us now see if this question can be easily answered using BRDFs. Given an incident light from the direction $\mathbf{k_i}$, the fraction that is reflected should be the ratio of outgoing irradiance (or radiance exitance) to the incoming irradiance. Therefore, the directional hemispherical reflectance for the $D(\mathbf{k_i})$ is given by the ratio of radiance exitance E to irradiance H as

$$D(\mathbf{k_i}) = \frac{E}{H}. \tag{9.13}$$

From Equation 9.12 we know that

$$L_s(\mathbf{k_o}) = H\rho(\mathbf{k_i}, \mathbf{k_o}) \tag{9.14}$$

Also, from the definition of radiance in Equation 9.8 we know that

$$L_s(\mathbf{k_o}) = \frac{\Delta E}{\Delta \sigma_o cos\theta_o}. \tag{9.15}$$

Therefore,

$$H\rho(\mathbf{k_i}, \mathbf{k_o}) = \frac{\Delta E}{\Delta \sigma_o cos\theta_o}. \tag{9.16}$$

Rearranging terms we get the contribution of E/h reflected in the direction of $\mathbf{k_o}$ as

$$\frac{\Delta E}{H} = \rho(\mathbf{k_i}, \mathbf{k_o})\Delta \sigma_o cos\theta_o \tag{9.17}$$

Therefore,

$$D(\mathbf{k_i}) = \frac{E}{H} = \int_{\forall \ \mathbf{k_o}} \rho(\mathbf{k_i}, \mathbf{k_o}) cos\theta_o d\sigma_o \qquad (9.18)$$

An ideal diffuse surface is called *Lambertian* and is considered to have a constant BRDF at any viewing direction. In other words, the appearance of the object is view-agnostic or view-independent. Though such surfaces are pratically non-existent, many objects with *matte* appearance are often modeled as a Lambertian surface. Let us consider such a Lambertian surface with $\rho = C$. Then the directional hemispherical reflectance of such a surface is given by

$$D(\mathbf{k_i}) = \int_{\forall \ \mathbf{k_o}} C\Delta\sigma_o cos\theta_o d\sigma_o \qquad (9.19)$$

$$= \int_{\phi_o=0}^{2\pi} \int_{\theta_o=0}^{\frac{\pi}{2}} C \ cos\theta_o \ sin\theta_o \ d\theta_o \ d\phi_o \qquad (9.20)$$

$$= \pi C \qquad (9.21)$$

Therefore, if we consider a perfectly reflecting Lambertian surface where $D(\mathbf{k_i}) = 1$, then its BRDF is $\frac{1}{\pi}$.

9.1.2   Light Transport Equation

Using the aforementioned equations, we can now write a simple light transport equation that defines how light is transported via surfaces or objects in the presence of lights from many different directions. If we consider radiance L_i from the direction $\mathbf{k_i}$ along a small solid angle $\Delta\sigma_i$, the irradiance due to this light is given by $L_i cos\theta_i \Delta\sigma_i$ where θ_i is the angle between $\mathbf{k_i}$ and $\mathbf{n}$. Therefore, the outgoing radiance ΔL_o in the direction $\mathbf{k_o}$ due to the radiance coming in from direction $\mathbf{k_i}$ is given by

$$\Delta L_o = \rho(\mathbf{k_i}, \mathbf{k_o}) L_i cos\theta_i \Delta\sigma_i \qquad (9.22)$$

Therefore, to consider the radiance from all the different directions (all different values of $\mathbf{k_i}$), the total irradiance in the direction of $\mathbf{k_o}$ is given by

$$L_s(\mathbf{k_o}) = \int_{\forall \ \mathbf{k_i}} \rho(\mathbf{k_i}, \mathbf{k_o}) L_f(\mathbf{k_i}) cos\theta_i d\sigma_i \qquad (9.23)$$

This is called the rendering equation or light transport equation and is the cornerstone of building illumination models, simple or complex.

9.2   Photometry and Color

For every radiometric property, there is a corresponding photometric property that intuitively measures "how much of it can the human observer make use of".

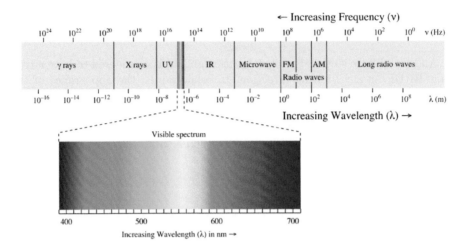

Figure 9.2. This figure shows the visible spectrum of light and its position with respect to the invisible spectrum.

Therefore, photometric quantities have an aspect of perception associated with them. Color is a photometric quantity that we use all the time. It is a part and parcel of our lives, so much so, that we probably cannot appreciate it unless we lose our perception of color. Mr. I, who lost color perception due to an accident exclaimed with anguish, "My dog looks gray, tomato juice is black and color TV is an hodge podge". Color not only adds beauty to our life, but serves important signaling functions. The natural world provide us with many signals to identify and classify objects. Many of these come in terms of color. For e.g. banana turns yellow when its ripe, the sky turns red when it is dawn and so on.

The *color stimuli* is the radiometric quantity (usually radiance), that reaches the human eye from any point in the world. The important parameter in the context of photometry, and therefore color vision, is the associated wavelength λ. The visible light spectrum has wavelength that varies between 400 nm and 700 nm. Figure 9.2 shows the visible spectrum of colors. A illumination or an object selectively emit or reflect respectively certain wavelengths more than other. Two things are responsible for our color vision and hence photometry. The first is the selective reflection of wavelengths by different objects. However, it is only one of the factors responsible for the color of an object. The second important factor is the eye's selective response to different wavelengths. This response, and hence the perception of color, can be different from species to species, and also shows a variance across individuals of the same species. That is the reason, color is often considered as a perception, and not reality!

Let us start from the illumination of a scene. A scene is usually lighted by

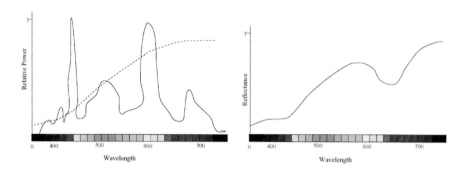

Figure 9.3. Left: The illumination spectrum ($I(\lambda)$) of a flouroscent (bold line) and tungsten lamp (dotted line). Right: The reflectance spectrum ($R(\lambda)$) of a red apple.

some light source. This light source emits light differently at different wavelengths λ. This function, denoted by $I(\lambda)$ gives the illumination spectrum. Similarly, for an object, its relative reflectance at different wavelengths define its reflectance spectrum $R(\lambda)$. Since the reflectance or reflectivity of an object is the ratio of outgoing to incoming power, it is between 0 and 1 for an object that is not a light source. These spectra for a couple of light sources and a red apple are illustrated in Figure 9.3.

When an object is illuminated by a light source, the amount of light that is reflected from that object at different wavelengths is given by the product of $I(\lambda)$ and $R(\lambda)$. Since this is the spectrum that stimulates the vision, this is called the *color stimuli* or color signal, denoted by $C(\lambda)$. Thus,

$$C(\lambda) = I(\lambda) \times R(\lambda) \tag{9.24}$$

as illustrated in Figure 9.4. The physical quantity we are dealing with here is power per unit area per unit solid angle per unit wavelength $(W\ m^{-1}\ (nm)^{-1}\ (sr)^{-1})$. Therefore, we are essentially thinking of radiance when we are defining the color stimuli.

Color stimuli can be of different types as shown in Figure 9.4. When it has light of only one wavelength, it is called *monochromatic*, e.g. a laser beam. When the relative amount of light from all wavelengths is equal, then it is called *achromatic*. The sunlight is close to achromatic in the day time. Finally, if the stimulus has different amounts of light from different wavelengths, it is called *polychromatic*. Most of the time we deal with polychromatic light. Most of the manmade and natural color stimuli are smooth spectra of polychromatic light.

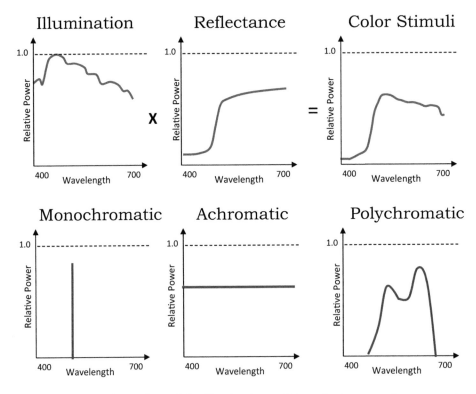

Figure 9.4. Top: The product of the illumination and the reflectance spectrum generate the color stimuli. Bottom: This shows different types of color stiumli - from left to right – monochromatic, achormatic and polychromatic.

9.2.1 CIE XYZ Color Space

Interestingly, the perceived color is different than the color stimuli. Human eye has three sensors (usually called cones in biology) which have differential sensitivities to different wavelengths. In 1939, the CIE (International Commission on Illumination) came up with standard spectral responses of these sensors based on earlier studies done by color scientists. Let us denote these by $\bar{x}(\lambda), \bar{y}(\lambda), \bar{z}(\lambda)$. Therefore, multiplying the color stimuli with these sensitivities give us the perceived spectrum as illustrated in Figure 9.5. Now, the strength of each of these perceived spectrum is the area under the curve and is computed by integrating each of the three curves. These provide three numbers quantifying the strength

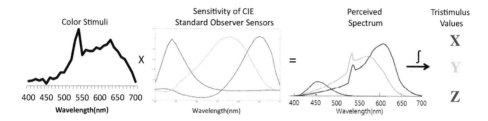

Figure 9.5. The color stimulus is multiplied by the sensitivities of the three cones of the CIE standard observer – $\bar{x}(\lambda)$ (red), $\bar{y}(\lambda)$ (green), $\bar{z}(\lambda)$ (blue) – to generate the perceived spectrum. The strength of each of these perceived spectrums is given by the area under these spectral curves given by their integration. This is also the process to find the XYZ tristimulus values of a color stimulus which quantifies how the human brain perceives this stimulus.

of the stimuli in each of the sensors called the XYZ tristimulus values

$$X = \int_\lambda C(\lambda)\bar{x}(\lambda)d\lambda = \sum_{\lambda=400}^{700} C(\lambda)\bar{x}(\lambda) \tag{9.25}$$

$$Y = \int_\lambda C(\lambda)\bar{y}(\lambda)d\lambda = \sum_{\lambda=400}^{700} C(\lambda)\bar{y}(\lambda) \tag{9.26}$$

$$Z = \int_\lambda C(\lambda)\bar{z}(\lambda)d\lambda = \sum_{\lambda=400}^{700} C(\lambda)\bar{z}(\lambda) \tag{9.27}$$

Let us now consider the units of the functions we are dealing with. The eye's response is measured as lumens per watt. Lumens (lm) is an estimate of the light produced. For example, think of a light bulbs. They are usually rated in terms of the power they consume (i.e. watt) and the useful light energy they produce (i.e. lumens). Therefore, the higher the lm/W, the more efficient the lamp. Similar is the case for the human response functions, but they also have a wavelength dependency. Therefore, the unit for X, Y or Z is $(lm/W)(W/(m^2\ sr) = lm/(m^2\ sr)$. Note that since we integrate over the wavelength, the nm disappears from the unit. One lumen per steridian of solid angle is defined as one cd (candela). Therefore, the units of the tristimulus values is cd/m^2.

It has been shown that the human brain usually works with these three numbers thus generated instead of the spectrum. It also provides us a better paradigm to study color stimuli and perception than working with the spectra directly. Note that the XYZ tristimulus values offer us a 3D space to define colors, just like coordinate systems for geometry. Any color can be plotted as a point in this 3D space. However, since XYZ values are essentially given by the area under a curve, there can be two different spectra which can provide the same XYZ value. This means that these two spectra will produce the same sensation in the human eye and hence will be plotted at the same point in the XYZ space. Multiple such spectra which produce the same XYZ values are called *metamers* and the phenomenon, *metamerism*. Metamerism is a boon in disguise. During color reproduction from a display, to say print, all we need to do is to create a metamer of the original color i.e. a color with the same XYZ value and not necessarily the same spectrum. Therefore, CIE XYZ space is always considered in the context of human perception and not in the context of the real color spectra.

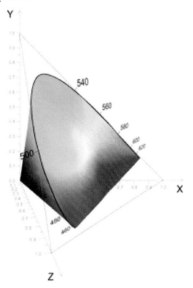

Now let us analyze this XYZ space. First, note that only the positive octant (where the values of X, Y and Z are always positive) of this space really makes sense since physically there is no negative light. Therefore, we are only concerned with the first octant of this space. Second, some XYZ tristimulus values, even in this octant, are not valid since they indicate an impossible spectrum. For example, there cannot be any spectrum that can result in XYZ value of $(1, 0, 0)$ i.e. evoking response in one of the sensors and no response in any other sensors. This is primarily due to the fact that the sensitivities of the three sensors overlap significantly. These XYZ values which do not correspond to a physical spectrum are called imaginary colors. Therefore, a part of the XYZ space are actually imaginary without having any corresponding real spectra to go with the tristimulus values. Therefore, the real colors or spectra only span a subset of the positive octant of the XYZ space. Munsell, the famous color scientist of 20th century, was instrumental in finding the shape of this subset of real colors as a conical volume as shown in Figure 9.6.

Figure 9.6. This shows the conical volume occupied by the real colors in the CIE XYZ space.

Put a Face to the Name

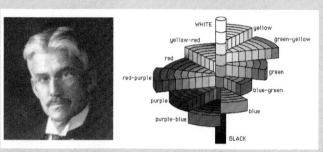

Albert Henry Munsell was an American painter and teacher of art who was the first to attempt to create a numerical system to accurately describe color by designing the Munsell Color System that was the precursor and inspiration for the development of the first scientific color order system, CIE XYZ color space. As an artist, Munsell found color names "foolish" and "misleading" and therefore strived to create a system that has meaningful notation of color. In order the achieve this, he invented the first photometer and also a patented device called the spinning top that helped him measure colors and how colors change. Munsell color system created a necessary bridge between art and science providing enough structure to allow scientists to expand upon and use it, while being simple enough for artists with no scientific background to use it for selecting and comparing colors. Munsell's System essentially created a way of communicating colors. He also coined the terms chrominance or chroma and lightness. Munsell also investigated the relationship between the color and the light source used for illumination to find that the light source used drastically effected the color perceived thereby leading him to eventually develop the standard for daylight viewing of colors for accurate color evaluation. Munsell lived from 1858 to 1918 and wrote three books, all of which considered the most fundamental readings in color science – *A Color Notation* (1905), *Atlas of the Munsell Color System* (1915) and one published posthumously, *A Grammar of Color: Arrangements of Strathmore Papers in a Variety of Printed Color Combinations According to The Munsell Color System* (1921). In 1971 he created the Munsell Color Company which is now called the Munsell Color Labs located in the Rochester Institute of Technology and is the premier research institution for color science.

9.2.2 Perceptual Organization of CIE XYZ Space

We now have an understanding of how the tristimulus values are derived from a color stimuli spectra. but we still do not have a sense of how colors are organized in this space. Where do the grays lie? How does the trajectory of a color move in this space if only its brightness is increased? Therefore, given the coordinates

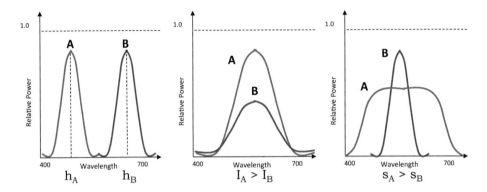

Figure 9.7. This figure illustrates how the properties of color relate to the properties of their spectra. Left: A and B have the same area (therefore same intensity) has different dominant wavelength and therefore different hues. Middle: A and B have the same hue but different intensities. Right: A and B have the same hue but different saturations.

of a color in this space, we cannot predict how this color will look.

In order to find the perceptual organization of color in this XYZ space, we should first study all the perceptual parameters that we use to describe color and relate them to mathematical properties of their spectra. We often find ourselves comparing one color to be more or less bright than the other. What does this mean? Intuitively, this can be thought of as the total energy of color perceived by the eye and is given by the area under the curve $C(\lambda)$ after it is weighted by the sensitivity functions of the eye. Therefore, $X + Y + Z$ is a good measure of this total energy reaching the eye. Although, there is no good term for this quantity in the color literature, this term is considered very important in color perception. So, for lack of a better term, we call this the intensity I of a color and $I = X + Y + Z$. The opponent theory of color, something we will not go into in this book, explains the perception of color in higher level processing in the brain that indeed adds up the tristimulus values to estimate the total energy of colors – a quantity that is not independent of the distribution of the energy across the wavelengths. The tristimulus theory of color, on the other hand, deals with the perception of color in the eye.

Hue, h, can be thought of as the colorfulness of a color and can be given by the weighted mean of the wavelengths present in the color spectra weighted by their relative power. This results in a wavelength, the color of which defines the dominant sensation the spectra will create in the human eye. Finally, most of us will agree that, for example, pink is a less vibrant version of red. Vibrancy, or saturation s, of a color can be thought of as the amount of white (or achromatic color) present in a color. The more white in a color, the less vibrant it is. Therefore, saturation is inversely proportional to the standard deviation of the color spectra from the hue. Therefore, monochromatic color, with zero standard deviation from the hue, has 100% saturation. Keep adding white to this you will get different levels of reduced saturation. When you get to a completely achromatic color, that is the most unsaturated color with saturation of 0%. Figure 9.7 illustrates these concepts using different color spectra.

In the XYZ space, we define the hue and saturation of a color using its chromaticity coordinates x and y as the proportion of X and Y in I respectively. Therefore,

$$x = \frac{X}{I} = \frac{X}{X+Y+Z} \qquad (9.28)$$

$$y = \frac{Y}{I} = \frac{Y}{X+Y+Z} \qquad (9.29)$$

Note that $z = 1 - x - y$ is the proportion of Z in I and is redundant since it can be computed from the chromaticity coordinates. Therefore, chromaticity coordinates are a way to remove one dimension from the XYZ space to create a 2D space defined by the chromaticity coordinates. This 2D space is called the chromaticity chart. Now, you will see when studying geometric transformations in this book, that Equation 9.29 defines a perspective projection of a point in XYZ space on to a plane given by the equation $X + Y + Z = k$. This plane has a normal of $(1, 1, 1)$. Perspective projection of a 3D point (X, Y, Z) is defined as the 2D point where a ray from origin to the 3D point intersects the plane of projection which is the plane with normal $(1, 1, 1)$ in this case. Note that multiple such planes

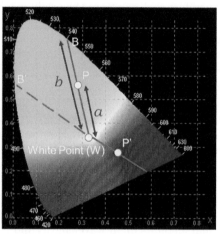

Figure 9.8. This shows the chromaticity chart and the placement of different colors on it. W is a color with equal proportions of X, Y and Z resulting in chromaticity coordinates of $(\frac{1}{3}, \frac{1}{3})$. The hue of color P is defined by the dominant wavelength is given by B, the point where the straight line WP meets the spectral periphery. For another color P', WP' meets the non-spectral periphery. Therefore, the hue of P' is defined by the complementary wavelength B', the point where WP' extended backwards meet the spectral periphery. The saturation is given as the ratio of the distance of the color from W to the distance from W to its dominant or complementary wavelength, B or B' respectively.

can be defined based on the value of k. But the location of the projection will be the same since the chromaticity coordinates define a normalized coordinate system that ranges between 0 to 1. One such plane is shown by the gray bordered triangle in Figure 9.6.

Now, consider a ray from the origin to a 3D point (X, Y, Z). Any point on this ray is given by a coordinate (kX, kY, kZ). Note that the chromaticity coordinates of (kX, kY, kZ) is identical irrespective of the value of k. Therefore, all points on this 3D ray project to the same point in the chromaticity chart — and therefore have the same chroma. However, what changes for each of these colors on this ray is their intensity I. Therefore, this projection allows removal of this intensity information providing us only the information about the chroma of the color. Therefore each ray from the origin is a iso-chroma trajectory in the XYZ space.

Although the 3D coordinate of a color in the XYZ space defines the color uniquely (upto metamerism), it does not provide us with an adequately good image in our mind. For example, even with the information of the 3D coordinates of a color – say, $(100, 75, 25)$ – we cannot imagine the chroma of the color. However, if we use the (Y, x, y) representation of $(75, 0.5, 0.28)$, we can immediately imagine the chromaticity chart and know that this must be a color in the red region of the chart. Therefore, most specification sheets for devices will follow the (Y, x, y) format. However, the (X, Y, Z) and (Y, x, y) representations are completely interchangeable, i.e. one can be computed from the other.

Let us discuss the expected response of the eyes when seeing an achromatic color. It is intuitive that the brain makes a decision on color based on relative difference in the firing of the three cones – therefore the relative difference in the tristimulus values. If all these values are identical, the brain would interpret it as equal amounts of all wavelengths and hence would perceive an achromatic color. In other words, for an achromatic color, $X = Y = Z$ and therefore $(x, y) = (\frac{1}{3}, \frac{1}{3})$. Thus, all the grays including black at origin $(X = Y = Z = 0)$ to white at infinity lie on the ray from the origin in the 3D XYZ which all map to the same chromaticity coordinate of $(\frac{1}{3}, \frac{1}{3})$ that is called the *white point* in the chromaticity chart and is denoted by W. Finally, note that we need to limit the space to some finite values. This is achieved by normalizing the maximum value of Y to be a well defined white – usually the luminance of a perfectly diffused reflector. This allows us to limit the space spanned by physical colors to the cone shown in Figure 9.6.

Similarly, when the chromaticity coordinates of physical colors are plotted on the chromaticity chart, the result it yields is shown in Figure 9.8 which is rather intuitive to follow. First, the projection of the conical volume on the triangular plane shown in Figure 9.6 would lead to a horseshoe shaped as shown in Figure 9.8. Second, note that the higher x means a much larger proportion of X which in turn means more intensity in the longer wavelengths and hence red. Similarly, the higher value of y indicates more intensity in middle wavelengths which is green. If both x and y are small - i.e. z is high – it means that Z is highest indicating more intensity in the lower wavelengths or blue.

Fun Facts

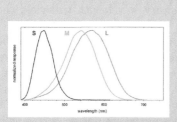

This plot shows the actual sensitivities of the S, M and L cones in the human eye where S, M and L stand for short, middle and long wavelengths. Note that the M and L plots are very close to each other. Further, the number of S cones is also much less than the M and L cones. Note that these plots are significantly different from the standard observer functions shown in Figure 9.5. This is due to the fact that these plots were physiologically measured only after mechanisms to do so were available to biological scientists and happened much later than the design of the standard observer functions. However, it can be shown that the LMS space can be related to the XYZ space by a linear transformation.

Note that in the chromaticity chart all the monochromatic colors are at the periphery. This is called the *spectral* boundary of the chromaticity chart. It is almost as if the wavelengths 400 to 700 have been placed around the boundary. There exists a straight line periphery of the chart that connects the two ends of the horseshoe. Which wavelengths do they represent?

To answer this question, let us go back to the visible spectrum of light (Figure 9.2). There are no wavelengths corresponding to purple which nevertheless is a color we perceive quite often. Also the colors change hue smoothly across the wavelengths i.e. blue changes slowly through cyan to green which then changes slowly through green-yellow to yellow which then changes through orange to red. So, cyan, which can be thought of as a combination of blue and green rests between blue and green. Similarly, orange rests between yellow and red. But, where are the purples? Shouldn't there be shades of purples between the high wavelengths red and lower wavelengths blue completing a circular representation of the visible colors of light? This is exactly the purples that show up as the straight line periphery of the chromaticity chart. There is no single wavelength to represent these colors and hence it is called the non-spectral boundary of the chromaticity chart.

Let us now consider a color P in this chart as in Figure 9.8. We connect W and P and extend the line backwards to meet the periphery of the chromaticity chart at B. The wavelength of the color at B is considered to be the *dominant wavelength* of P. Dominant wavelength is the sensation of the monochromatic wavelength evoked by P and is an estimate of the perceptual property of hue. Note that instead of P, if we consider a color P' and try to find its dominant hue, we will end up in the non-spectral part of the boundary which does not have a wavelength attached to it. In such cases (for purples), a dominant hue is undefined. Instead, we connect W to P' and extend the line backwards to B'. This is called the *complimentary wavelength*, i.e. if a dominant wavelength for P' existed, its superposition with the complimentary wavelength will yield the neutral W.

Saturation of any color in the chromaticity chart is defined as the ratio of the distance of the color P from the white point to the distance of the line that passes through P from W to the periphery. In Figure 9.8, saturation is thus given by $\frac{a}{b}$. Note that when the P is a monochromatic color, it coincides with B. Therefore $a = b$, leads to a saturation of 1 or 100% as is expected for a monochromatic color. On the other hand, if $P = W$, then $a = 0$ and the saturation is 0%, as is expected for an achromatic color.

Finally, let us consider one more property of color called *luminance* which is Y. This is defined as the *perceived brightness* of a color. For this, consider the following experiment. Consider two different colors of the same intensity – a blue and a green. Therefore, blue will have higher Z while green will have higher Y. Though these two colors have the same intensity, almost all humans will find the green to be brighter than the blue. This is due to a preferential importance given to the middle wavelengths, indicated by Y, than the others. This stems from the evolutionary reasons that man had to be extra sensitive to green all around him to survive on land. This is why Y is called the luminance and is often considered to be very important in perceptual tasks. For example, when compressing images, Y is maintained at full resolution while the other two channels are heavily sub-sampled.

9.2.3 Perceptually Uniform Color Spaces

CIE XYZ color space is perfectly suited for color matching applications. For example, if you want to superimpose two projectors and want to match their colors — you just have to make sure that they are projecting colors with the same XYZ values at the two overlapping pixels. However, there are applications where perceptual distances between colors are more important.

What do we mean by perceptual distance? This means how much distance we have to move from one color before the difference becomes visible. To understand the importance of perceptual distance, let us consider an application of image compression. When an image is compressed, we may want to change the colors slightly to aid the compression. This would mean moving the colors from their original location in the color space. But, we would like to move it just enough so that the change is not visible and the compressed image looks very close to the original image. The distance between the original and compressed image colors can give us an idea of how close perceptually these images are and hence use them to evaluate different compression techniques. In such applications, distance between colors become very important.

Unfortunately, CIE XYZ color space is perceptually non-uniform. This means

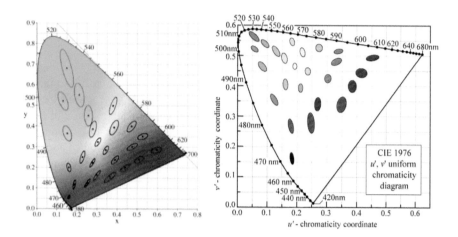

Figure 9.9. This figure shows the MacAdam ellipses plotted on the CIE 1939 chromaticity chart (left) and the CIE 1976 $u'v'$ chart (right). The colors in each of ellipse are indistinguishable to the human eye.

that equal distance at different regions of the color space does not signify equal perceptual difference. A perceptually uniform color space would signify that if we want to draw a geometric shape around a color P in the chromaticity chart to show us the set of all colors close to P that are indistinguishable from P, this shape would be a circle and the size of the circles will be the same irrespective of the position of P. However, this is not true when considering the chromaticity chart devised from the CIE 1939 XYZ color space. Scientist MacAdam plotted this geometric shape for different colors in the chromaticity chart and what transpired is shown in Figure 9.9. Note that these are all ellipses and the shape and size of the ellipses change with the position of the color. This shows that our ability to distinguish between different shades of green is much worse than our ability to distinguish between purples or yellows. Our ability to distinguish between blues is probably the best.

Therefore in 1964 and subsequently in 1976 efforts were made to design a perceptually uniform color space via non-linear transformation of the chromaticity coordinates. There are several such spaces designed like the CIE LUV or CIE Lab. Figure 9.9 shows one such color space designed in 1976 called the CIELUV space. Note that the perceptual distances are much more uniform than the chromaticity chart of 1939, but still not ideal. The most popular space in this context is the CIELab space designed much later and is derived from the CIE XYZ space. The Euclidian distance of 3 in the CIE Lab space is considered to be *just noticeable difference*. However, if you have color matching applications nothing can beat the simple old CIE XYZ space and the chromaticity chart devised thereof.

Fun Facts

Did you know that the concept of perceptual distance is well studied in the perception literature? People always wondered how much difference in stimulus we can tolerate without noticing it. For example, if you are carrying a heavy book and a 20 page thin book is added to your load, you will most likely not notice it. However, if your load is a thin 20 page book, you will definitely notice the change. Therefore the more important thing to consider is not the absolute change in the stimuli, but rather the relative change in the stimuli. While in the former case your stimuli of weight is changing by a few percent; in the latter case, it is almost doubling. In fact, Weber's law, named after its discoverer, a well-known law in perception literature says that our ability to perceive a difference (what is more formally called the difference threshold) is directly proportional to the amount of stimuli. The constant of this proportionality changes across different perceptions, but 10% has been found to be a reasonably good approximation empirically. We see the same thing when perceiving the range of grays. We are more sensitive to differences at dimmer gray values than at brighter gray values. This is also another reason why our display $\gamma > 1.0$. This helps us to provide greater resolution at the lower channel values than at the higher ones.

9.3   Conclusion

The rendering equation was introduced first in two seminal works in computer graphics in 1986 – [Immel et al. 86] and [Kajiya 86]. Its use in image synthesis was popularized by the seminal work of Dr. James Arvo in [Arvo and Kirk 90]. Color is one of the most confusing topics in the domain of visual computing primarily due to the long history of color and its wide use in various ways in many diverse domains starting from art, painting, physics, vision, human perception, video processing and compression and then lately in image processing, computer vision and graphics. [Stone 03] is an excellent practical handbook to understand these diverse viewpoints. [Reinhard et al. 08] provides a detailed formal treatise.

Bibliography

[Arvo and Kirk 90] James Arvo and David Kirk. "Particle transport and image synthesis." *SIGGRAPH Computer Graphics*, pp. 63–66.

[Immel et al. 86] David S. Immel, Michael F. Cohen, and Donald P. Greenberg. "A Radiosity Method for Non-diffuse Environments." *SIGGRAPH Computer Graphics* 20:4 (1986), 133–142.

[Kajiya 86] James T. Kajiya. "The Rendering Equation." *SIGGRAPH Computer Graphics* 20:4 (1986), 143–150.

[Reinhard et al. 08] Erik Reinhard, Erum Arif Khan, Ahmet Oguz Akyz, and Garrett M. Johnson. *Color Imaging: Fundamentals and Applications.* A. K. Peters, Ltd., 2008.

[Stone 03] Maureen C. Stone. *A Field Guide to Digital Color.* A K Peters, 2003.

Summary: Do you know these concepts?

- Radiometry and Photometry

- Radiance and Irradiance

- Bidirectional Reflectance Distribution Function (BRDF)

- Diffused Illumination

- Specular Illumination

- Phong Illumination Model

- Visible Spectrum of Light

- Color Stimuli

- Metamerism

- Tristimulus Values

- CIE XYZ Space

- Chromaticity Coordinates and Chart

- Intensity, Hue and Saturation

- Perceptual Distance

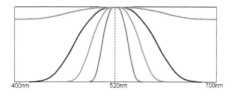 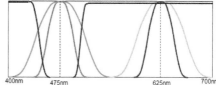

400nm   520nm   700nm   400nm   475nm   625nm   700nm

Exercises

1. The spectra of color $C_1 = (X_1, Y_1, Z_1)$ and $C_2 = (X_2, Y_2, Z_2)$ are given by $s_1(\lambda)$ and $s_2(\lambda)$ respectively. Let the color formed by multiplications of the spectra s_1 and s_2 be s_3, i.e. $s_3(\lambda) = s_1(\lambda) \times s_2(\lambda)$. Is it true that the XYZ coordinate corresponding to s_3, denoted by C_3, is $(X_1 X_2, Y_1 Y_2, Z_1 Z_2)$? Justify your answer with calculations.

2. Consider the four spectra in the left image of the above picture, their color is not related to their visible colors, but used for visualization.

   (a) What is the relationship between the dominant wavelengths of all these colors?

   (b) What is the relationship between the saturation of all these colors?

   (c) What is the relationship of the distances of these colors from the white point on the chromaticity chart?

   (d) What is the relationship between the $I = X + Y + Z$ of these colors?

   (e) The chromaticity coordinates of all these colors will lie on a single geometric entity (e.g. circle, parabola). What is that geometric entity?

   (f) The CIE XYZ coordinates of all these colors will lie on a single geometric entity (e.g. circle, parabola). What is that geometric entity?

3. Consider the spectra in the right image of the above picture, their color not related to their visible colors, but used for visualization.

   (a) The blue spectra is the most likely complementary color to which spectra?

   (b) The chromaticity coordinates of which of the spectra would lie in the same line?

   (c) If the chromaticity coordinates of the orange and pink spectra are $(0.1, 0.1)$ and $(0.6, 0.3)$ respectively, what is the most likely chromaticity coordinates of a color formed by their addition?

4. In the figure below match the objects on the right with their most probable color spectra on the left.

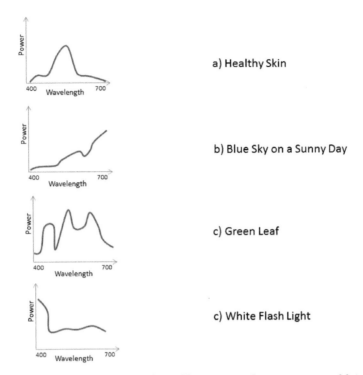

a) Healthy Skin

b) Blue Sky on a Sunny Day

c) Green Leaf

c) White Flash Light

5. Consider a Lambertian surface. How many dimensions would its BRDF have? Briefly describe an simple hardware setup and an algorithm that would allow you to measure the BRDF of a Lambertian surface?

6. You are measuring a surface patch with center P using a spectro-radiometer. The radius of the patch is $2mm$. The light is coming from a 45 degree angle and has an angular extent of 20 degrees. The measured energy is 200 Watts per nm. Find the irradiance and radiance at P.

7. When you switch on the projector in the class you see that it is projecting predominantly blacks and purples. You figure out that one of the wires connecting to the primaries R, G and B may be malfunctioning. Which one is it and why?

10

Color Reproduction

Maybe after the treatise on color in the previous chapter you are wondering what is the use of the XYZ color space? To understand this, we have to take a look at what is called color reproduction. When you think of images created by any device – for example, a digital camera capturing an image or a projector projecting an image or a printer printing one – these are all reproduction of colors from the physical scene (e.g. camera) or from another device (e.g. printer printing a camera captured image). The quality of a color reproduction is evaluated by how close the reproduced image is to that of the original image or scene. The term 'close' can be measured both quantitatively and qualitatively.

Color reproduction systems can be of two types – additive and subtractive – depending on the way two or more colors are mixed to create a new color in the color reproducing system. When learning painting, children are taught that the primary colors are red, blue and yellow. Yet in the field of image processing, we are taught that the primary colors are red, blue and green. So, wherein lies the contradiction? Apparently, both the art teachers and the image processing books are right. The difference stems from the fact that there are two ways to mix colors – *additive* and *subtractive*. While red, green and blue are primary colors of the former, the primary colors for the latter are cyan, magenta and yellow which are often referred to as blue, red and yellow for simplicity.

In subtractive color mixture, the color of a surface depends on the capacity of the surface to reflect some wavelengths and absorb others. When a surface is painted with a pigment or dye, a new reflectance characteristic is developed based on the capacity of the pigment or dye to reflect and absorb the different wavelengths of light. Consider a surface painted with *yellow* pigment which reflects wavelengths $570 - 580nm$ and another surface painted with *cyan* pigment which reflects $440 - 540nm$. If we mix both the pigments, only the wavelengths that are not absorbed by either of these pigments will be reflected, thus resulting in the color *green*. The yellow absorbs the wavelengths evoking the sensation of blue while the cyan absorbs the wavelengths evoking the sensation of yellow. Hence, what is left behind is a sensation of green. This is called *subtractive color mixtures* since bands of wavelengths are subtracted or canceled by the combination of *light absorbing materials*. And the resulting color, as you have probably noticed, is given by the *intersection* of the two spectrums. The yellow, cyan and magenta are termed as the color primaries of the subtractive color mixtures, because they form the smallest set of pigments required to produce all other colors. Dyes and inks usually follow subtractive color theory and hence images generated using these media are the result of subtractive color reproduction.

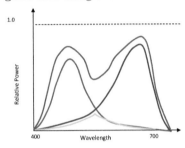

Figure 10.1. The blue and red show two different color spectra whose additive and subtractive mixtures are shown by the purple and green spectra respectively.

In additive color mixture systems, colors are mixed such that bands of wavelengths are added to each other. This is called *additive mixture of colors*. Thus, the spectrum of the color formed by superposition of multiple colors is given by the *addition* of their respective spectra. This is similar to how the human eye visualizes color. Devices like cameras and projectors follow additive color mixture.

Let us look at additive mixing of colors a little more formally. Let S_1 and S_2 be the spectra of two different color stimuli in Figure 10.1 shown in red and blue respectively. When they are combined additively, the resulting spectrum $S(\lambda)$ is given by the *addition* of the relative powers of each of S_1 and S_2 at each wavelength resulting in the purple spectrum. Therefore, $S(\lambda) = S_1(\lambda) + S_2(\lambda)$. However, while representing a spectrum for subtractive color mixture such as paint, a value of x at a particular wavelength means $x\%$ is reflected by the paint while $(1 - x)\%$ is absorbed. This curve is a spectral reflectance curve (values between $[0, 1]$) – fraction of the incident spectral value reflected by the material. Therefore, when two paints are superimposed, only the part that is not absorbed by either is reflected and therefore the resulting spectrum becomes the multiplication of the two spectral reflectances and the incident illumination spectrum creating the green spectrum in Figure 10.1.

10.1   Modeling Additive Color Mixtures

Modeling additive color space and color mixtures is easy in the XYZ color space. When two colors are mixed additively, the XYZ values of the resulting color are just the addition of the XYZ values of the individual colors in the mixture. In other words, the color resulting from an additive mixture of two colors (X_1, Y_1, Z_1) and (X_2, Y_2, Z_2) is given by their vector addition $(X_1 + X_2, Y_1 + Y_2, Z_1 + Z_2)$ and so forth.

Let us consider two colors $C_1 = (Y_1, x_1, y_1)$ and $C_2 = (Y_2, x_2, y_2)$. The easiest way to add these two colors would be to convert each of these to (X, Y, Z) format and add providing $C_s = (X_s, Y_s, Z_s) = (X_1 + X_2, Y_1 + Y_2, Z_1 + Z_2)$. Now converting this back to (Y, x, y) format, we get

$$Y_s = Y_1 + Y_2 \qquad (10.1)$$

$$x_s = \frac{X_1 + X_2}{X_1 + X_2 + Y_1 + Y_2 + Z_1 + Z_2} \qquad (10.2)$$

$$y_s = \frac{Y_1 + Y_2}{X_1 + X_2 + Y_1 + Y_2 + Z_1 + Z_2} \qquad (10.3)$$

Let us consider Equation 10.3 of x_s

$$x_s = \frac{X_1 + X_2}{I_1 + I_2} = \frac{x_1 I_1}{I_1 + I_2} + \frac{x_2 I_2}{I_1 + I_2} = x_1 \frac{I_1}{I_1 + I_2} + x_2 \frac{I_2}{I_1 + I_2}. \qquad (10.4)$$

Using the same concept to y_s, we find that

$$(x_s, y_s) = (x_1, y_1) \frac{I_1}{I_1 + I_2} + (x_2, y_2) \frac{I_2}{I_1 + I_2} \qquad (10.5)$$

Note that the above equation gives you lot more information than the equations 10.3. It says that the chromaticity coordinate of C_s is a convex combination of those of C_1 and C_2. Therefore the new color C_s can only lie on the straight line segment between (x_1, y_1) and (x_2, y_2) in the chromaticity chart. It also says that the location of C_s on this line will be solely dictated by the proportion of its intensity. So, if C_1 is blue and C_2 is red, C_s will be a purple and if I_1 is much higher than I_2, it will be a bluish purple landing closer to C_1 on the straight line between (x_1, y_1) and (x_2, y_2). If I_2 is larger, then it will be reddish purple. Note that this also provides us an alternate way of doing the addition of colors in the (Y, x, y) representation without going to the (X, Y, Z) representation — add the luminance and find the convex combination of the chromaticity coordinates weighted by the proportions of the intensities of each color. When considering addition of n different colors, the formulae are given by

$$Y_s = \sum_{i=1}^{n} Y_i \qquad (10.6)$$

$$(x_s, y_s) = \sum_{i=1}^{n} (x_i, y_i) \frac{I_i}{\sum_{i=1}^{n} I_i} \qquad (10.7)$$

Therefore, the chromaticity coordinates of the new color are given by the proportion of the intensities of C_1 and C_2 and not their luminance. Most color science literature makes this mistake and says that the chroma needs to be combined in proportion of their luminance and not the total intensity. This fundamental mistake makes it impossible to match colors by combining one or more additive colors in an experimental set up, and you may often think of moving to more complicated perceptually uniform color spaces. However, we can show with correct derivation of model parameters as above, that we can do perfect matching of colors just working with the XYZ color space.

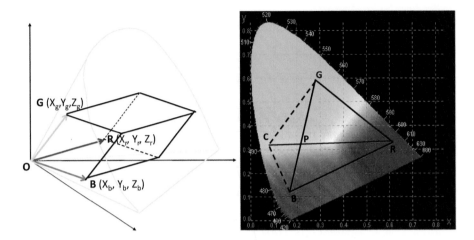

Figure 10.2. The 3D gamut (left) and the 2D gamut (right) of a linear three-primary device. In the 3D gamut on left, $R = (X_r, Y_r, Z_r)$, $G = (X_g, Y_g, Z_g)$ and $B = (X_b, Y_b, Z_b)$. The 2D gamut is shown by the black triangle RGB. Any color in the triangle RGB is reproduced by a convex combination of R, G and B using unique weights given by its barycentric coordinates with respect to R, G and B. If a fourth primary C is added, the 2D color gamut is now given by the polygon RGCB, the convex hull of R, G, B and C. However, note that in this case, the color P inside this gamut RGCB can be reproduced by non-unique combinations of different primaries – one with G and B and another with R and C.

10.1.1 Color Gamut of a Device

Equation 10.7 provides an interesting insight. This equation shows that a large number of colors are generated by a convex combination of a few colors. Therefore, in order to reproduce a reasonable area of the chromaticity chart we will need at least three colors, the convex combination of which creates colors with chromaticity coordinates in the triangle formed by the chromaticity coordinates of these three colors. These three given colors are called the *primaries* of the device and the triangle formed by their chromaticity coordinates is called the *2D color gamut* of the device. Typically, these three primaries lie in the regions of blue, red and green to cover a reasonably large area of the chromaticity chart. That is why most devices we see today have red, green, and blue primaries as shown in Figure 10.2. Nowadays, devices with more than three primaries are also designed to increase the color gamut.

Let us now consider a device with three primaries, usually red, green and blue. Let the input value for each channel, given by i_r, i_g, and i_b, be normalized and therefore range between 0 and 1. Suppose that the XYZ coordinates of each of the primaries at maximum intensity is given by $R = (X_r, Y_r, Z_r)$, $G =$

(X_g, Y_g, Z_g) and $B = (X_b, Y_b, Z_b)$ as illustrated in Figure 10.2. This means if we change the input of only one of the channels – say red – keeping the other two at zero, the XYZ values of the reproduced colors will travel along vector OR starting from O at $i_r = 0$ and reach R at $i_r = 1$. Also, when we change the inputs of multiple channels we get a vector addition of the vectors OR, OG and OB scaled by their respective input values. In other words, the color $C = (X, Y, Z)$ produced for input $I_p = (i_r, i_g, i_b)$ is computed as

$$C = (X, Y, Z) = O + i_r(R - O) + i_g(G - O) + i_b(B - O) \tag{10.8}$$
$$= i_r(X_r, Y_r, Z_r) + i_g(X_g, Y_g, Z_g) + i_r(X_b, Y_b, Z_b) \tag{10.9}$$

The space spanned by C as i_r, i_g and i_b changes from 0 to 1 is given by the parallelepiped shown in Figure 10.2 (left). This is the entire gamut of colors that the device can reproduce and hence is called the *3D color gamut* of the device. Typically, the tristimulus values of the primaries of practical devices are well within the visible colors, the parallellopiped formed by these primaries will usually be strictly inside the visible color gamut. Hence, usually our devices typically reproduce only a subset of the colors human can see. How big a subset they can produce depends on the properties of their primaries given by the coordinates of R, G and B. Note that Equation 10.9 can be written in the form of a 3×3 matrix as

$$\begin{pmatrix} X \\ Y \\ Z \end{pmatrix} = \begin{pmatrix} X_r & X_g & X_b \\ Y_r & Y_g & Y_b \\ Z_r & Z_g & Z_b \end{pmatrix} \begin{pmatrix} i_r \\ i_g \\ i_b \end{pmatrix} \tag{10.10}$$
$$C = MI_p \tag{10.11}$$

When the entire color gamut (or characteristics) of a device can be represented by such a matrix M, it is called a linear device. Note that the above matrix M indeed tells us everything about the color characteristics of a device. Also, if we know the desired color C to be reproduced, we can find the unique input that would create it by $I_p = M^{-1}C$. Therefore, every color within the gamut is created by a unique combination of input values.

The intersection of the vectors OR, OG and OB with the chromaticity chart provides us with the chromaticity coordinates of these three primaries which will define the set of all chromaticities that the device can reproduce (without considering the intensity). Since there are three vectors, their intersection in the chromaticity chart will create a triangle as illustrated in Figure 10.2 by the black triangle RGB. This is called the 2D color gamut of a device.

From Figure 10.2 it is obvious that adding an additional primary, like C, outside the current color gamut, would help us cover a larger area of the chromaticity chart and thereby increase the 2D gamut to the polygon $CBRG$. This is of course true and is often used as a technique to increase color gamut by

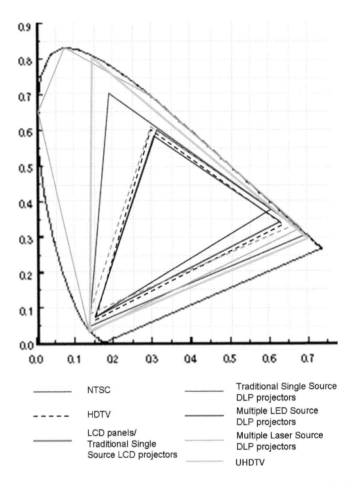

Figure 10.3. This figure shows a few standard and empirically measured color gamuts of current display devices.

the TV manufacturers. However, there is a downside to it. Consider this four primary system and the color P. Notice that it can be generated in more than one way — either by combining G and B or by combining C and R. Therefore, unlike a three-primary system which provides a unique way to create a color by combination of the primaries, here there are multiple ways to create a color.

In Equation 10.9, we assume that O is at the origin $(0,0,0)$. This means that the black produced by the device (output for $I_p = (0,0,0)$) indeed generates zero light. Unfortunately, in some display devices today, especially projectors, some constant leakage light is always present, even for input $(0,0,0)$, which is often called the *black offset*. If this black offset is characterized by the XYZ coordinates (X_l, Y_l, Z_l), then the equation 10.9 becomes

$$(X, Y, Z) = O + i_r(R - O) + i_g(G - O) + i_b(B - O) \tag{10.12}$$
$$= (X_l, Y_l, Z_l) \tag{10.13}$$
$$+ i_r(X_r - X_l, Y_r - Y_l, Z_r - Z_l) \tag{10.14}$$
$$+ i_g(X_g - X_l, Y_g - Y_l, Z_g - Z_l) \tag{10.15}$$
$$+ i_r(X_b - X_l, Y_b - Y_l, Z_b - Z_l) \tag{10.16}$$

The above equation represented as a matrix becomes

$$\begin{pmatrix} X \\ Y \\ Z \end{pmatrix} = \begin{pmatrix} X_r - X_l & X_g - X_l & X_b - X_l & X_l \\ Y_r - Y_l & Y_g - Y_l & Y_b - Y_l & Y_l \\ Z_r - Z_l & Z_g - Z_l & Z_b - Z_l & Z_l \end{pmatrix} \begin{pmatrix} i_r \\ i_g \\ i_b \\ 1 \end{pmatrix} \tag{10.17}$$

Fun Facts

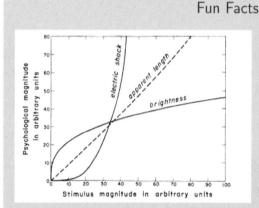

Interestingly, not only human perception of light, but all human perceptions follow a power law. This is called the Steven's power law based on the scientist who first made the observation. This law says that response R of any human perception to the input stimuli I follows the equation $R = KI^\gamma$. If $\gamma > 1.0$, perception is expansive. An example is electric shock. If $\gamma < 1.0$, perception is compressive. An example is our perception of brightness. γ is seldom 1.0. The expansive or compressive nature of our perception is shaped by evolution! The compressive nature of our perception to brightness saves our eyes from getting burned regularly by sunlight. The expansive nature of our perception to electric shock helps to put us on guard when the stimulus is not too large to cause damage.

However, note that the parameters of the matrix are not available to the user directly from the specification sheet of the devices. In general, they can be derived from other parameters that are specified. The first of these is a standardized 2D gamut. The 2D gamut of a device should usually conform to a predefined standard gamut, like sRBG, HD and NTSC (Figure 10.3). However, the 2D gamut does not contain any information about the minimum and maximum luminance that can be reproduced by the display. It only signifies the hue and saturation of the primaries, each of which can have different minimum and maximum brightness. So, to complete the description, we need the information about the *white point* and the *dynamic range*(more commonly called contrast). The white point gives the chromaticity coordinate of the white and the dynamic range is given by the ratio of the brightest and dimmest gray, i.e. white and black. The white point is specified as conforming to a predefined standard. Whites can be of different tints — purplish, bluish or reddish. And different cultures have been seen to prefer different white points. So, such standard whites have been defined like D65 ($x = 0.31271, y = 0.32902$) or D85. However, note that dynamic range can provide you the color gamut only up to a scale factor and hence the display's maximum brightness (usually produced at white) is required to get the absolute color gamut. The intensity of the white is usually specified as a measure of the brightness of the device. All these parameters together define a 3D color gamut in the XYZ space and the matrix M.

10.1.2 Tone Mapping Operator

Let us now discuss another important property of a color reproducing device, the tone mapping operator. As we change the input of one channel from 0 to 1 as the other two remain zero, the output will travel on the vectors OR, OG or OB (Figure 10.2). Now, the way the resulting output moves on these vectors may not be linear. In fact, most of the times it is a non-linear function trying to adjust for the response of human eye. Let us call this function h. For example,

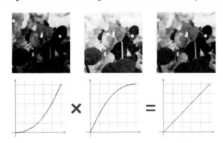

the human eye has a non-linear response to light which is compressive in nature i.e. if the eye is stimulated with k times the intensity, the perceived intensity is less than k times. For example in cameras, this function $h(i_r)$ is modeled by i_r^γ where $\gamma < 1.0$. In displays, $\gamma > 1.0$, most commonly $\gamma = 2.0$ to compensate for the compressive gamma imparted by cameras. Although we have assumed the same h for all channels here, these functions can be different for different channels. In fact, predefined γ was in vogue in the pre-digital age when film cameras usually had a $\gamma = 0.5$ which

Figure 10.4. On the left is a typical display gamma of 2.2. In the middle is the gamma of a capture device set to $\frac{1}{2.2}$ to mimic the human eye. The result of putting this image on to the display is a linear response of input to output as seen in the right.

was compensated by displays with a $\gamma = 2.0$. Hence, the name gamma function or gamma correction. This is illustrated in Figure 10.4.

Let us consider a more general representation of the gamma function which need not be a power function in a color reproducing device and can be thought of as a parameter of a color producing device that can be controlled to change the appearance of the device. This more general representation is called the *tone mapping operator* or *transfer function.* When color devices were non-existent and people were used to only black and white devices (a better way to call them would be gray), transfer function was the only function that dictated the picture quality. All the terminology for user interfaces to control picture quality originated at this time and hence they have a direct relationship with the tone mapping operator.

The basic assumption was that the tone mapping operator would be a smooth monotonically increasing function. There used to be two controls — usually called brightness and contrast or picture. Brightness used to act like an offset on the transfer function moving the function up or down. Contrast or picture used to change the gain of the transfer function. This is illustrated in Figure 10.6 along with its effect on an example image.

When color displays came into being, the natural thing to do was to have an independent transfer function for each channel. This provided a much greater control on image appearance. Changing the tone mapping operator for one of the channels to be different than others created different effects like producing unique color tints, as shown in Figure 10.7. This is often described as changing the *color balance.* Changing the tone mapping operator is the only one way to change the color balance of the display. The same effect can be achieved by changing the relative intensities of the different primaries using the offset control.

Figure 10.5. Notice the contouring in the flat colored region around the portrait. This is the tell-tale quantization artifact due to insufficient intensity resolution.

In the displays menu from the control panel on a standard Windows desktop, all these three transfer functions, one for each channel, can be seen. Therefore today, γ has become a way for users to have control on the device to create a different look and feel for images. In any laptop, in properties and settings, one can change the h for different channels differently which need not be even an exponential function. To get the best reproduction and to take full advantage of the dynamic range of the medium, the image should have pixels with range of colors that span throughout the entire tone range.

10.1.3 Intensity Resolution

Intensity resolution ideally means the number of visible intensity steps. However, this visibility depends on the viewing environment such as ambient light color

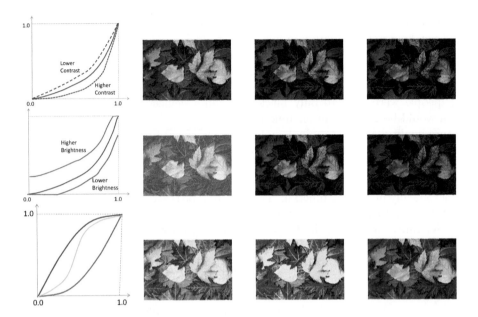

Figure 10.6. This figure shows the effect of brightness and contrast control on the transfer function. Top Row: Increasing contrast increases the slope of the red curve. Three different contrasts are shown on the left followed by their effect on the image produced on the right of the plot of the curves – top, middle and bottom curves from left to right. Middle Row: On the other hand, the brightness control moves the transfer function up and down. Increasing the brightness function means saturating higher values with the same value and vice versa. The effect of all these changes to a picture is shown on the right of the plot of the curves – top, middle and bottom curves from left to right. As the brightness is reduced notice the clamping near 0 and as a result the rightmost image has most of its parts darkened to zero. When the brightness is increased the higher values are clamped to 1 giving it a washed out appearance as is seen in the leftmost image. Bottom Row: This shows the effects of changing the tone mapping operator altogether to have more general functions – the images on the right show the effect of the top, middle and bottom tone mapping operators from left to right. For each of these figures, all of the three channels have the same transfer function.

and absolute brightness. So, in order to make it simple, intensity resolution is defined by the number of digital steps used to define the intensity of each channel. Thus, for an 8 bit display, intensity resolution is 256. The distribution of these intensities across different input values depends on the transfer function. Insufficient intensity resolution introduces quantization artifacts in the form of contour lines as illustrated in Figure 10.5. In practice, a perceptually uniform brightness distribution rarely shows contouring with 8 bits per pixel. However, perceptually uniform distribution indicated non-uniform steps in output value

Figure 10.7. The figure shows the effect of having different gamma functions for different channels. The green channel is kept the same in all the pictures (the red curve in the third row of Figure 10.6) while the red and blue channels are changed similarly for the left, middle and right picture respectively using respectively the red, green and blue curves in the third row of Figure 10.6. This results in different color tints in the different pictures – the right two are much warmer than the left one while the rightmost one is much more purplish than the other two.

for equal steps in input values. But, if we want to achieve a linear encoding on such displays, that provide uniform steps in output value for equal steps in input value, we would need a much larger number of bits – around 10-12 bits per pixel.

10.1.4   Example Displays

Displays are one of the most common devices where the effects of these different properties of color gamut, tone mapping operators and intensity resolution can be readily observed. So, let us study a few common display technologies from this perspective.

Cathode Ray Tube (CRT) Displays: CRT monitors excite phosphors with rays of electrons from an electron gun. Different types of phosphors are used to emit red, green and blue light. The phosphor colors match the sRGB color gamut, but they age easily becoming progressively less bright. In addition, the blue phosphor often degenerates faster leading to a change in color balance, giving the monitor a yellowish appearance. Note that though the hue and saturation of the primaries remain the same in the chromaticity diagram, the color of the display changes just due to deterioration of their brightness.

The CRT's transfer function is a non-linear power function which is defined by the physics of the electron gun exciting the phosphors. In the simplest form, it is approximated as

$$I = V^\gamma$$

where I is the measured intensity and V is the input voltage corresponding to channel inputs. If the intensity produced at V_0 (black) is non-zero, then this equation is modified to

$$I = (V + V_0)^\gamma$$

Finally, the entire curve can be scaled by a constant k to give the most general

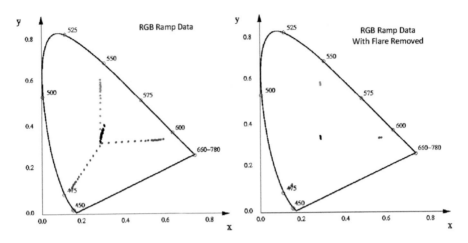

Figure 10.8. The effects of flare light on the chromaticity of the primaries of an LCD display. The XYZ chromaticity coordinates of each channel are plotted (with red, green and blue color) as the input values are ramped from 0 to 1 while the other channels are left at zero. The black plots show the chromaticity coordinates of the grays as they ramp from 0 to 1. Left: With flare. Right: With flare subtracted. Note that with flare, each of these plots starts from the black offset chromaticity coordinates and move to the channel chromaticity value on a straight line as the input value increases thereby decreasing the effect of the black offset in the additive combination. If the black offset or flare is removed, the chromaticity coordinates are constant (since only the intensity changes with the change in input values).

form of the equation

$$I = k(V + V_0)^\gamma$$

The brightness and contrast controls in CRTs therefore change k and γ respectively.

Liquid Crystal Displays (LCD): LCD displays are a spatial array of red, green and blue segments, each of which is a colored filter over a cell of liquid crystal material that can be made variably transparent. A backlight shines through the LCD array so that the resulting color is a function of both the backlight and the filters. However, colored filters are significantly different than colored phosphors. The more intensely colored (saturated) the filter is, the less light it will pass making the display dim. The less saturated the filters, the brighter the display but it is also a lot less colorful. Thus, to get a highly bright and colorful display, very bright backlights are needed in addition to saturated color filters. But a trade-off needs to be made due to the huge power consumption. When compared to the gamut of the CRT displays, the blues of the LCD displays are often much less saturated.

The transfer function of LCDs is usually linear. But usually LCDs include electronics by which this can be changed to match the traditional CRT displays. Also, usually LCDs project some light even at input zero due to leaking of the backlight through the front of the display. Though it is called flare, it is the same as the black offset. Thus, the chromaticities of the primaries shift towards the white point for the lower intensities. Even with flare subtracted, the LCDs deviate measurably from the ideal RGB model since the chromaticity is still not constant at all brightness levels.

Projection Displays: A digital projector contains a digital imaging element like an LCD panel or an array of digital micromirrors (DMD) that modulate the light coming from a high intensity light bulb. Most LCD projectors and the larger DMD projectors have three imaging elements and a dichroic mirror that splits the white light from the bulb into its red, green and blue components. These are recombined and displayed through a single lens. The smaller projectors use a single imaging element with a wheel of filters, so that the separations are displayed sequentially in time. Some DLP (digital light processing) projectors have a fourth filter called the clear filter which are used while projecting the grays to achieve a higher brightness for grays. However, note that this is not equivalent to using more than three primaries since the chromaticity coordinate of the fourth filter lies inside the color gamut formed by the red, green and blue filters.

10.2   Color Management

So far we have been discussing a single device. Let us now consider multiple devices. When considering multiple devices, even if they are of the same type and same brand, the primaries in them can differ significantly. Especially, when we consider a complete imaging system including acquisition (using cameras), monitors, displays and printing, we need to make sure that the color in one device looks similar to that in another. You may have had the situation that your picture in the camera display looked nice and vibrant, but looked dull when projected in a slide projector for an event. Or, the printed picture looked washed out while the same picture looked perfectly fine in your monitor!

Color management entails modifying the input going to each device so that the output coming from each device matches. Since each device has different primaries, it is evident that they will need different inputs to create the same color. Since our goal is to maintain the same color across multiple devices, the only way to achieve this is to change the input from one device to another. We will consider two fundamental techniques of color management in this section.

10.2.1 Gamut Transformation

Let us consider two devices with linear gamma and color gamut defined by two different matrices M_1 and M_2. Let us assume that the input (R_1, G_1, B_1) creates the color (X, Y, Z) in the first device. Our goal is to find the input (R_2, G_2, B_2) in the second device that would create the same color. Note that, as per Equation 10.10

$$\begin{pmatrix} X \\ Y \\ Z \end{pmatrix} = M_1 \begin{pmatrix} R_1 \\ G_1 \\ B_1 \end{pmatrix} = M_2 \begin{pmatrix} R_2 \\ G_2 \\ B_2 \end{pmatrix} \tag{10.18}$$

From the above equation, we find that

$$\begin{pmatrix} R_2 \\ G_2 \\ B_2 \end{pmatrix} = M_2^{-1} M_1 \begin{pmatrix} R_1 \\ G_1 \\ B_1 \end{pmatrix} \tag{10.19}$$

Therefore, if we multiply the input to the first device with the matrix $M_2^{-1} M_1$, we will get the appropriate input to create the same color in the second device. This is called gamut transformation.

However, there are some issues with this technique as illustrated in Figure 10.9. Here we show the parallelepiped gamut of two devices, one in black and the other in gray. Let us now consider the color marked by the blue dot which is within the gamut of the first device. Once we apply our gamut transformation to find the corresponding input in the second device to produce the same color, the values

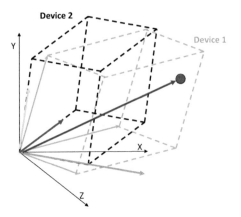

Figure 10.9. This figure illustrates the problem caused by out of gamut colors during gamut transformation.

we get are marked in the orange, magenta and cyan vectors. Note that the required color is outside the color gamut of the second device, and to reproduce that color one of the primaries has to be scaled by a value greater than 1.0. This means that the inputs generated will be out of the range. This indicates that the color to be generated is out of the device gamut and cannot be generated using convex combination of the primaries. Such colors are call *out-of-gamut* colors and cause a problem in any gamut transformation.

There are multiple ways to deal with out-of-gamut colors and they can be chosen based on the application. One option is to use an in-gamut color on the planar boundaries of the gamut that is closest to the out-of-gamut colors.

This can be achieved by clamping the out-of-bound input to 1 or 0. Such a clamping creates a local movement of only the out-of-gamut colors retaining the appearance of all in-gamut colors. This method can yield rather effective results for GUI applications where flat colors are used for buttons and slides.

However, when dealing with natural images, multiple out-of-gamut colors can land on the same in-gamut colors creating color blotches. Another option is to scale the position of all the colors to fit the out-of-gamut colors within the gamut. This can be achieved by scaling the input by an appropriate factor to move the out-of-range value to 1. This leads to a global movement of colors yielding better result for images though vibrancy and brightness of all the colors are sacrificed.

10.2.2 Gamut Matching

Gamut matching is a technique that tries to eliminate out-of-gamut colors altogether. The mainstay of this method is to find a common gamut that all devices can reproduce. The method is illustrated in Figure 10.10. Let us consider two devices with gamuts G_1 and G_2 – shown in red and blue respectively. Let the linear matrices representing these two devices be M_1 and M_2 respectively. First, we find the intersection of G_1 and G_2. $G_1 \cap G_2$ is shown in green. There is no guarantee that $G_1 \cap G_2$ is a parallelepiped. However, in order to express the common gamut as a matrix, we find the largest parallelepiped, G_c, inside $G_1 \cap G_2$. We desire to find the largest such parallelepiped to increase the gamut of colors that can be reproduced by both the devices. Let the matrix that represents this black gamut be M_c.

Next, we consider any input to be an input (R_c, G_c, B_c) in this common gamut G_c and find the corresponding input in the ith device from Equation

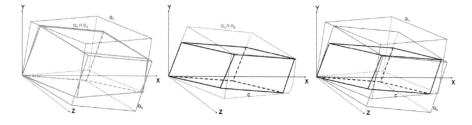

Figure 10.10. This figure illustrates the process of gamut matching for two devices whose gamuts are denoted by the red and blue parallelepiped. The intersection of these two gamut is shown by the green volume (left). The biggest parallelepiped inscribed in this intersection is shown in black (middle). Finally, the red and blue parallelepipeds are transformed to the black one via appropriate linear transformations (right).

10.19 as

$$
\begin{pmatrix} R_1 \\ G_1 \\ B_1 \end{pmatrix} = M_i^{-1} M_c \begin{pmatrix} R_c \\ G_c \\ B_c \end{pmatrix}
\tag{10.20}
$$

This method can be easily generalized to n devices. However, finding the intersection of multiple parallelepipeds and the largest inscribed parallelepiped in it are time consuming computations.

We have discussed here only two very basic color management techniques that only apply to linear devices. Non-linear devices (for e.g. ones with more than three primaries) have more complex shaped gamuts. Complex geometric entities like Bezier patches or splines may be used to handle such non-linearities. Further, we also only considered methods that are content-independent (i.e. that does not depend on the particular content). There are other gamut matching techniques that take content into consideration. For example, if you are dealing with fall images where you know you will have predominantly reds, oranges and yellows, you can do lot better by adapting the method to these colors so that they are maintained at higher fidelity while larger movements occur for colors which are sparse in the image.

10.3   Modeling Subtractive Color Mixture

We have discussed additive color mixtures so far. Though this is the system which we will use mostly, paint based systems (e.g. printers) still use subtractive color mixtures. We will discuss some of the basic issues about subtractive color mixtures in this section.

Cyan, magenta and yellow are considered to be the primaries of subtractive color systems. Cyan absorbs red, magenta absorbs green and yellow absorbs blue. The ideal response of such paints or filters is shown by the bold lines in Figure 10.11). When we say that yellow has an input of 0.5, it essentially means that 50% of blue is absorbed. Or, a input of 0.75 magenta means 75% of green is absorbed. Considering this, a very simplistic model of CMY systems can be

$$
(C, M, Y) = (1, 1, 1) - (R, G, B).
\tag{10.21}
$$

Therefore, it is easy to find the RGB inputs of a device given the input to a subtractive CMY device using the above equation. However, the problem is that the real CMY filters rarely behave as ideal block filters. They show lot of cross talk due to ink impurities. This also causes gray imbalance which means that equal amount of the different primaries do not lead to a neutral gray color. Therefore, this simplistic model rarely holds.

In addition, when dealing with dyes on paper, several other issues come into play. For example, it is evident that depositing C, M and Y on top of each

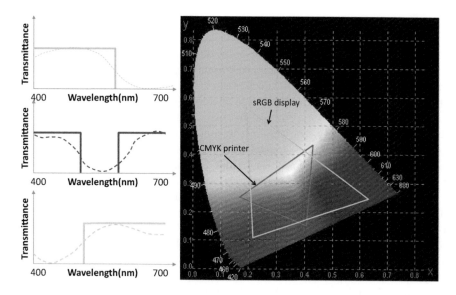

Figure 10.11. Left: This figure illustrates the transmittance profile of cyan, magenta and yellow filters in a subtractive color system. The ideal responses are given in bold lines and the real responses in dotted line. Right: This shows a subtractive 2D color gamut of a printer (CMYK) when compared to an additive 2D color gamut of a display (standard RGB - often referred to as sRGB).

other in layers should create a dark black. But due to the cross talk, the black thus created usually does not provide a good contrast. Sometimes, layering of so many different primaries causes the paper to get wet and tear off. So, almost all subtractive color devices use an inexpensive high contrast black dye to avoid tearing of paper and to reduce cost by reducing the usage of the ink of other primary colors. This creates a 4-primary $CMYK$ system where K stands for black (traditionally K stood for key color — this was during the age of text printing when black was considered a very important dye for printing books; also since B is used for Blue, the convention is to use K for black). However, this means that the primaries are no longer independent of each other and hence each color can be produced in more than one way. Thus, such devices undergo careful factory calibration to decide what amount of which primaries will be used to generate a particular color and it is difficult to reverse engineer due to the non-uniqueness of the process. Figure 10.11 shows the comparison between a common subtractive gamut of a printer to additive gamuts of displays. Note that the subtractive gamut is usually much smaller than an additive gamut.

10.4   Limitations

At this point, it is very important to point out that a single image from a device cannot really reproduce exactly how the humans perceive the scene. The limitations in any color reproduction mechanism stem from two fundamental reasons – the range of brightness and the range of colors (chromaticity coordinates) that a device can capture or reproduce are usually smaller than what is found in nature. We have already seen this in the context of chromaticity coordinates. It is evident that no 3-primary device can reproduce the entire color gamut that is perceived by a human since any triangular 2D color gamut will leave some part of the chromaticity chart uncovered.

The same phenomenon happens for brightness. Let us consider Figure 10.12(top). The brightness in any scene can range from 10^{-2} lumens (in shadows of trees on a moon-lit night) to 10^{10} (in sky on a bright sunlit day). This is a huge variation and is hence plotted in log scale. The response of the human eye to these brightness range is *not* as shown by the black dotted line. Instead, at any particular time, the human eye can only perceive a smaller subset of this large range of brightness – maybe only 3-4 orders of magnitude. This means the human eye can perceive a contrast ratio or *dynamic range* of about 1:10,000. In contrast, the dynamic range that can be reproduced by 8-bit device is usually of the order of 1:100.

Further, based on the brightness of different parts of the scene our eye can quickly adapt to the most appropriate range that should be sensed to gather the maximum information. We have all experienced this, especially when our eye goes through a drastic adaptation, like when we are blinded for some time when we come out to brightly lit outdoors from a dark room, only to get adapted to the new condition quickly. This flexible adaptation capability of the human eye at different brightness level is illustrated by the different colored curves in Figure 10.12(top). Each curve is linear only within a range of illumination beyond which it saturates. Given a particular illumination level, the curve which provides a linear response around that range of illumination provides the curve in which the eye is responding to the scene. Therefore, when given a scene as Figure 10.12(bottom), the eye adapts to a higher range of brightness to extract information of the appearance of the sky and then to a lower range of brightness to extract the information of the objects like houses, roads and cars. Then it combines all this information to create a mental picture similar to the combination of these two. In this section, we discuss a few ways by which we address the limited capability of usual devices when compared to the human eye.

10.4.1   High Dynamic Range Imaging

High dynamic range imaging is a technique by which we mimic the dynamic range of the human eye and create an image that can have a contrast similar to the human eye or even as high as what is present in nature. Here images are captured at different camera settings so that different amounts of light illuminate the sensor. When more light illuminates the sensor, we capture the darker parts of the scene while the brighter parts of the scene over-saturate the sensor creating 'burned out' or over-exposed regions. When less light illuminates the sensor, we capture the brighter part of the scene while the darker parts of the scene under-saturate the sensor creating under-exposed regions. Therefore, by capturing different images and allowing different levels of light to impinge on the sensor, different parts of the scene get captured while others are either under or over exposed. But, by combining the information from all these pictures, information in all parts of the scene can be gathered creating one high dynamic range image that has a much higher dynamic range than what is available to us via any standard 8-bit image capture.

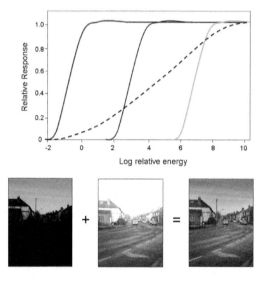

Figure 10.12. Top: This figure illustrates the adaptive dynamic range of the eye. Each colored curve shows how the response of the eye spans a range of brightness much smaller than the total range of brightness available in nature. At any instant the response of the eye is linear within this small range. The black dotted lines show the hypothetical response if there was no adaptation to different ranges of brightness and the eye had a response that spanned the entire dynamic range available in nature. Bottom: This shows how the eye processes information from capturing the different parts of the scene in an appropriate dynamic range.

Therefore, the obvious question is, how do we control the light that illuminates the sensor to create a high dynamic range image? This can be done by changing the exposure of the camera which can, in turn, be achieved either by changing the aperture size of the camera or the shutter speed. Shutter speed dictates the amount of time the shutter is open to expose the camera sensor to light. Usually the latter control is chosen for a reason that is completely device dependent. As the aperture of the camera changes, due to the complex lens system, the light that reaches the center of the sensor is not the same as that which reaches the periphery. In fact, the amount of light reaching different pixels of the sensor shows a steady fall-off from center to periphery. This is often called the *vignetting effect* of a camera. The vignetting effect in a camera is minimal at low apertures ($f/8$ or below) when the camera is closest to the pinhole model. At other aperture settings the vignetting effect is considerable thereby influencing the accuracy of the light sensed at each pixel. Therefore, it is better to use the shutter speed (and not change aperture) to change the exposure to minimize the inaccuracies due to the different vignetting effects at different aperture settings.

Let us now consider a static scene captured by a camera using n different shutter speeds. For the jth shutter speed, $1 \leq j \leq n$, let the time for which the shutter is open be given by t_j, the number of pixels in the image captured by the camera be m, the gray scale value captured at ith pixel, $1 \leq i \leq m$, be Z_{ij}, the scene irradiance at the ith pixel be E_i, and the camera transfer function be f. Therefore, for any i and j, we can model the imaging process as

$$Z_{ij} = f(E_i t_j). \tag{10.22}$$

Assuming that f is monotonic and invertible, we can write the above equation as

$$f^{-1}(Z_{ij}) = E_i t_j. \tag{10.23}$$

Assuming a natural logarithm of both sides, we can write this equation as

$$g(Z_{ij}) = ln \; E_i + ln \; t_j \tag{10.24}$$

where $g = ln \; f^{-1}$. In this equation, t_j and Z_{ij} are known and E_i and g are unknown. For every pixel i, for every shutter speed j, we can write one such equation leading to a system of mn linear equations. g is a function with 256 values for a 8-bit device. Therefore, we will need to solve $m + 255$ unknowns from set of mn over-constrained linear equations using linear regression. To constrain the g to be monotonic, we can add additional constraints to the system of equations. We can even add curvature constraints to assure the g is smooth. The solutions of these equations will provide a high dynamic range image as shown in Figure 10.13.

However, a high dynamic range image brings forth the obvious question of how can we display it? Its range of brightness and contrast are many orders of magnitude beyond what a traditional 8-bit display can do. Seeking the solution to this problem has created a large body of research literature on designing complex, often spatially varying, tone mapping operators that can take this huge range of contrast and successfully compress it within the range (usually 0-255) that can be reproduced. The basic idea behind the tone mapping operators is to create the effect of spatially varying exposure such that every region is exposed the right amount to reveal information (Figure 10.13) without getting under or over saturated. Though such an image may not look photo-realistic since we are

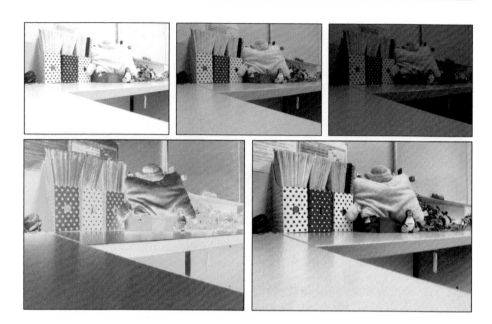

Figure 10.13. Top Row: Three different images captured at different shutter speeds. Bottom Left: The recovered high dynamic range radiance map. Note that this image cannot be displayed in the regular display since it has a much higher contrast. Therefore we show a heat map visualization of the radiance, blue indicating low and red indicating high. Right bottom: This shows the same image being shown on a regular 8-bit display using a tone-mapping operator.

not used to seeing such photos from a real camera where both very bright and very dark regions are well-exposed, it nevertheless is perfect for conveying the information content in every region of the picture.

10.4.2 Multi-Spectral Imaging

Multi-spectral imaging is a technique that addresses the issue of a limited 2D color gamut of any 3-primary color reproduction system. It is evident that the triangular 2D gamut formed by 3-primaries can never capture the entire 2D color gamut of the human eye. In fact, as shown in Figure 10.3, most color reproduction systems do not have highly saturated primaries thereby restricting the 2D gamut of 3-primary systems even further. This is due to a fundamental physical limitation. Saturated primaries are achieved by narrow band primaries which are light inefficient since they filter out most of the light in a scene retaining only a very narrow band. Therefore, saturated primaries require addressing this fundamental trade off between light efficiency and a larger 2D color gamut. An obvious solution to this problem is to choose more than 3-primaries and various

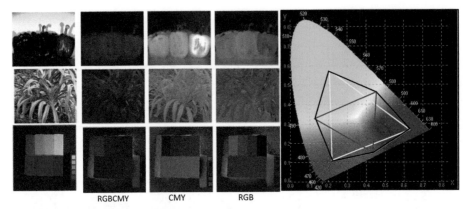

RGBCMY   CMY   RGB

Figure 10.14. Consider a scene captured by a hyperspectral camera that captures the accurate spectrum at every pixel from which we compute the XYZ values at every pixel. Left: This image shows the RGB image from a standard camera that the spectra captured by the hyperspectral camera would produce. Middle: Consider the same scene being captured by a 6 primary camera (RGBCMY), a standard RGB camera and a CMY camera - the primaries are shown on the chromaticity chart on the right. The Euclidian distance between the reconstructed spectrum from these three cameras and the ground truth spectra captured by the hyperspectral camera are normalized and represented as a gray scale value at every pixel (brighter means greater distance). As expected, a 6-primary camera captures a spectra closer to the ground truth than the 3-primary cameras.

systems have been designed over the years using 4-6 primaries. More than six primaries lead to compromising other properties like spatial resolution to offset the gain in the 2D color gamut.

Referring to Figure 10.14, consider a scene captured by a hyperspectral camera that captures the accurate spectrum at every pixel. The XYZ tristimulus value can be computed at every pixel from the captured spectra. Next, the same scene is captured by a 6 primary camera (RGBCMY), a standard RGB camera and a CMY camera. The inaccurate spectra captured via these primaries are reconstructed using a linear combination of the sensitivities of the primary weighted by the values captured. The Euclidian distance between the inaccurately captured spectra from the six or three primary cameras and the ground truth spectra captured by the hyperspectral camera are normalized and represented as a gray scale value at every pixel. Brighter values indicate greater distances and therefore greater errors in capture. Note that, as expected, as the number of primaries increases, the errors reduce. Also, note that the errors for the CMY camera is much greater than that from the RGB camera. This is evident from examining the coverage of these two 2D gamuts in Figure 10.14 where the area covered by the CMY gamut is clearly much smaller than the area covered by the RGB gamut.

This light inefficiency becomes the primary inhibitor for creating spectrally accurate reproduction of colors in displays. Spectrally accurate displays demand highly saturated (and therefore narrow band) primaries which are very bright to create the light efficiency demanded by a display. Creating such super-bright saturated primaries for displays was practically impossible till date. But, the promised gamut of the 6-primary laser projectors of the future, shown by the cyan polygon in Figure 10.3 shows promise to achieve this, hopefully sometime soon.

Fun Facts

The world of color standards can be a confusing aspect to deal with for consumers. What do the terms NTSC, HDTV, UHDTV mean and what are their consequences on the picture quality? The color standard definitions have their origin in the 1940s when the TV industry felt the need to define some standards for transmission of the video signals. Since this was the era of black and white television, the only property that was related to the transmission was the spatial resolution of the imagery which was 640 × 480 at a 4:3 aspect ratio (ratio of the screen width to height). In 1953, color was added to the standards after the advent of color televisions. This included specifying a standard 2D color gamut and white point (as shown in Figure 10.3). This was the main standard until around 2010 when the advent of digital content led to the development of the HDTV (high-definition TV) standard. This changed the resolution to 1920 × 1080 and expanded the color gamut slightly (as shown in Figure 10.3). This also introduced the concept of widescreen TV with aspect ratio of 16:9. Recently, there is a new standard called UHDTV (Ultra high-resolution TV) which doubles the resolution to 3840 × 2160. But we have reached the limits of our ability to perceive resolution in standard TV size with HDTV. Therefore UHDTV does not promise a huge difference in quality in terms of resolution. However, the color gamut has also expanded significantly promising much more vibrant displays. Also, UHDTV standard now allows high dynamic range imagery which together with the expanded color gamut should improve the color quality of TVs greatly.

10.5 Conclusion

Color reproduction needs to combine the precision of hard mathematics along with the limitations and/or imprecision of human perception – often involving human cognition – making it a very difficult science or application to succeed in.

Deep knowledge of human perception, as offered in books like [Palmer 99], can be very useful in this domain. Texts that focus on engineering aspects of color reproduction include [Hunt 95, Berns et al. 00]. The involvement of people from diverse domains, like biology, art, science and engineering, in the development of color models has been the boon and the curse resulting in color reproduction being a complex area to explore. Therefore, it is often useful to see color from an alternate perspective as in [Livingstone and Hubel 02].

Much work was done in high dynamic range images in the late 1990s and early 2000s. This was stimulated by the initial work on capturing HDR image in [Debevec and Malik 97]. This led to a plethora of work in appropriate tone mapping operators to display such images on traditional displays [Tumblin and Turk 99, Larson et al. 97, Gallo et al. 09]. New HDR cameras were designed [Yasuma et al. 10]. Displays that can truly display such a high dynamic range were also designed [Seetzen et al. 04]. Today, such displays are becoming more mainstream and slowly emerging in the market. HDR imaging continues to be an active area of research even today [Gupta et al. 13]. A comprehensive reference for HDR imaging pipeline is Reinhard et al.'s recent book on this topic [Reinhard et al. 05]. Multi-spectral cameras [Yasuma et al. 10, Susanu 09, Shogenji et al. 04] and displays [Li et al. 15] have been explored before, but they have not made their way to the consumer devices yet. The only successful case is that of the RGBW projector that uses a clear filter in addition to red, green and blue filters only during the projection of grays to increase the brightness rating. However, this does not help in increasing the 2D color gamut since the white lies within the gamut created by R, G and B.

Bibliography

[Berns et al. 00] Roy S. Berns, Fred W. Billmeyer, and Max Saltzman. *Billmeyer and Saltzman's Principles of Color Technology.* Wiley Interscience, 2000.

[Debevec and Malik 97] Paul E. Debevec and Jitendra Malik. "Recovering High Dynamic Range Radiance Maps from Photographs." In *Proceedings of the 24th Annual Conference on Computer Graphics and Interactive Techniques, SIGGRAPH '97*, pp. 369–378, 1997.

[Gallo et al. 09] O. Gallo, N. Gelfand, W. Chen, M. Tico, and K. Pulli. "Artifact-free High Dynamic Range Imaging." *IEEE International Conference on Computational Photography (ICCP).*

[Gupta et al. 13] M. Gupta, D. Iso, and S.K. Nayar. "Fibonacci Exposure Bracketing for High Dynamic Range Imaging." pp. 1–8.

[Hunt 95] R. W. G. Hunt. *The Reproduction of Color.* Fountain Press, 1995.

[Larson et al. 97] G. W. Larson, H. Rushmeier, and C. Piatko. "A Visibility Matching Tone Reproduction Operator for High Dynamic Range Scenes." *IEEE Transactions on Visualization and Computer Graphics* 3:4.

[Li et al. 15] Yuqi Li, Aditi Majumder, Dongming Lu, and Meenakshisundaram Gopi. "Content-Independent Multi-Spectral Display Using Superimposed Projections." *Computer Graphics Forum*.

[Livingstone and Hubel 02] Margaret Livingstone and David H. Hubel. *Vision and Art : The Biology of Seeing*. Harry N Abrams, 2002.

[Palmer 99] Stephen E. Palmer. *Vision Science*. MIT Press, 1999.

[Reinhard et al. 05] Erik Reinhard, Greg Ward, Sumanta Pattanaik, and Paul Debevec. *High Dynamic Range Imaging: Acquisition, Display, and Image-Based Lighting (The Morgan Kaufmann Series in Computer Graphics)*. Morgan Kaufmann Publishers Inc., 2005.

[Seetzen et al. 04] Helge Seetzen, W. Heidrich, W. Stuezlinger, G. Ward, L. Whitehead, M. Trentacoste, A. Ghosh, and A. Vorozcovs. "High Dynamic Range Display Systems." *ACM Transactions on Graphics (special issue SIGGRAPH)*.

[Shogenji et al. 04] R. Shogenji, Y. Kitamura, K. Yamada, S. Miyatake, and J. Tanida. "Multispectral imaging using compact compound optics." *Opt. Exp.*, p. 16431655.

[Susanu 09] Peterescu S. Nanu F. Capata A. Corcoran P. Susanu, G. "RGBW Sensor Array." *US Patent 2009/0,167,893*.

[Tumblin and Turk 99] J. Tumblin and G. Turk. "Low Curvature Image Simplifiers (LCIS): A Boundary Hierarchy for Detail-Preserving Contrast Reduction." pp. 83–90.

[Yasuma et al. 10] F. Yasuma, T. Mitsunaga, D. Iso, and S.K. Nayar. "Generalized Assorted Pixel Camera: Post-Capture Control of Resolution, Dynamic Range and Spectrum." *IEEE Transactions on Image Processing* 99.

Summary: Do you know these concepts?

- 3D or 2D Color Gamut of a Device
- Gamma Function
- Color Management
- Gamut Transformation and Matching
- Dynamic Range
- Adaptation
- High Dynamic Range Imaging
- Multi-spectral Imaging

Exercises

1. C_1 and C_2 are colors with chromaticity coordinates $(0.33, 0.12)$ and $(0.6, 0.3)$ respectively. In what proportions should these colors be mixed to generate a color C_3 of chromaticity coordinates $(0.5, 0.24)$? If the intensity of C_3 is 90, what are the intensities of C_1 and C_2?

2. Consider a linear display whose red, green and blue primaries have chromaticity coordinates of $(0.5, 0.4)$, $(0.2, 0.6)$ and $(0.1, 0.2)$ respectively. The maximum brightnesses of the red, green and blue channels are 100, 200 and $80 cd/m^2$ respectively. Generate the matrix that converts the RGB coordinates for this device to the XYZ coordinates. What are the XYZ coordinates of the color generated by the RGB input $(0.5, 0.75, 0.2)$ on this device?

3. Consider two colors $C_1 = (X_1, Y_1, Z_1)$ and $C_2 = (X_2, Y_2, Z_2)$ in the CIE XYZ space. Let their chromaticity coordinates be (x_1, y_1) and (x_2, y_2) respectively.

   (a) If C_1 is a pure achromatic color, what constraint will hold on its tristimulus values and its chromaticity coordinates? In that case, would black and white lie on the ray connecting the origin to C_1 in XYZ space? Justify.

   (b) If $C_2 = (50, 100, 50)$, then what is the value of (x_2, y_2)?

   (c) What is the dominant wavelength of C_2?

   (d) To create a color of chromaticity coordinates $(7/24, 10/24)$, in what proportions should be C_1 and C_2 be mixed? What are the intensity and luminance of C_1 required for this mixture?

4. When we mix blue paint with yellow paint we get green. But when we project blue light on yellow light, we get brown. How do you explain this contradiction?

5. Consider a grayscale image with linear gamma function. How would you change the gamma function to make the image have higher contrast?

6. Another name for the gamma function is the *tone mapping operator*. Consider a $8 - bit$ display (each channel is represnted with $8 - bit$ integers) with tone mapping operator of i^2 across all channels where i is the channel input. Consider the properties of brightness, contrast, color resolution, white point and tint. Which of these properties will change in each of the following scenarios?

   (a) The tone mapping operator is made i^3 across all the channels.

(b) The tone mapping operator is made i^3 across the green channel only.

(c) The number of bits are changed to 10 bits across all channels.

(d) Which property would remain unchanged across all the above changes?

7. Gamut matching addresses the problems caused by out-of-gamut colors in gamut transformation. But what do you expect to be the negative effect of gamut matching?

8. Consider a projector with sRGB color gamut (RGB chromaticity coordinates given by $(0.64, 0.33)$, $(0.3, 0.6)$ and $(0.15, 0.06)$). The luminance of the red, green and blue channels is given by 100, 400 and 50 lumens. Since this is a projector, it has a black offset with chromaticity $(0.02, 0.02)$ and luminance of 10 lumens.

(a) Find the matrix to convert the input values of this projector to the XYZ space.

(b) Consider another projector whose all color properties are the same except the luminance of green which is 200 lumens. Find the same matrix for this projector.

(c) Find an input in the first projector that will be out of gamut for the second projector.

(d) Is there any color in the second projector that will be out of gamut for the first projector? Justify your answer.

9. Consider a display with the following specs. The 2D gamut of the display, given by the chromaticity coordinates of its blue, green and red primaries respectively, are $(0.1, 0.1)$, $(0.2, 0.6)$ and $(0.6, 0.2)$. The intensity of the white is 1000 lumens. The white point is $(0.35, 0.35)$. Find the matrix M that defines the color property of this display.

11

Photometric Processing

In the previous chapter, we learned about different ways to represent color. Though device dependent, RGB representation is still the most common representation of color images. In this chapter, we will learn about some fundamental image processing techniques that deal with processing the colors of the image. There are two ways to process color images. (a) In the first approach, the processing techniques are applied similarly to the red, green and blue color channels assuming each of them to be an independent 2D image. (b) In the second approach, RGB images are converted to 1D luminance (Y) and 2D chrominance (via some linear or non-linear color space transformation), which are processed separately and the processed image is subsequently transformed back to RGB.

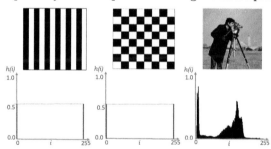

Figure 11.1. This figure shows different images (top row) and their corresponding histograms (bottom row). Since these are single channel gray scale images, the gray values i range from 0 to 255. The probability of their occurance, $h(i)$ ranges from 0 to 1. Note that the left and middle image have same histogram though they are very different in appearance.

When applying tone mapping operators, the first approach is commonly used since this does not change the relative differences between the colors. But, a good example of the second approach is *contrast enhancement* where the contrast of the luminance but not of the chrominance is enhanced. As opposed to enhancing the contrast of each of the red, blue and green channel independently, enhancing only the luminance while keeping the chrominance unchanged helps to preserve the hue which can be important in many applications. Similarly, in other applications like image compression, changing to a luminance and chrominance representation is justified since human perception gives more importance to luminance which can be leveraged by compressing the chrominance channels more aggressively than the luminance. Which of the two methods is used to process color images often depends on several criteria like application, network bandwidth and processing power. We will discuss different techniques in this chapter assuming they will be applied to a single channel of an image, be it be luminance or red or green or blue channel. It is left to the application developer to decompose an image to appropriate channels.

11.1 Histogram Processing

Histogram is defined as the probability density function of the values in an image. Let us consider a single channel of an image of size $m \times n$ providing a total of $N = m \times n$ pixels. The value at each pixel i can take p discrete values normalized between zero (black) to one (white). p is given by the number of bits used for representing the grays. Let us consider 8-bit gray values leading to $p = 256$. Therefore, i can take k different values, i_k, $1 \leq k \leq 256$. Let the number of pixels in the image with value i_k be N_k. Therefore, the histogram $h(i_k)$ is defined as

$$h(i_k) = \frac{N_k}{N} \qquad (11.1)$$

Figure 11.2. This figure shows how the exposure (the measure of how much light is let into the camera to capture the picture) affects the histogram of an image. When the image is underexposed (left) the histogram h shifts left with higher values of h for lower values of i. When the image is overexposed (right) the histogram h shifts right. For correctly exposed image, the values of h at 0 and 255 are not outliers.

i.e. the probability of a pixel to have a value i_k. Since $h(i_k)$ is a probability, $0 \leq h(i_k) \leq 1$, and therefore h is a probability density function.

Figure 11.1 shows some example of histogram of images. Note that two images can have entirely different appearances but have very similar histograms. Figure 11.2 shows the effect of the exposure on the histogram of an image. Exposure decides how much light comes into the camera when capturing the image. If too little light is let in, darker regions of the image are underexposed having the value 0 therefore creating a high value for $h(0)$. If too much light is let in, brighter regions of the image are overexposed having the value of 1 thereby creating a spike at $h(1)$. Figure 11.3 shows the effect of contrast on the histogram of an image. Low contrast images usually do not have the near black and near white grays. Therefore, $h(i_k)$ is non-zero only in a small middle ranges of values of i_k.

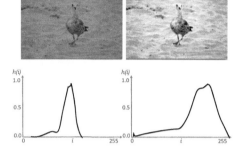

$h(i)$ ... $h(i)$

One common application of histogram processing is to *enhance the contrast* of the image, using a process called *histogram stretching*. As we saw earlier, the range of values for which $h(i_k) \neq 0$ is small for a low contrast image. The goal of histogram stretching is to map every input value i_k to a new input j_k such that the range of the new histogram $h(j_k)$ has positive values for all values of k.

Figure 11.3. This figure shows the effect of change of contrast on the histogram of an image. Note that low contrast image (left) means very dark grays and very bright grays are both absent and therefore the range of values for which $h(i_k)$ is positive is small and in the middle range.

The first step of histogram stretching is to find the *cumulative probability distribution function*, H, where

$$H(i_k) = \sum_{0}^{k} h(i_k) = H(i_k - 1) + h(i_k).$$

(11.2)

$H(i_k)$ is a monotonically non-decreasing function that goes from $H(0) = 0$ to $H(1) = 1$. Let us examine the function H for a low contrast image (Figure 11.4). Let the range of values with non-zero values in h be from d to u where $d \leq u$, $d >> 0$ and $u << 255$. Therefore, $H = 0$ from $0 \leq i_k \leq d$ and $H = 1$ for all $u \leq i_k \leq 1$.

In order to stretch the histogram to increase the contrast of the image i_k is mapped to $j_k = H(i_k)$. Therefore, $j_k = 0$ for all $0 \leq i_k \leq d$ and $j_k = 1$ for $u \leq i_k \leq 1$. All the values in between d and u are mapped between the entire range of 0 and 1. Therefore the range of the values in the new image now spans the entire range 0 to 1 instead of d to u resulting in an improved contrast. However, j_k depends entirely on the slope of H at i_k. Also, since j_k can only have $u - d + 1$ values

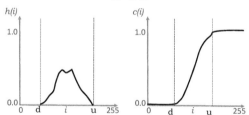

$h(i)$ ... $c(i)$

Figure 11.4. This figure shows a histogram $h(i)$ on the left and the cumulative probability distribution $c(i)$ corresponding to $h(i)$ on the right. Note that the cumulative probability distribution function is a monotonically increasing function ranging between 0 to 1.

between 0 to 1, the histogram of the contrast enhanced image will be non-zero only at $u - d + 1$ input values and not all p values between 0 and 1. Figure 11.5 shows the results of this contrast enhancement via histogram stretching.

However, the mapping function used to map i_k to j_k in histogram stretching is spatially invariant across the image. Therefore, this method is called *global* histogram stretching. Histogram stretching is often also called histogram equalization.

Applying the same mapping at every pixel, as is done in global histogram stretching, inherently assumes that the contrast is similar across the entire image. Therefore when the image has spatially varying contrast, global histogram stretching leads to an artifact of *burn and dodge*. In other words, some parts of the image get over-exposed while other parts are under-exposed. This is illustrated in Figure 11.6. To avoid this artifact, a variant of the global method, called adaptive histogram stretching, is used.

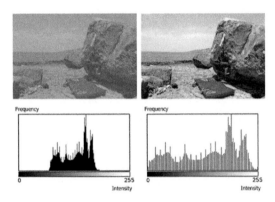

Figure 11.5. This figure shows contrast enhancement of the image on the left using global histogram stretching to produce the image on the right. The bottom row shows the corresponding histograms of these two images.

Let us consider an image whose contrast is varying spatially. In such cases, a global histogram stretching technique should be applied in a local $P \times P$ neighborhood window around a particular pixel (u, v) to compute the mapping of the input value at (u, v). Applying this technique at every pixel results in adaptive histogram stretching where the mapping at every pixel is different and is guided by the local contrast given by the local histogram in its $P \times P$ neighborhood. However, in this case, the same value i_k will get mapped to different values at different pixel locations in the image based on the local contrast. Therefore, even if the original image had only $u - d + 1$ gray values in the range d to u, they could be mapped to more than $u - d + 1$

Figure 11.6. This left image is one with spatially varying contrast – note that the top right part of the image has much better contrast than the rest of it while the left bottom part has much worse contrast than the rest of the image. The right image shows the result after global histogram equalization which shows that the former region is now over-exposed while the latter is under-exposed.

gray values after adaptive histogram stretching. Therefore, the histogram of the enhanced image will not be sparse as in global histogram stretching.

As is evident, the quality of the result from this adaptive histogram stretching will depend on the value of P. This is illustrated in Figure 11.7. If P is too small,

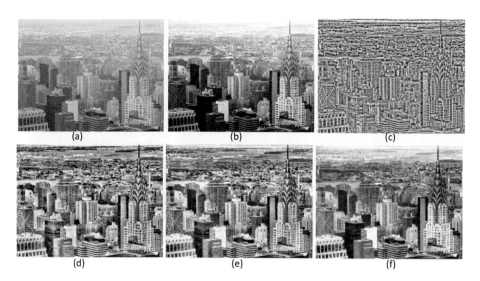

Figure 11.7. The original image is shown in (a) and the result of global histogram stretching in (b). The result of adaptive histogram stretching with $P = 12, 60, 100, 250$ is shown in (c), (d), (e), (f) respectively. Note that the noise is much more than (b) in (c), (d) and (e). However, the window size in (f) is optimal and it provides a much better contrast enhancement than (b), especially in the horizon at the back whose contrast is significantly lower than that of the city. Also note that each buildings contrast is differently enhanced in (f). However, in some places in (f), burning still occurs due to over exposure.

the contrast is evaluated at a small granularity leading to a tremendous amount of noise. As the window size increases, the noise reduces. But if P is too high, local burns and dodges will start to appear. Therefore, choosing an optimal window size is important for adaptive histogram stretching.

11.1.1 Handling Color Images

To apply histogram processing to color images, the treatment differs based on the application. One option is to apply the same method to the three channels independently. But this does not ensure that the hue is preserved. Therefore, most of the time, the RGB image is first converted to luminance and chrominance using standard RGB to XYZ linear transformation followed by finding the chromaticity coordinates computation. Then, contrast enhancement is applied only to Y while the chromaticity coordinates are left unchanged. Following the enhancement in Y, the image is transformed back to RGB format using inverse transformations. What results is called *hue preserving contrast enhancement*. This is illustrated in Figure 11.8. When hue is not preserved, blotches of pink and green show up in different parts of the scene during contrast enhancement.

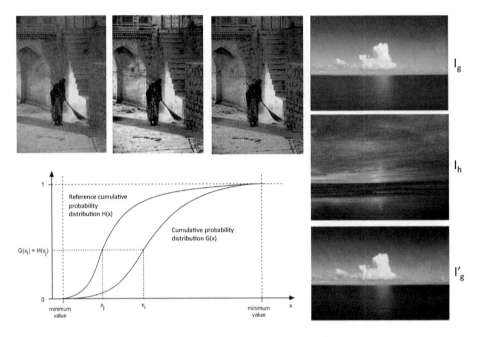

Figure 11.8. The left three images on the top shows two different ways to handle color during adaptive histogram stretching. Left: Original image; Middle: Adaptive histogram stretching of each of the channels independently. Therefore hue is not preserved as is shown by the shift towards green on the door on top of the stairs and the shift to purple on the left side of the shadow and on the wall. Also notice that the shawl of the lady now has more saturated pink. Right: Adaptive histogram stretching applied only on the luminance channel resulting in a hue preserving contrast enhancement. However, whether it is more realistic or pleasant to look at may still be an arguable issue. For some, the more saturated pink and the higher contrast between sunlight and shadows may make the second picture aesthetically more appealing. The bottom figure shows the process of histogram matching. On the right, the histogram of I_g is matched with that of I_h creating I_g' that has the same look and feel as I_h.

However, in some cases hue-preservation may not be applicable. One such example is that of histogram matching. Histogram matching is a technique that allows us to impart the look and feel of one image onto another which is only possible by modification of both hue and luminance. Consider two images I_h and I_g with two different histograms h and g respectively. The goal of histogram matching is to make these two histograms identical. To achieve this, we first find the corresponding cumulative distribution functions H and G respectively. Next, for an input x_i in I_g, we map it to an input x_j such that $G(x_i) = H(x_j)$. Following this mapping h becomes identical to g and I_g' looks similar to I_h. These are illustrated in Figure 11.8. Here the color tone of a sunset scene is imparted into an oceanside scene via histogram matching.

11.2 Image Composition

Compositing images is another important application of color image processing. In this section we will discuss a number of methods to achieve this. Image composition is quite commonly used in the entertainment industry where often virtual objects or characters from past or artistic effects have to be worked in the images/videos captured.

The simplest technique for image composition is by using *sprites* – currently available in Photoshop as a feature called intelligent scissors. Sprites are parts cut out from an image I given by $S \times I$ where S is a binary image called the *sprite mask* with 1 for the pixels included in the sprite, and$\times$ indicates pixel-wise multiplication. A sprite is defined for each image to be used in the composition. These sprites are then appropriately translated and scaled and placed in layers on top of each other in a specific order. Figure 11.9 shows an example where I_1 is a picture of downtown Seattle and I_2 is a picture of Bill Gates. Two sprites S_1 and S_2 are defined each scissoring out the foreground from I_1 and I_2 respectively. Therefore, three layers I'_1, I'_2 and I'_3 are defined as

$$I'_1 = I_1 \times S_1 \tag{11.3}$$
$$I'_2 = I_2 \times S_2 \tag{11.4}$$
$$I'_3 = I_1 \times S'_1 \tag{11.5}$$

The combined image I is created by overlaying translated and scaled I'_1, I'_2 and I'_3. I'_3 provides the first layer of the background made by the sky in I. This is then overlaid by the the image of Bill Gates formed by I'_2 which is in turn overlaid by the image of the city I'_1. Note that overlaying means replacing the pixels in I' wherever the sprite is 1. However, such overlaying of pixels hardly work. For example in Figure 11.10, I_1 multiplied with sprite S_1 is overlaid on I_2 to create I. Mathematically, the operation can be expressed as $I = I_1 S_1 + I_2(1 - S_1)$. But the pawn does not look like it is placed on the chess board, rather it looks pasted on the chess board. This problem is addressed by the image blending operation.

11.2.1 Image Blending

Sprite is binary and only allows for complete retention or removal of a pixel of the source image in the composited image. So the sprites do not work while compositing images with translucent objects where the background is partially visible through the foreground, or when compositing images with furry objects

Figure 11.9. This figure shows the layer based composition of two images I_1 and I_2 to form the new image I using sprites S_1 and S_2 respectively.

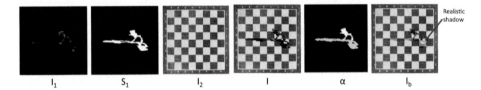

Figure 11.10. I_1 multiplied with binary sprite S_1 is overlaid on I_2 to create I mathematically expressed as $I = I_1 S_1 + I_2(1 - S_1)$. When S_1 is modified to an alpha mask α, the same operation of $I = I_1 \alpha + I_2(1 - \alpha)$ results in a nice result where the pawn looks as if it is sitting on the chess board due to the more realistic shadow.

for which generating a precise binary mask separating the foreground from the background is extremely hard.

This brings us to the concept of a more general sprite mask that does not need to be binary. In fact, as a more general concept, a sprite mask can have any fractional value between 0 to 1 and therefore can be used to attenuate colors of different parts of the image differently on multiplication. Such a sprite is termed as an *alpha mask* (Figure 11.10). Using the same operation $I_1 \alpha + I_2(1 - \alpha)$ with the alpha mask now creates the image I_b where the shadow of the pawn is weighted by a smaller value than 1.0, resulting in an image I_b where the pawn looks like it is placed on the chessboard and not pasted on it due to a more realistic shadow. This process is called *alpha blending*.

Now, let us explore some application of such alpha masks. Figure 11.11 shows the compositing of two images I_1 and I_2 (a and b) which have a horizontal common region. In the first image I_1, most of the bottom part of the image is dark (a) while in the second one, I_2, most of the top part is dark (b). These are composited using different alpha masks using the equation $I_1 \alpha + I_2(1 - \alpha)$. The first alpha mask (c) is a binary mask, similar to a sprite mask, that uses a central seam and assigns all the pixels on one side of it to I_1 and all the pixels on the other side to I_2 resulting in the image in (d) that shows a clear seam along the central line of division. In the second alpha mask (e), all the pixels in the non-overlapping region below the central seam are assigned to one image and those above the seam to the other image and all the pixels in the overlapping region are allowed to have equal contribution from I_2 and I_2 by assigning a weight of 0.5. The result is an image (f) where the seam is smooth but still visible. Finally, in the third mask (g), the assigned alpha value is dependent on the distance from the two boundaries created by the overlapping region of I_1 and I_2 with each of I_1 and I_2 respectively. In order to see how this distance can contribute to the alpha value, let us consider a pixel in the overlap region between I_1 and I_2. Let the distance of this pixel from the boundary of this overlap with I_1 and I_2 be d_1 and d_2 respectively. Note that as d_1 increases d_2 decreases and vice-versa. The weight assigned to this pixel in the alpha mask is $\alpha = \frac{d_2}{d_1+d_2}$. Therefore, when d_1 is 0, i.e. the pixel is close to the boundary with I_1, $\alpha = 1$ and therefore all the contribution is from I_1. $(1-\alpha)$ is zero signifying no contribution from I_2. But, as $d_2 = 0$ near the boundary with I_2, $\alpha = 0$ signifying no contribution from I_1 but $(1 - \alpha) = 1$ signifying complete contribution from I_2. For pixels in between, the α is in between 0 and 1 based on the relative distances from the boundaries with I_1 and I_2. Note that the resulting image with this alpha mask (h) based blending is truly seamless.Blending of images using such an alpha mask will result in a smooth change in the contribution of the individual images in the final image and will appear seamless.

(a)   (b)   (c)   (d)

(e)   (f)   (g)   (h)

Figure 11.11. This shows the compositing of two images in (a) and (b) using three different alpha masks shown in (c), (e) and (g) with the resulting images shown in (d), (f) and (h).

The same process of alpha blending can be used to blend two images placed adjacent to each other. This is a common process in applications like panoramic image generation and image synthesis. We show such an example in Figure 11.12. The goal is to composite I_1 and I_2 to create an image whose left and right half will look like I_1 and I_2 respectively with a seamless transition in between. For this, we will resort to alpha blending using the function $I_1\alpha + I_2(1 - \alpha)$ where α is the mask. Consider a step function α which is black (0) in the left half and white (1) in the right half. Let $I_1\alpha$ and $I_2(1 - \alpha)$ result in I_l and I_r respectively. $I_l + I_r$ does not show a smooth transition between I_1 and I_2 at the center due to a step blending

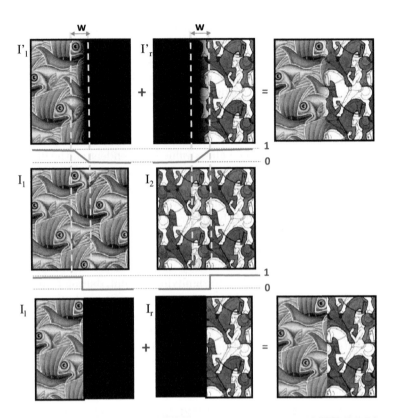

Figure 11.12. This figure shows blending or feathering technique being used to composite two images. The goal is to composite I_1 and I_2 to create an image whose left and right half will look like I_1 and I_2 respectively with a seamless transition in between. On the bottom we show a step blending function (to compute alpha mask) which steps down from 1 to 0 exactly at the middle giving I_l. The complementary blending function, given by $(1 - \alpha)$, steps up from 0 to 1 in the middle providing I_r which is then added to I_l to create the composite image on bottom right. The step function creates a drastic seam. On the top we show how the blending function is changed to transition in a smooth manner from 0 to 1 across the width w to create a much more seamless composite image on the top right.

function. A better way to achieve a smooth transition would be to choose a width w around the central line of the image and smoothly transition the alpha mask from 0 to 1 through the w pixels as shown in I'_l and I'_r. This process is called *feathering* and creates a much smoother blending.

Note that for such blending operations the most appropriate way to handle color images would be to apply the same blending functions across the three channels independently. Using a luminance and chrominance does not make much sense since the same pixel from two images in the overlap region may not have the exact same luminance and chrominance to start with.

Feathering effects depend on two parameters: the blending width and the blending function, as shown in Figure 11.13. Too large blending width results in ghosting while too small blending width results in a visible seam. The optimal blending width is dependent on the size of the features in the blending region. In fact, if

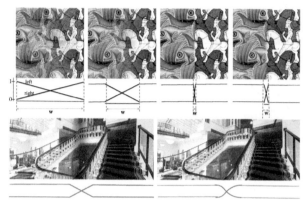

Figure 11.13. This figure shows the effect of blending width (top) and blending function (bottom) on feathering.

this problem is cast into the fourier domain, it can be shown that if the size of the features span one octave (should be between 2^i and 2^{i+1} pixels), an optimal blending width of 2^{i+1} will not show any ghosting artifacts, but will result in a smooth seam between the two images. In terms of the blending function, so far we have only seen functions that ramp down or up linearly. But such functions lead to a gradient discontinuity when they transition from the flat part to the linear part which results in visible artifacts called *Mach bands*.

Mach bands, as illustrated in Figure 11.14, are caused by the phenomenon of lateral inhibition in the human eye which is the perception of any gradient discontinuity as higher than the actual value at one end and lower than the actual value at the other. Figure 11.14(a) shows a figure created by a step function of different gray levels. Yet at the boundary, instead of being perceived as a clean step, it is perceived as a gradual change of gray that goes higher and then lower before achieving the next gray level. This phenomenon is just an illusion called Mach bands that is broken in (b) when one of the bands is removed. This is explained by how our human perception distorts the perception of a gradient discontinuity. In the context of blending, the same problem arises at the gradient discontinuities of a linear ramp as shown in (d). The human perception distortion is illustrated in (e). A more conducive function for blending, therefore, is one whose gradient is continuous like a cosine function or a spline function, as shown in Figure 11.13. Since these functions do not show any gradient discontinuity, they do not lead to Mach bands.

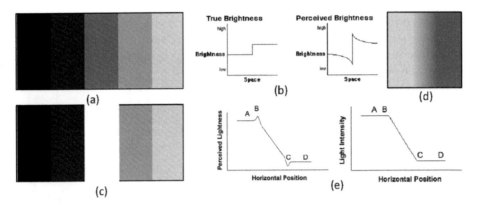

Figure 11.14. Mach band effect: Note that close to the boundary between bands the gray scale values are perceived to be darker or lighter than at pixels slightly farther than the boundary making it look almost like a 'curtain'. However, the actual values of the bands are defined by step functions that move from one gray level to another as shown in (b)-left. The 'curtain' illusion occurs due to the phenomenon of lateral inhibition in the human eye that makes our perception of a gradient discontinuity as shown in (b)-right and therefore the 'curtain' illusion is removed when we remove one of the bands in (a) as shown in (c). The same effect is also shown even with the presence of a smaller gradient discontinuity as in (d) whose effect on perception is illustrated in (e).

Fun Facts

The largest non-digital photographs in the world are made by stitching smaller images together. The largest single non-digital photograph captured in the world is of a control tower and runways at the US Marine Corps Air Station in El Toro, Orange County, California. It measures 32 feet high and 11 feet wide. It was taken in a decommissioned jet hangar, which was turned into a giant pinhole camera. The film was a $32' \times 111'$ piece of white fabric covered in 20 gallons of light-sensitive emulsion. The fabric was exposed to the outside image for 35 minutes. Print washing the image was done with fire hoses connected to two fire hydrants.

The longest photographic negative in the world is 129 feet and was created by Esteban Pastorino Diaz in March, 2015. The negative is of a panorama of major streets in Buenos Aires, Argentina. The images were captured by the slit camera (a camera that captures 1 pixel wide columns at a time while panning from left to right), which was mounted to the roof of a moving car.

Combining the constraints on feature size and smoothness of blending function, the ideal way to blend images is to blend the image in multiple resolutions. Such a multi-resolution decomposition is created by the Laplacian pyramid where each level of the pyramid provides a different resolution and the levels can be combined to create the image back. Let us consider two images I_1 and I_2, each of size $2^N \times 2^N$, to be blended to create the new image I. Let the respective Laplacian pyramids be denoted by L_1 and L_2 – each with $ln(N)$ levels. Using k as an index to the levels of the pyramid, each level of the Laplacian pyramids of I_1 and I_2 are given by L_1^k and L_2^k respectively, where $0 \leq k \leq ln(N)$. To achieve a smooth blending, a different blending function b_k with width w_k should be used for each level k. Most commonly, b_k is a spline that provides a smooth function whose resolution (i.e. how fast it ramps down or up) can be changed to provide the different w_k. Recall that size of the image at each level of the Laplacian pyramid is different, the size of the image at level k being $2^{(N-k)}$. The blended images at each level create a new Laplacian pyramid L. The images in L are then combined to provide the blended image I. To learn more about this, refer to the seminal work by Burt and Adelson [Burt and Adelson 83]. An illustration is presented in Figure 11.15.

Blending is a common technique used in an application called panoramic image generation, illustrated in Figure 11.16. The goal here is to capture multiple narrow field of view images from consumer camera to create a single wide field of view panoramic image. Multiple contiguous images of a location are captured with adequate overlap between spatially adjacent images. Since the camera moves between two adjacent images, the first step is to register adjacent images geometrically. This is achieved applying a geometric transformation to the image (e.g. scaling, translation or rotation) so that the overlapping areas can be spatially overlapped to match exactly. We will learn more about such geometrical transformations in the next chapter of the book. The overlapping area of two adjacent images are blended together to create a seamless transition between the images. This results in a panorama or an image that covers a much wider angle of view.

11.2.2 Image Cuts

Blending is not always the best way to achieve a nice transition between two images. This is especially true if there are moving objects in the common or overlapping areas of the image. In these cases, a blending operation will blend between two time instances of the same scene creating a ghosting artifact, almost similar to what we see in motion blur. In such cases, instead of blending, we apply a cut operation. This is a complementary operation to blending. In blending, a pixel in the composite image can have contributions from multiple source images. But, in an image cut, every pixel in the composite image comes from only one of the source images. Therefore, when stitching two adjacent images as in Figure

Final Composite Image

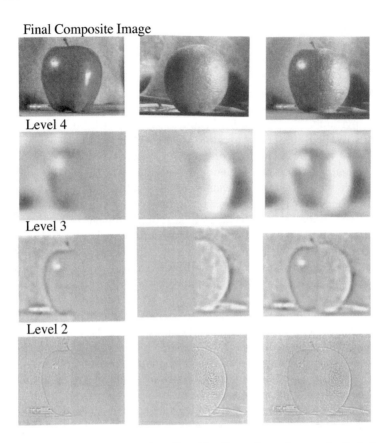

Level 4

Level 3

Level 2

Figure 11.15. This figure shows the effect blending of two image via their laplacian pyramid. The left two rows shows the different levels of the Laplacian pyramid blended using different spline functions achieving the effect of a wider blending region of lower resolution levels and narrower blending regions of higher resolution levels. The images are composited to create the image on the right to create each level of the laplacian pyramid of the composited image. These are then composited to create the image of the apple-orange on top right.

11.17, the contribution should switch from the blue image to the red image as we move from left to right. If this switch is done at a point where the pixel to the left coming from the blue image has very similar color to the pixel on the right coming from the red image, then the composition will be seamless. The problem is formulated as a minimal error boundary cut problem. The goal is to find a cut through the overlapping area such that the neighboring pixels at any point has the minimum energy transition across the cut. To learn more about this, please refer to the work from Efros and Freeman [Efros and Freeman 01].

Figure 11.16. Multiple contiguous images of a location are captured (top) with adequate overlap between adjacent images (shown by similarly colored rectangles). These images are then registered geometrically (bottom left) placing the overlap region exactly on top of each other. Finally, the regions coming from two adjacent images are blended using a blending function and the image cropped to form a rectangular panorama (bottom right).

Figure 11.17. On the left you see two images (shown by red and blue rectangle) with a large overlap in between that are composited using blending. Since the people and the truck moved between the capture of these two images, this results in severe ghosting of the moving objects. On the right, the same two images are composited using an image cut operation resulting in a artifact free composition.

11.3 Photometric Stereo

The final photometric processing application that we are going to discuss in this chapter is of computing shape of objects from illumination changes in images. Illumination contributes to the photometric properties of an object. Photometric stereo is a different kind of photometric processing that uses some knowledge on geometric processing you have been exposed to earlier in this book. We use a very

Input Images

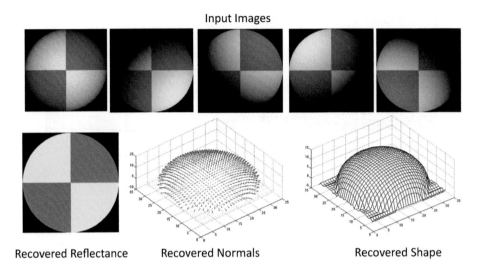

Recovered Reflectance   Recovered Normals   Recovered Shape

Figure 11.18. This figure shows the recovery of shape from illumination under the assumption that the object is Lambertian. The top row shows the five input images taken by changing only the light position. The bottom row shows the recovered reflectance function (left), the normals (middle) and the reconstructed shape (right).

simple illumination model considering a point light source from the direction L illuminating an object point P with normal N. More light is incident per unit area of the object if L and N are coincident i.e. the light is illuminating P head-on. As the angle between L and N increases, the amount of incident light decreases. It can be shown that the fall off of incident light per unit area is proportional to $cos\theta$ where θ is the angle between N and L. Therefore, the illumination at P is given by $N \cdot L$. To compute the amount of the illumination that is reflected from P, we assume the object to be Lambertian i.e. reflects light equally in all directions. Let fraction of light reflected be ρ. Therefore, the amount of reflected light is given by $\rho \cdot N \cdot L$.

Now, for photometric stereo, let us consider a set up where a Lambertian object is being seen by a camera. The locations of both the object and the camera do not change. However, it is lighted by n lights located at different locations surrounding the object. These light directions are known and given by $L_1, L_2, \ldots, L_n$, where L_i is a vector given by (L_i^x, L_i^y, L_i^z). Let the light intensities be unity. Let the point P with normal $N = (n^x, n^y, n^z)$ and reflectance ρ on the object be seen at a camera location (p, q). Let the reflected illumination recorded by the camera at (p, q) with only light L_i illuminating the surface be $R_i(p, q)$. Therefore at each pixel (p, q)

$$R_i = \rho N \cdot L_i. \tag{11.6}$$

Expanding this, we get the following equation at each pixel (p, q) for each light L_i.

$$R_i - \rho \begin{pmatrix} L_i^x & L_i^y & L_i^z \end{pmatrix} \begin{pmatrix} N_x \\ N_y \\ N_z \end{pmatrix} = 0 \tag{11.7}$$

Taking into account all the n lights we get

$$\begin{pmatrix} R_1 \\ R_2 \\ \cdot \\ \cdot \\ \cdot \\ R_n \end{pmatrix} - \rho \begin{pmatrix} L_1^x & L_1^y & L_1^z \\ L_2^x & L_2^y & L_2^z \\ & \cdots & \\ & \cdots & \\ & \cdots & \\ L_n^x & L_n^y & L_n^z \end{pmatrix} \begin{pmatrix} N_x \\ N_y \\ N_z \end{pmatrix} = 0 \tag{11.8}$$

Rearranging the terms, we get

$$\begin{pmatrix} L_1^x & L_1^y & L_1^z \\ L_2^x & L_2^y & L_2^z \\ & \cdots & \\ & \cdots & \\ & \cdots & \\ L_n^x & L_n^y & L_n^z \end{pmatrix}^{-1} \begin{pmatrix} R_1 \\ R_2 \\ \cdot \\ \cdot \\ \cdot \\ R_n \end{pmatrix} = \rho \begin{pmatrix} N_x \\ N_y \\ N_z \end{pmatrix} \tag{11.9}$$

$$or \quad \mathbf{L}^{-1}\mathbf{R} = \rho N \tag{11.10}$$

where $\mathbf{L}^{-1}$ is the pseudo-inverse of the non-square matrix L. Note that, at every pixel (p, q) the light directions and the intensity recorded are known. Each term in the left hand side of the Equation 11.10 is known. Therefore, we compute the right hand side of the equation and the vector obtained is the normal vector at the point on the object scaled by ρ. The magnitude of this vector provide reflectance ρ at (p, q) and the corresponding unit vector provide the normal N. This is illustrated in Figure 11.18.

The above process gives us the surface normal at any pixel (p, q) and its reflectance, but not its depth. So, in the next step we need to find the depth of the surface with respect to the camera. Let us consider the camera's coordinate system and the surface as shown in Figure 11.19. Let the depth of points at (p, q), $(p, q+1)$, and $(p+1, q)$ given by $z_{p,q}$, $z_{p+1,q}$ and $z_{p,q+1}$ respectively. Note that the tangent plane to the normal $N = (N_x, N_y, N_y)$ recovered by the photometric stereo can be approximated by the vectors V_1 and V_2, assuming a smoothly varying normal, and is given by

$$V_1 = (p+1, q, z_{p+1,q}) - (p, q, z_{p,q}) = (1, 0, z_{p+1,q} - z_{p,q}) \tag{11.11}$$

$$V_2 = (p, q+1, z_{p,q+1}) - (p, q, z_{p,q}) = (0, 1, z_{p,q+1} - z_{p,q}) \tag{11.12}$$

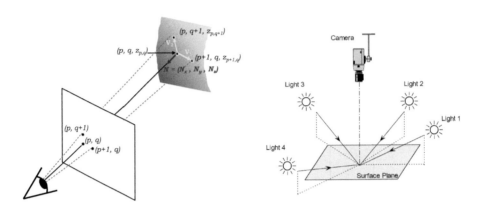

Figure 11.19. Left: This figure shows the depth of points at (p,q), $(p,q+1)$, and $(p+1,q)$ given by $z_{p,q}$, $z_{p+1,q}$ and $z_{p,q+1}$ respectively. The normal $N = (N_x, N_y, N_y)$ recovered by the photometric stereo is shown as well. Right: This figure shows the assumptions on the light and camera set up for photometric stereo given by three or more non-coplanar lights and camera distant to the surface.

Note that since V_1 and V_2 are both perpendicular to N. Therefore, we can find two constraints, first of which is as follows.

$$0 = N \cdot V_1 \tag{11.13}$$

$$= (N_x, N_y, N_z) \cdot (1, 0, z_{p+1,q} - z_{p,q}) \tag{11.14}$$

$$= N_x + N_z(z_{p+1,q} - z_{p,q}) \tag{11.15}$$

Similarly, the second constraint is given by

$$0 = N \cdot V_2 \tag{11.16}$$

$$= (N_x, N_y, N_z) \cdot (0, 1, z_{p,q+1} - z_{p,q}) \tag{11.17}$$

$$= N_y + N_z(z_{p,q+1} - z_{p,q}) \tag{11.18}$$

There the two constraints can be summarized as

$$\frac{N_x}{N_z} = z_{p,q} - z_{p+1,q} \tag{11.19}$$

$$\frac{N_y}{N_z} = z_{p,q} - z_{p,q+1} \tag{11.20}$$

where the depth values are the only unknown.

Every pixel thus contributes to constraints involving the depths of its neighboring pixels. Considering the camera image of $P \times Q$ size, let us assume that

Figure 11.20. From left: One of the input images; the recovered surface normals encoded by a RGB vector; the surface albedo; the reconstucted depth where depth is inversely proportional to the gray value; the same object relighted from a different virtual direction.

the total number of all such constraints is C. The value of C would have been $2PQ$ but for the pixels in the boundary for whom only one such constraint can be designed. Therefore, $C < 2PQ$ but still sufficiently larger than PQ to create an over-constrained system of equations given by

$$MZ = B \qquad (11.21)$$

where M is a $C \times PQ$ matrix whose entries can be either 1 or -1, Z is a $PQ \times 1$ column vector containing the unknown depth and B is a same sized column vector of known scalar values. Therefore, Z can now be solved using a linear regression or singular value decomposition. The reconstructed depth is shown in Figure 11.18.

There are a few things to note from the above equations. We need to solve for three unknowns. Therefore, we need at least three lights (i.e. $n = 3$) to find the shape of the object. However, if the light directions are coplanar, L is rank-deficient and hence cannot be solved. Hence, we need at least three noncoplanar light directions. Other assumptions include that the camera image plane is parallel to the XY plane in the global coordinate system and the camera and the lights are distant from the object. Figure 11.19 shows that photometric stereo can give accurate surface normals but the recovered depth is inaccurate if the camera or the lights are placed very close to the object. Note that we do not need to find any correspondence unlike most stereo methods we have studied in Chapter 8 since the camera location does not change. Before processing the images in photometric stereo you have to undo the effect of the camera transfer function. Finally, once the surface normals and reflectance are computed, the object can be relighted from a light direction that was not available originally, as shown in Figure 11.20

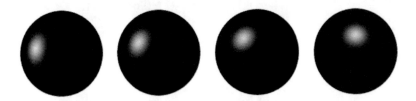

Figure 11.21. This figure shows the specular highlight on the chrome sphere in the scene when lighted from four different light directions.

Figure 11.22. This figure shows the results of photometric stereo on color images in the presence of shadows. On the left we see the input images and on the right the reconstructed face geometry is shown from different views.

11.3.1 Handling Shadows

One limitation of photometric stereo stems from the shadows. If any pixel (p, q) is in shadow in the image i, its importance should be undermined. To achieve a formulation where such confidence can be imparted to the accuracy of the recorded illumination, we can weight each equation in Equation 11.8 by the pixel intensity recorded. The pixels in shadow will therefore be given less weight due to their lower intensity. This gives us the equations

$$
\begin{pmatrix} I_1 \\ I_2 \\ . \\ . \\ . \\ I_n \end{pmatrix} - \rho \begin{pmatrix} I_1 L_1^x & I_1 L_1^y & I_1 L_1^z \\ I_2 L_2^x & I_2 L_2^y & I_2 L_2^z \\ & \cdots & \\ & \cdots & \\ & \cdots & \\ I_n L_n^x & I_n L_n^y & I_n L_n^z \end{pmatrix} \begin{pmatrix} N_x \\ N_y \\ N_z \end{pmatrix} = 0 \tag{11.22}
$$

that can be solved as before for more accurate normals.

11.3.2 Computing Illumination Directions

Photometric stereo computes the light directions also in addition to the surface geometry. To achieve this easily, one option is to put a chrome sphere of known radius r in the scene. This sphere will show specular highlights at different places in different images from which the light directions can be computed. An example of the chrome sphere in the scene when lighted by the same lights are shown in Figure 11.21.

Let us assume that we can detect the center of the highlight at (h_p, h_q) and the center of the sphere (which appears in the same place in all camera images) as (c_p, c_q). Let us assume that the depth of the center is 0 and the depth of the center of the highlight is h_z. Therefore, we can find h_z using

$$h_z = \sqrt{r^2 - (h_p - c_p)^2 - (h_q - c_q)^2} \qquad (11.23)$$

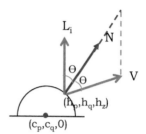

Now we know the 3D location of the highlight (h_x, h_y, h_z) and the center $(c_x, c_y, 0)$.

From this, we can compute the normal N to the sphere at the highlight. The view direction is given by $V = (0, 0, 1)$. The light vector L_i was reflected about N and then seen along the view direction to create the highlight. Therefore, reflecting V about N will give the light direction. This is illustrated in Figure 11.23. Note that N is given by the sum of V and L which are equal sized vectors. Their projection along N is given by $N.V$. Therefore L_i is given by the vector addition

Figure 11.23. This figure illustrates the computation of light directions from the specular highlight on a chrome sphere introduced in the scene.

$$L_i + V = 2(N.V)N \qquad (11.24)$$

Therefore, we can easily compute the light direction from the above equation for each of the images.

11.3.3 Handling Color

There are two ways to handle color in photometric stereo. The first option is to generate three sets of equations, one for each channel.

$$\mathbf{L^{-1}I_R} = \rho_R N \qquad (11.25)$$

$$\mathbf{L^{-1}I_G} = \rho_G N \qquad (11.26)$$

$$\mathbf{L^{-1}I_B} = \rho_B N \qquad (11.27)$$

In this case, first N can be solved using only one of the channels. Then it can be substituted in the above set of equations to find ρ_R, ρ_G and ρ_B. The other option is to combine the three channels assuming a channel-independent ρ where $I = \sqrt{I_R + I_G + I_B}$. Figure 11.22 shows an example which takes the latter route in the presence of shadows in the input images.

11.4   Conclusion

In this chapter we learned a large number of techniques that start from one or more color images and combine them to create a new image or generate new information about the scene. Here are some pointers to follow advanced topics. To learn more on contrast enhancement, you can check out [Majumder and Irani 07]. Instead of generating all the levels of Laplacian pyramid during blending, [Brown and Lowe 03] shows a way to do a two band blending that achieves comparable results more efficiently. Blending inherently assumes that the colors of the input images are similar. Therefore blending does not yield a good result if the colors of the objects in the images are vastly different – e.g. oceans with differently colored waters. In such cases, better composition is achieved by blending the gradients of the images instead of their values. To learn more about gradient domain blending, refer to [Pérez et al. 03]. To learn more about complex image cutting for texture synthesis, refer to the use of graph cuts in [Kwatra et al. 03]. To learn more about human perception phenomena and Mach bands, please refer to [Goldstein 10]. In case of photometric stereo, the two big limitations are the assumptions of known light vectors and a Lambertian object. [Basri et al. 07] presents a method for photometric stereo where the light directions are also unknown. [Wu et al. 11] presents a method that can achieve surface reconstruction from photometric stereo even in the presence of specular reflections.

Bibliography

[Basri et al. 07] Ronen Basri, David Jacobs, and Ira Kemelmacher. "Photometric Stereo with General, Unknown Lighting." *International Journal of Computer Vision* 72 (2007), 239–257.

[Brown and Lowe 03] M. Brown and D. G. Lowe. "Recognising Panoramas." In *Proceedings of the Ninth IEEE International Conference on Computer Vision - Volume 2*, 2003.

[Burt and Adelson 83] Peter J. Burt and Edward H. Adelson. "A Multiresolution Spline with Application to Image Mosaics." *ACM Trans. Graph.* 2:4.

[Efros and Freeman 01] Alexei A. Efros and William T. Freeman. "Image Quilting for Texture Synthesis and Transfer." In *Proceedings of the 28th Annual Conference on Computer Graphics and Interactive Techniques, SIGGRAPH '01*, 2001.

[Goldstein 10] Bruce E. Goldstein. *Sensation and Perception*. Thomas Wadsworth, 2010.

[Kwatra et al. 03] Vivek Kwatra, Arno Schdl, Irfan Essa, Greg Turk, and Aaron Bobick. "Graphcut Textures: Image and Video Synthesis Using Graph Cuts." *ACM Transactions on Graphics, SIGGRAPH 2003* 22:3 (2003), 277–286.

[Majumder and Irani 07] Aditi Majumder and Sandy Irani. "Perception-based Contrast Enhancement of Images." *ACM Trans. Appl. Percept.* 4:3.

[Pérez et al. 03] Patrick Pérez, Michel Gangnet, and Andrew Blake. "Poisson Image Editing." *ACM Trans. Graph.* 22:3 (2003), 313–318.

[Wu et al. 11] Lun Wu, Arvind Ganesh, Boxin Shi, Yasuyuki Matsushita, Yongtian Wang, and Yi Ma. "Robust Photometric Stereo via Low-rank Matrix Completion and Recovery." pp. 703–717.

Summary: Do you know these concepts?

- Histogram
- Histogram Stretching or Equalization
- Histogram Matching
- Contrast Enhancement
- Image Composition
- Image Blending
- Image Cuts
- Mach Bands
- Panoramic Image Generation
- Shape from Illumination
- Photometric Stereo
- Reflectance and Normal Reconstruction

Exercises

1. An image has a linear histogram $p(r) = r$. We want to transform this image so that its histogram becomes quadratic, $p(z) = z2$. Assume continuous images and find out the equation for this transformation.

2.

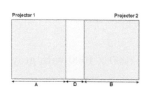

Two projectors overlap partially to create a bright overlap region as shown in Figure 2. (a) We would like to reduce the brightness in the overlap region using a blending operation. What should be width of the blending function? (b) Consider a linear blending function. Draw the blending function of the projector 1 (in blue) and projector 2 (in red). (c) Are linear blending functions continuous in terms of gradient? Justify your answer. (d) What are the artifacts caused by linear blending functions? What property should a blending function have to alleviate these artifacts? Name one or two blending functions that will alleviate these artifacts.

3. Consider the 16 pixel 1D image $I = \{4, 2, 3, 6, 2, 3, 4, 5, 2, 3, 4, 5, 5, 1, 5\}$. Assume for padding the last element of the 1D image are repeated on both sides.

   (a) Represent its histogram.
   (b) Show the array for the output of the low pass filtering of this image with the filter $[1/3 1/3 1/3]$.
   (c) Show the array for the output of the high pass filtering of this image with the filter $[-1 0 1]$.
   (d) Show the output of applying a 1x3 median filter to this image.

4. Consider a 10×10 checker board image whose checkers are white and gray (value $= 128$) instead of white and black. Each checker is 10×10 pixels in size. Draw the histogram of this image. Can you think of another image that will have the same histogram? Justify your answer.

5. Let us consider two images with histograms A and C shown in the above figure. Which of A and C has a lower contrast? If we apply a global histogram stretching to A which of the histograms shown will be the most likely resulting histogram? If we take a cumulative sum of A, which will be the most likely histogram? What kind of artifacts can global histogram stretching result in? What is the cause of these artifacts? What method can be used to alleviate this artifact?

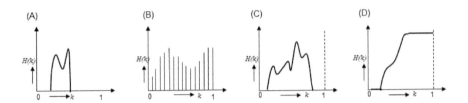

6. Consider two flat square images each of size 1000×1000 with gray values 200 and 100.

   (a) Consider a blending width of 10 pixels. What will be the resulting size of the blended image?

   (b) Consider a linear blending which causes Mach bands. One solution to this problem would be to use cosine blending. Can the same problem be alleviated by using a blending region of 300 pixels? Justify your answer.

   (c) When using this wider blending region, what is the resulting size of the image?

   (d) Which solution - the linear ramp of 300 pixels or cosine ramp of 10 pixels - would yield better blending? Justify your answer.

   (e) Do you anticipate choosing the lower blending width for any other reason than blending quality?

7. Both image blending and image cuts work best if applied on images with similar color temperature. In the absence of such a precondition, what technique should you apply to the images before compositing to assure a better quality result?

8. One way to handle shadows in photometric stereo is to use weighted light vector. What inaccuracies in the depth reconstruction do you expect if this is not done? Justify your answer.

Part V

Visual Content Synthesis

12

The Diverse Domain

So far we have explored concepts where we start with inputs captured from devices and systems and try to reverse engineer the properties of the scene. For example, spectral analysis techniques allow us to analyze information about the color at any point on a surface from a camera captured image. Or; feature detection techniques allow us to detect lines and corners in an image which can be used subsequently for object detection or image segmentation. Epipolar geometry helps us to find correspondences between stereo pair of images from which we can find the 3D geometry of an object. Therefore, these can be thought of as image/scene analysis techniques.

In the next section of this book, we are going to explore the inverse problem of generating a digital image similar to one generated by a device (e.g. camera) with the 3D scene of the world as the input. This process is that of image/scene synthesis. Synthesis therefore takes as input (a) *a scene* described as a collection of objects, lights, materials and textures represented using precise formal digital representation; (b) *a view set up* specifying the constraints on the eye/viewer that is viewing the scene; and outputs a 2D image similar to one captured by a device (e.g. camera, photometer) or seen by a viewer. Synthesis can be broken down into three steps: modeling, processing and rendering. Modeling deals with computer representation of objects and associated processes. Processing involves computation on the models for some specified outcomes and goals. Rendering involves drawing an image to convey the appearance of the model/processes to a human user.

12.1 Modeling

Modeling is the process of digitally representing an object or a phenomena so that it can be interpreted and processed by the computer. For example, there can be multiple ways to model an object – a dense collection of points, or a large number of planar triangles each of which approximates a small almost planar region of the object, or a number of curved patches to represent the object. All the above surface representations are used to represent only the *surface* properties

Figure 12.1. This figure shows multiple ways to model an object. From left to right: We show a teapot modeled by a set of points, a mesh of quadrilateral polygons, a set of surface patches (each patch shown in a different color) and finally we show a volume representation of an object with tissue density values at every volume location which can be used to visualize the tissue density as transparency values (mapping least dense to transparent and most dense to opaque).

Figure 12.2. This figure shows modeling of different phenomena. (a) creates beautiful cloth rendering by micro-scale simulation of how every fibre of the cloth interacts with light.(b) and (c) show the physically based modeling of sub-surface scattering for accurate appearance modeling of translucent objects. (d) show the effects of modeling illumination accurately.

of the object (e.g. geometric appearance given by gradients or curvature, color appearance given by textures or RGB colors). Alternatively, we may want a representation for the volume occupied by the object and its associated properties (e.g. density of the material in a human body part). Therefore, the primitives we choose for modeling inherently depends on what we would like to model (e.g. 3D volume or 2D surface) and the operations we would like to perform. For example, in aeronautical simulations, representation using surface patches may provide a more accurate simulation of fluid/air on the surface than triangular mesh. Therefore, one may want to use patch based representation for running mathematical simulation while using mesh based representation for rendering using an interactive graphics renderer. Figure 12.1 illustrates different representations for the same object.

Modeling need not be of objects alone. We can even model different natural phenomena as illustrated in Figure 12.2. For example, we can model the phenomenon of subsurface scattering that can be used to render translucent objects

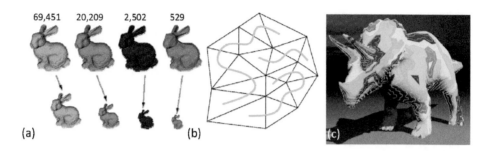

Figure 12.3. (a) shows model simplification and how the appearance is acceptable from larger distances even when the model is rendered at lower level of details. (b) shows the stripification of a small simple model. (c) shows stripification of a complex model. Same colored triangles belong to the same strip.

(Figure 12.2b and c), or light transport from the emitter through reflector, absorbers and refractors resulting in the realistic lighting of a scene (Figure 12.2d). We can model how every micro-fibre of a fabric interacts with the environment to create beautiful cloth renderings (Figure 12.2a). In fact, modeling need not be physically accurate also. At times, the objectives that drive the modeling can be entirely different. For example, modeling of movements of ocean water in the animation movies typically are not physically realistic or accurate — but it is artistically appealing for the purpose of story telling, as in rendering of water in the movie 'Finding Nemo'.

12.2 Processing

Processing refers to methods or techniques that are used on models of objects or phenomena usually driven by objectives like accuracy, performance (usually faster rendering) or application dependent efficiencies. Examples of such processing includes model simplification or stripification. Model simplification is the process where an object is stored at different *levels of details* that use different number of primitives to represent the same object. Objects need larger number of primitives when represented at higher level of details and fewer primitives when represented at lower level of details. When rendering, the right level of detail to be rendered is chosen based on how far the object is from the viewer. When the object is farther away, a lower level of detail would suffice for acceptable appearance but can be rendered much more efficiently in much less time due to fewer primitives. This is illustrated in Figure 12.3.

Similarly, let us take the example of streaming a 3D mesh, a popular application today for e-commerce. The goal here is to stream the 3D mesh to a remote location for rendering. Streaming would entail sending three vertices per triangle

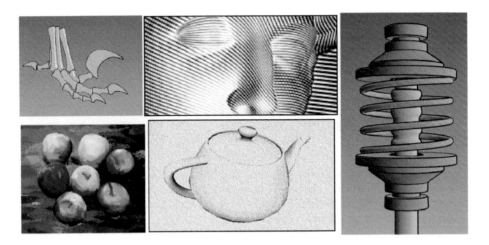

Figure 12.4. This figure shows several non-photorealistic rendering that can imitate painterly rendering or charcoal sketches, mechanical illustrations, and newsprint.

of the mesh to send the geometry information and then send the connectivity information (refer to Chapter 1). However, if we can stream one triangle after another such that every triangle is adjacent to the previous one streamed, we will need to send three vertices for the first triangle and one vertex per triangle for every subsequent triangle, thereby reducing the amount of data to be sent by almost 66%. Such a set of triangles is said to form a *triangle strip*. Therefore, a common kind of processing, called stripification, is to take a triangle mesh and represent it as multiple triangle strips (Figure 12.3).

Unlike processing such as compression and strip generation that may change the object models, processes such as collision detection just use objects to generate other application specific results. A collision detection operation computes the locations of the moving objects and detect if any of the triangles in the object intersect with any triangle of another object thereby causing a collision. Processing can also be motivated via higher performance as in organizing the model in a special data structure, like octree or BSP trees, that would enable fast access and retrieval of objects using a spatial index. Such data structures are useful in ray-tracing, culling of objects that are outside the observer's field of view, collision detection, etc.

12.3   Rendering

Rendering is the process of taking as input a 3D scene, a view set up and creating the 2D image of the 3D scene that will be seen from the particular viewpoint.

The two main aspects of rendering are the quality of appearance of the 2D image generated and the time it takes.

Quality: It is easy to assume that quality means accuracy of rendering. Instead, quality is an application dependent notion. Quality is determined by what is acceptable based on the goal of the application. For example, when playing a game where players are moving fast and are focused on specific tasks (e.g. picking up treasures or killing adversaries), they may not notice if the scene is lighted very realistically. On the other hand, when watching an animation movie, unrealistic lighting would most likely get noticed. When rendering fluid simulation results to evaluate and improve the design of a car before it is built, accuracy is probably of the highest priority, irrespective of the appearance. But, when creating special effects for movie, quality is guided by how close the digital content matches the real. Further, the style of the appearance need not always be photorealistic (i.e. like a photo) though it has been the focus of mainstream computer graphics historically. More recently, we have discovered an immense opportunity in creating non-photorealistic renderings. For example, a student of mechanical engineering would not like to see the photo of a greasy part of a motor to learn about it. He would rather want an illustration of the 3D parts which abstracts the different components and their functionality better. A student of medicine would not want to study the human digestive system from its photo. He would rather want a colored highlighted illustration of the same to learn more about the anatomy. Such renderings that are specifically designed to be not like photos are called non-photorealistic renderings, a few examples of which are shown in Figure 12.4. These kind of effects were also used in animation movies. For example, specific artistic effects of fur and grass was extensively used in animation movies like *Monsters Incorporated* or *Lorax*.

Speed: The speed of rendering a scene is always dependent on how much complexity is modeled and rendered. The most fundamental parameter that dictates speed is the number of primitives since it is inversely proportional to the speed of rendering. Complex phenomena like caustic effects or realistic illumination effects can make the rendering very slow. In popular terms, if the rendering can be achieved at a video rate, i.e. 30 frames per second (fps), it is called an interactive rendering. But, it should be kept in mind that the term interactive is also application dependent. For example, a game application may need to be rendered at 30 fps to be termed as interactive, but a sketch application can respond at 10 fps and the user may still feel that the system is responding appropriately to the sketch strokes. However, more often than not, rendering a frame for minutes or hours is termed as non-interactive. Most complex phenomena like subsurface scattering or cloth appearance modeling are usually associated with non-interactive rendering. Therefore, almost all the renderings in Figure 12.2 have taken multiple machines and many hours to render a single frame.

Quality vs Interactivity Trade-off: In the domain of image synthesis, there is an omnipresent issue of the tradeoff between quality and interactivity. The choice is often purely based on an application. A game application favors speed over quality while an animation movie favors quality over speed. The available computing resource, whether it is a mobile device or a farm of powerful machines, is allocated to the appropriate need, namely to enhance speed or quality.

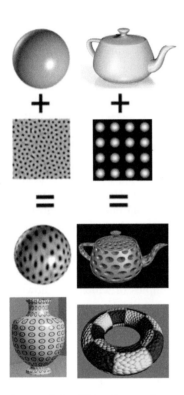

The next question is how much can the quality be compromised when speed is of concern. It is imperative that if the quality goes down beyond a certain 'acceptable' level, it will affect the user experience even if rendering speed is at its best. Therefore, in interactive graphics we see several techniques, which can be thought of 'tricks' that try to mimic complex visual phenomenon so that they do not stand out to be jarringly wrong. For example, the technique of texture mapping pastes images on geometric primitives such as triangles to provide visual complexity to the scene without increasing the number of primitives. Similarly, bump mapping simulates the effects of small bumps by perturbing normals thereby creating lighting effects visible in bumps, again without increasing the number of primitives (Figure 12.5). Environment mapping pastes an image of the environment on the object to simulate a shiny object in the scene.

The aforementioned discussion may bring forth an idea that realism is always good and you cannot go wrong with non-interactive realistic image synthesis. This impression is also not accurate. Note that the complexity of a realistic image is phenomenal and it is not true that a rendering as close to realistic as possible is always desired. The phenomenon of 'uncanny valley' is well-known among artists. If the realistic replication is very close but not absolutely correct, it can create discomfort in users despite being very realistic. In fact, more often than not it creates a disturbing experience. In fact, the uncanny valley has been attributed to the failure of very expensively made animation movies like *The Polar Express* or robots like Cubo girl (Figure 12.6).

Figure 12.5. This shows that a plain 3D model can be used in conjunction with a richer image to create texture mapped objects (left) and bump mapped objects (right).

12.4   Application

The domain of visual content synthesis is literally innumerable and diverse. In this section we will discuss some popular applications, specifically focusing towards 3D content synthesis.

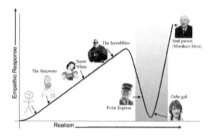

Figure 12.6. This shows the user evaluated empathy towards several digital characters in animation movies.

The rise of the field of computer graphics was motivated from its inception by a concept that is well-defined today as *virtual reality* or VR. The goal here was to simulate environments and experiences that are virtual and yet so real that people can use them for training. Examples of such environments are flight simulators for pilots, training environments for Army, Navy and Airforce. The name 'virtual reality' stems from the fact that the virtual environment would be the reality for the users for some time and at some space. The basic concept of virtual reality lies in having an immersive display on which a computer generated scene is presented to create a sense of immersion in a virtual environment. In addition, users can have different interactive devices (e.g. joystick) for navigating the 3D world or interacting with it. The immersive display can be instrumented by a surround seamless large area display made of multiple projectors. The perception of depth can be achieved by active stereo glasses that switch synchronously with the projectors time multiplexing between the views for the two different eyes. It can also be achieved via passive stereo glasses where superimposed projectors of different polarity are used to project the views of two eyes. The glasses are equipped with identical polarizers the allow the different projections to reach two eyes. The display can also be a head mounted display (HMD) where two different views of the scene are generated in real time and presented to the two eyes synchronously (e.g. Occulus Rift, Google cardboard). The head of the user is usually tracked (e.g. using cameras) which is used to determine the viewpoint from which the scene will be rendered. Today, we are seeing the advent of retinal displays where light is shone into the retina for the user to experience the same images as projected on a HMD. Virtual reality, even today, is one of biggest consumer for computer graphics. These are now routinely being used for 3D gaming experiences. Further, with the advent of high-speed networking, it is now also being used for applications like teleconferencing. Some of such applications are shown in Figure 12.7.

Figure 12.7. This figure shows some use of computer generated scenes in virtual reality environments - a military training simulator (left) and a flight simulator (middle). More recently real content is being streamed on such systems for immersive teleconferencing (right).

Fun Facts

Though it may seem that concepts of VR and AR are very futuristic, these concepts were envisioned by visionaries more than 50 years ago. Morton Hellig built the first VR environment in 1962 to enable his vision of the "experience theatre" or "cinema of the future". This was called the sensorama (left), a VR environment, fully equipped with 3D wide-vision moving color images along with stereo sound, aromas, wind and vibrations – much of which is not available today in our 3D cinema experience. In 1968, Ivan Sutherland, a professor at the University of Utah, built the first head-tracked head-mounted display which he called "Sword of Damocles". Later on, he founded the first computer graphics company in the US, Evans and Sutherland, with his colleague David Evans. They were the only name in flight simulators in the early days and are still in business building projection environments for planetariums.

In the past decade, we have seen merging of the virtual and real world in what is called augmented reality (AR). Figure 12.8 shows an example of augmented reality using both tablets or see through displays. In the former, the real world is captured in the tablet and augmented with the virtual world. Therefore the augmentation happens in the virtual space. In the latter, the augmented world is presented on see through displays and is automatically merged with the real world when seen through (e.g. Microsoft Hololens). Note that in this case, it is not a mere composition of images. The 3D virtual models are merged with the 3D real world which is much more challenging than the image composition.

Figure 12.8. This figure shows some tablet based and other see-through-display based augmented reality systems.

For example, light from the real world should interact with the virtual objects to make the experience believable. Also, it is worth appreciating that AR is a domain which sees a hand-in-hand functioning of computer vision and image synthesis. The real world needs to be reconstructed at some level to decide the location where virtual objects should be placed or merged.

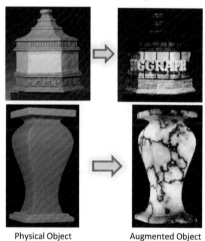

Physical Object   Augmented Object

Figure 12.9. Real objects (left) augmented with light projected from three different projectors to create spatially augmented reality objects (right).

Currently, we are seeing the advent of spatially augmented reality where real objects are augmented with projected light to change their appearance without encumbering the human being with a device. For example, a white model of Taj Mahal can be augmented with projected light to show the intricately detailed artwork in different illumination conditions starting from rosy dawn to white moonlight. Similarly, other cultural heritage artifacts can be restored to their original appearance using projected light, and movements or bumps can be simulated on static models using projected illumination (Figure 12.9).

VR, AR and spatially augmented reality all have many applications in many different domains. Augmenting a patient with previously captured 2D or 3D images that registered accurately with his body can make surgeries minimally obtrusive (e.g. extraction of tissue for biopsy). Inexpensive VR training environments for law enforcement departments can provide extensive training for officers before they face tough real situations. Visualization of large 3D data like seismic data or weather data is extremely crucial to predict and prepare for natural disasters.

Animation and special effects offer many applications of visual content synthesis. The animation industry strives to discover methods to render different styles and feelings to their characters, provide realistic body motions, provide realistic draping of clothes and dresses consistent with the animation movements etc. Special effects, on the other hand, strive to match the realism of the virtual world with that of the real world similar to augmented reality. For example, in a special effect scene with both virtual and real characters, the real lights should affect the virtual scene accurately and vice versa – a great challenge in the industry.

Geometric modeling and processing has been the keystone for computer aided design and modeling (CAD, CAM) for a long time. With the advent of 3D printing, this domain of computer graphics is again coming back to the limelight via varied applications in the domain of 3D manufacturing. How do we print 3D objects with minimum material wastage? Can we design a geometric model this is printable, stable and functional and yet stackable? Can we design objects in pieces in such a manner so that their assembly instructions will have simpler or a reduced number of steps?

Designing novel graphics hardware is also a very vibrant area related to computer graphics. Until the mid-1990s, interactive graphics rendering was only possible on super-expensive mammoth machines (e.g. SGI Onyx, SGI infinite reality) which used to be hosted in large rooms under very controlled environment. Even then, rendering a few million triangles at 30fps was a major achievement. Radical changes in the graphics architecture borrowing heavily from parallel processing architectures has resulted in todays inexpensive graphics processing units or GPUs which can be put in any laptop or desktop and can perform a few orders of magnitude better than the mammoth machines of yesteryear. GPUs today are so powerful that they are being used as resources for even general purpose computing and scientific computing. Therefore GPU design and programming is also a very attractive domain of computer graphics.

Fun Facts

Toy Story was a big milestone for computer graphics being the first ever feature length computer animated movie. It was produced by Pixar Animation Studios and released by Walt Disney Pictures. Pixar was approached by Disney to make toy story following their success of the short film 'Tin Toy' in 1988. Pixar used to make such short films from time to time to promote their computers. They started as part of the graphics group in the computer division of Lucasfilm in 1979. They spun out a corporation in 1986 when they were funded by Steve Jobs. Though produced under some financial constraints, *Toy Story* was the top grossing film on its opening weekend and went on to earn over \$361 million worldwide. It is still widely considered by many critics as the best animated film ever made. In 2005 it won the Special Achievement Academy Award and was inducted in the National Film Registry as being "cuturally, historically or aesthetically significant".

12.5   Conclusion

The material covered in this book is not comprehensive by any measure. The goal of this book is to provide you with fundamentals to get you interested in exploring one or more of these domains in depth. For this purpose we will be focusing primarily on interactive graphics techniques which will provide you with all the fundamentals needed to move from 3D to 2D, the main objective of any image synthesis pipeline. We will be providing you pointers for more advanced readings on non-interactive, often physically-based, rendering. We will cover a number of processing and modeling techniques in the context of interactive rendering. The sections are organized based on the well-known graphics pipeline. We will refrain from limiting the content to a particular GPU hardware or a particular programming language. This section of the book will introduce you to the basic mathematical concepts of visual content synthesis which can be implemented on any GPUs using any programming language once you learn each of them respectively.

To learn more details of non-interactive processes, please refer to [Shirley and Marschner 09, Foley et al. 90]. OpenGL is still the most flexible and popular cross-language corss-platform API for graphics programming and interacting with GPUs. To learn more of how to implement graphics techniques in OpenGL, refer to [Angel 08, Hearn and Baker 10]. For extensive details on CUDA programming for GPUs, refer to [Cook 12, Cheng et al. 14]. To understand how to use GPUs as general purpose computing for massively parallelizing your general purpose application (e.g. massive sparse matrix multiplications), refer to [Kirk and mei W. Hwu 12].

Bibliography

[Angel 08] Edward Angel. *Interactive Computer Graphics: A Top-Down Approach Using OpenGL*. Addison Wesley, 2008.

[Cheng et al. 14] John Cheng, Max Grossman, and Ty McKercher. *Professional CUDA C Programming*. Wrox, 2014.

[Cook 12] Shane Cook. *CUDA Programming: A Developer's Guide to Parallel Computing with GPUs.* Morgan Kaufmann, 2012.

[Foley et al. 90] James D. Foley, Andries van Dam, Steven K. Feiner, and John F. Hughes. *Computer Graphics: Principles and Practice (2Nd Ed.).* Addison-Wesley Longman Publishing Co., Inc., 1990.

[Hearn and Baker 10] Donald D. Hearn and M. Pauline Baker. *Computer Graphics with OpenGL.* Pearson, 2010.

[Kirk and mei W. Hwu 12] David B. Kirk and Wen mei W. Hwu. *Programming Massively Parallel Processors: A Hands-on Approach.* Morgan Kaufmann, 2012.

[Shirley and Marschner 09] Peter Shirley and Steve Marschner. *Fundamentals of Computer Graphics.* A. K. Peters, Ltd., 2009.

Interactive Graphics Pipeline

The goal of the visual content synthesis pipeline is to take a 3D scene and a view point setup as input and generate a 2D image. We assume that this 3D scene is represented as a triangular mesh, as discussed in Chapter 1. Therefore the input data is given as a set of triangles whose vertices are defined with their position, at least one attribute (e.g. normal vectors, RGB color) and their connectivity. Let us consider Figure 13.1 in which the 3D scene consists of a very simple model of a pyramid defined by the vertices A, B, C and D. At every vertex, at least one attribute, namely its 3D location is defined. Other attributes can be color, normal and so on. The connectivity or topological information is given via four triangles ABC, DBC, DAC and ABD that define the way the vertices are connected by edges. The eye or view point or the position of the camera is defined by E and the image plane (or screen) by I. Let us describe the steps in the graphics pipeline that would draw a given object from the point of view of the eye E on the image plane I.

1. *Geometric Transformation of Vertices:* The first aspect of the image synthesis pipeline is to find the 2D location of the vertices in the 3D scene on the image plane after transforming the vertices based on their perspective projection from the eye E. These 2D locations are shown by the vertices a, b, c and d found by perspectively projecting the 3D object on I. This entails connecting the vertices A, B, C and D to the eye E via straight line rays and finding their intersections with the plane I.

2. *Clipping and Vertex Interpolation of Attributes:* Due the finite extent of I, all the vertices may not fall within the image plane I. We are only concerned with drawing the part of the scene that is inside I. Therefore, the projected triangles should be clipped by the boundaries of the image introducing new vertices at the intersection point of the image boundary and the triangle edges. Let us describe the steps in the graphics pipeline that would draw a given object from the point of view of the eye on the screen. For example, in Figure 13.1, new vertices f, e, h and g are introduced to clip the projected triangles shown in orange. The required attributes at the new vertices are

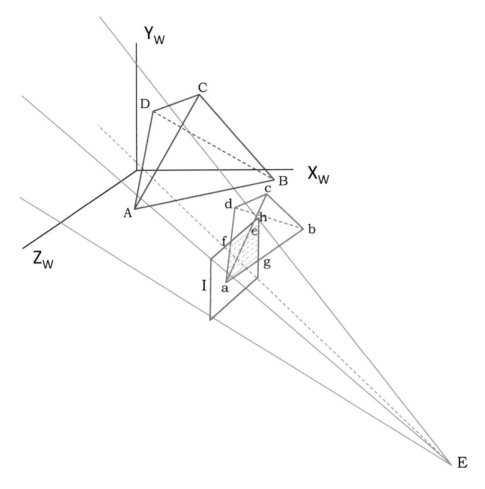

Figure 13.1. This figure shows the different steps of the graphics pipeline when rendering a simple 3D model of a pyramid defined by the vertices A, B, C and D.

computed via bilinear interpolation (see Chapter 2) from the other vertices of the clipped triangle. This is called the vertex interpolation of attributes.

3. *Rasterization and Pixel Interpolation of Attributes:* Finally, we have to paint the polygons formed by the clipped and/or unclipped vertices. This is done by computing all the discrete pixel locations on the screen that is covered by the rendered polygon, and computing the color of these pixels. First the pixels representing the edges of the polygon are computed, and the colors at these pixels are computed by linear interpolation of the colors at the end vertices representing the edge. Then each row of pixels is scanned

to compute the range of pixels in that row that are inside the polygon. The colors at these pixels are interpolated from the colors at either extreme of the range. This process of painting the triangles traversing the pixels in scanline order is called rasterization (shown by green pixels in Figure 13.1) and computing the attributes at the pixels is called pixel interpolation of the attributes.

13.1 Geometric Transformation of Vertices

The geometric pipeline consists of a sequence of transformations applied to every vertex of every triangle to find its corresponding 2D coordinate in the final image output. Chapter 6 discusses various geometric transformations that would be used in the current chapter.

The first step is the *model transformation*. This is the transformation of the objects from the object specific 3D coordinates to the single reference world coordinates. Second, the view setup describes the position of the eye and the orientation of the head which are used to apply a *view transformation* that represents a scene in a canonical view coordinate system. Third, in the *perspective projection* step 2D projections of the 3D vertices are computed. This step also includes making preparations to resolve occlusions. The final step in this process is the *window coordinate transformation* which maps all the vertices on to the exact window on the display in which the 3D scene will be rendered.

13.1.1 Model Transformation

The model transformation aids significantly in scene building. Consider every object defined in its own *object specific coordinate system*. When geometric models of various objects are downloaded from different places to populate a scene, it is rather natural that each of these will be defined in a different coordinate system. The goal of model transformation is to place these objects in different places in the scene in different forms (maybe scaled differently or oriented differently). Therefore, model transformation is the transformation M from the object coordinate system to one global world coordinate system of the entire scene where all the objects are placed.

The model transformation step also allows multiple instantiation of the same object in different position, orientation and scale. For example, if we are trying to create a 3D scene of a classroom, instead of storing the model of 100 chairs in the classroom, we can store one instance of the 3D chair in an object coordinate system and then instantiate 100 of them at different positions when building the scene. Figure 13.2 shows an example. Here there are three objects – a pyramid, a cylinder and a cube – defined in their own object coordinate system $X_o Y_o Z_o$. These are then converted to the global world coordinate system $X_w Y_w Z_w$ to

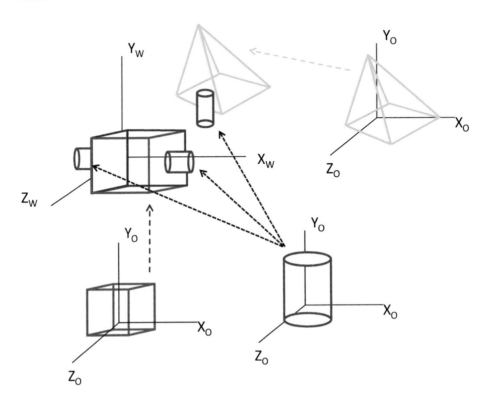

Figure 13.2. This figure shows the model transformation. Every object (e.g. cylinder, pyramid and cube) is defined in its own object coordinate system i.e. each of their vertices is defined in its own coordinate system. Then using model transformation they are instantiated multiple times and transformed differently to form the completely scene.

create the scene. For example, the pyramid has been translated, while the cube has been scaled and translated. The cylinder has been instantiated three times, each time with a different scaling, rotation and translation. Therefore, for the pyramid $M_p = T_p$ while for the cube $M_c = T_c S_c$ and for the cylinders there are three different transformations given by $M_1 = T_1 R_1 S_1$, $M_2 = T_2 R_2 S_2$ and $M_3 = T_3 R_3 S_3$. One important point to note here is that all the matrices – M_p, M_c, M_1, M_2 and M_3 – are of size 4×4 since we are transforming 4D homogeneous coordinates.

13.1.2 View Transformation

The input defining the view setup allows the rendering of the 3D scene for the defined view. The view setup is defined by the 3D location of the eye E, an image plane usually defined by a normal vector to the image plane N – also called the principle axis, and a view-up vector V which provides the up direction of the head. Ideally N and V should be perpendicular to each other. But when providing a view-up vector, it is often difficult on the part of the application programmer to provide a vector that is exactly orthogonal to N. Therefore, most graphics application programming interfaces (APIs) allow defining V as a vector close to the view-up vector from which the actual view-up vector that is perpendicular to N is computed.

The output of the graphics pipeline is a 2D image rendered from the 3D scene, that needs to be updated every time the view setup or part of the 3D scene is changed. Note that a change of the view set-up can be expressed as a change in the entire 3D scene. For example, if the eye moves to the right, it is equivalent to the scene moving to the left. The advantage of this approach is two fold: (a) All the transformations due to view point change can be applied to the model, and so the model transformation which is already applied to the models can be combined with the view transformation and this composition of transformation can be applied once to the model/scene; (b) since the view set-up and hence the image plane does not change, the perspective projection transformation remains the same. Therefore, most graphics APIs define a *default view* so that the scene is transformed in such a way that the view set-up remains at the default view. The most common default view is to have the eye at origin, the normal to the image plane to be the Z-axis and the view-up vector to be the Y-axis. Therefore, the default view setup can be defined as $E = (0, 0, 0)$, $V = (0, 1, 0)$ and $N = (0, 0, 1)$. This is illustrated in Figure 13.3.

The goal of view transformation is to convert an arbitrary view setup given by an arbitrary E, N and V to the default view. There are two steps to achieve this. First, the eye should be moved to the origin which is achieved by a translation $T(-E)$. Second, N should be aligned with the Z axis which is achieved by a rotation R. The rotation matrix R can be computed by defining a view coordinate system and aligning the view coordinate system with the standard coordinate system (X,Y,Z axes). Let the unit vectors defining the coordinate axes of the view coordinate system be u_x, u_y, and u_z given by

$$u_z = \frac{N}{||N||} \tag{13.1}$$

$$u_x = \frac{N \times V}{||N \times V||} \tag{13.2}$$

$$u_y = u_z \times u_x \tag{13.3}$$

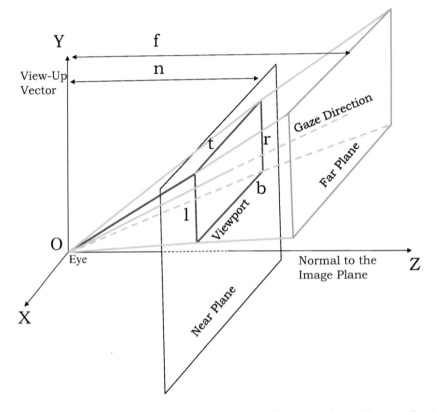

Figure 13.3. This shows the default view setup with eye at the origin, the Z axis perpendicular to the image plane (often called near plane in the computer graphics context), and the Y axis as the view-up vector.

Therefore R is given by

$$R = \begin{pmatrix} & u_x & & 0 \\ & u_y & & 0 \\ & u_z & & 0 \\ 0 & 0 & 0 & 1 \end{pmatrix}. \tag{13.4}$$

Since R is a function of N and V, we often denote it as $R(N,V)$. Thus, the final view transformation is given by $R(N,V)T(-E)$. Combining the model and view trnasformation, the transformation applied to a model vertex P via matrix pre-multiplication is given by $R(N,V)T(-E)M$. This 4×4 matrix $R(N,V)T(-E)M$ is identical to the 3×3 extrinsic parameter matrix of the cameras discussed in Chapter 7 except for the last row which is $(0,0,0,1)$ that is used to maintain the transformed 3D point to be in 4×1 homogeneous coordinates.

13.1.3 Perspective Projection Transformation

The perspective transformation matrix does the final transformation from the 3D scene to the 2D projection on the image plane. We now define the parameters to limit the extent of 2D image plane in order to define a field-of-view (FOV). The geometry on which the perspective projection transformation is defined is shown in Figure 13.3.

The eye or camera is viewing in the direction of the Z axis after the view transformation. In order to limit the data that is being processed, two planes, the *near plane* and the *far plane*, parallel to XY planes, are defined along the Z axes. Objects that are closer than the near plane and those that are farther than the far plane are not projected and drawn. These two planes are defined by their Z coordinates n and f, $n < f$. The near plane also serves as the image plane on which the 3D objects are projected. The axis-aligned rectangular window on the near plane through which we see the 3D scene from the eye point (origin) is called the *viewport*. Viewport is defined by the x coordinates of left and right vertical edges, and y coordinates of top and bottom horizontal edges. A rectangular viewport in graphics is more to mimic the rectangular sensor in the camera than to mimic the circular retinal image of the human eye. The four edges of the viewport – $x = l$, $x = r$, $y = t$ and $y = b$ – along with the origin (eye) define four planes. The truncated pyramidal structure formed by these four planes, the near plane and far plane is called the *view frustum* and the volume enclosed by the view frustum is called the *view volume*. Only the objects inside the view volume are rendered on the image plane. In the context of the human eye, the depth between n and f is usually termed as the *depth of field* and defines the range of depth in which objects form a focused and sharp image on the retina.

However, the viewport need not be centered around the Z axis. While E, V and N of the view set up describes the head position and orientation, the viewport describes the *gaze* or the orientation of the eye, when the head is fixed. The ray from the eye $(0, 0, 0)$ to the center of the viewport in the near plane $(\frac{l+r}{2}, \frac{t+b}{2}, n)$ is called the *gaze direction*. This effect of moving the gaze is very different from moving the entire head. Try the following experiment. Stand in front of a tiled wall. With head fixed look at and observe the tiles 30-45 degrees above or below. Then just rotate your head (but not change the position) to observe the same tiles straight ahead. The former effect keeps the image plane the same but changes gaze. The latter effect is created by the tilting of the image plane since the normal to the image plane changes with the rotation of the your head. Notice the effect of these two perspectives is very different from each other. When changing the gaze, the tiles will look stretched, but when rotating your head they will be not.

The primary function of the perspective projection is to project the 3D scene in the view volume onto the viewport. In addition, to simplify further computations of window coordinates and resolve occlusions, we need the perspective projection transformation to also convert the truncated pyramidal view frustum into a cuboid that extends in each of the X, Y, and Z direction from -1 to $+1$. This is illustrated in Figure 13.4. In order to illustrate this transformation, consider the point $P_v = R(N, V)T(-E)MP = (X, Y, Z)^T$ where P_v denotes the vertex after model and view transformation.

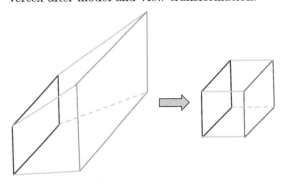

The first step is to coincide the general gaze direction to the default gaze direction that coincides with the normal to the image plane when the viewport is such that $l = -r$ and $b = -t$. This is achieved by a shear that brings the point $(\frac{l+r}{2}, \frac{t+b}{2}, n)$ to $(0, 0, n)$. Since the z coordinate remains unchanged, this transformation is a Z shear. Let the parameters of this shear be (a, b). Therefore,

Figure 13.4. This shows the transformation from the truncated conical frustum to the cuboid shaped frustum as part of the perspective projection.

$$\begin{pmatrix} 0 \\ 0 \\ n \\ 1 \end{pmatrix} = \begin{pmatrix} 1 & 0 & a & 0 \\ 0 & 1 & b & 0 \\ 0 & 0 & 1 & 0 \\ 0 & 0 & 0 & 1 \end{pmatrix} \begin{pmatrix} \frac{l+r}{2} \\ \frac{t+b}{2} \\ n \\ 1 \end{pmatrix} \qquad (13.5)$$

From this we can find the parameters a and b of shear to be

$$a = -\frac{l+r}{2n} \qquad (13.6)$$

$$b = -\frac{t+b}{2n} \qquad (13.7)$$

Therefore, the first matrix for perspective projection of P_v is $Sh_z(-\frac{l+r}{2n}, -\frac{t+b}{2n})$ and it provides the transformation to account for the off-axis viewport.

The next step is to transform the view frustum from a truncated pyramid to a cuboid. Let us assume, for the moment, that the z coordinate does not matter since after projecting to the image plane, all the vertices will have the depth n. We will revisit the depth issue later. Therefore, if z coordinate is ignored, the goal is to map the viewport which extends from l to r in the x direction to -1 to $+1$ in the x direction, and from t to b in the y direction to -1 to $+1$ in y direction. This means that lengths of $r - l$ horizontally and $t - b$ vertically should be mapped to 2. Since the center is already at $(0, 0)$ following the shear, this can be achieved by a scaling transformation $S(\frac{2}{r-l}, \frac{2}{t-b}, 1)$. Therefore, the complete transformation until this step for a vertex P_v is given by $P_s = S(\frac{2}{r-l}, \frac{2}{t-b}, 1)Sh_z(-\frac{l+r}{2n}, -\frac{t+b}{2n})P_v$.

Now, finally let us consider the perspective projection. We know from our camera calibration model in Chapter 7 that the perspective projection (x_p, y_p) of 3D point (X, Y, Z) is given by

$$x_p = \frac{Xn}{Z} \tag{13.8}$$

$$y_p = \frac{Yn}{Z} \tag{13.9}$$

which can be expressed as

$$\begin{pmatrix} x_p \\ y_p \\ n \\ 1 \end{pmatrix} = \begin{pmatrix} n & 0 & 0 & 0 \\ 0 & n & 0 & 0 \\ 0 & 0 & n & 0 \\ 0 & 0 & 1 & 0 \end{pmatrix} \begin{pmatrix} X \\ Y \\ Z \\ 1 \end{pmatrix} \tag{13.10}$$

$$= D(n) \begin{pmatrix} X \\ Y \\ Z \\ 1 \end{pmatrix} \tag{13.11}$$

In our case, the 3D point is what we achieve after cuboid transformation. Therefore, the complete transformation is given by $D(n)S(\frac{2}{r-l}, \frac{2}{t-b}, 1)Sh_z(-\frac{l+r}{2n}, -\frac{t+b}{2n})$ P_v. Since this matrix depends only on the view frustum parameters – r, l, t, b, n and f – let us call this $L(n, r, l, t, b)$. Therefore,

$$L(n, r, l, t, b) = D(n)S(\frac{2}{r-l}, \frac{2}{t-b}, 1)Sh_z(-\frac{l+r}{2n}, -\frac{t+b}{2n}), \tag{13.12}$$

and let $LP_v = P_l$.

13.1.4 Occlusion Resolution

The z coordinate P_l will be n always. This is expected since all the vertices are projected on the image plane at a depth of n. However, during image synthesis, the information about the depth of the projected vertex from the image plane or eye is very important to resolve occlusion and visibility.

Consider the triangles T_1 and T_2 in Figure 13.5 that intersect in 3D and when viewed from the view direction different parts of these two triangles are visible. Therefore, during scan conversion at each pixel only one of T_1 and T_2 should be drawn accurately based on this visibility from the view direction. Therefore, the depth information at the vertices of the triangles is retained as an attribute of the projected vertex to be used later on to interpolate the depth of interior pixels.

Let us consider this case of z-interpolation illustrated in Figure 13.6 in the projection of a 2D line between the points $A = (X_0, Z_0)$ and $B = (X_1, Z_1)$

where Z_0 and Z_1 are the depths of A and B respectively from the view point. The image plane is represented by the red line. The projection of the line on this image plane would be given by 1D coordinates s_0 and s_1. These are called the screen space coordinates of the 3D primitive. Let us assume that we have stored the depth of these two projected vertices - Z_0 and Z_1.

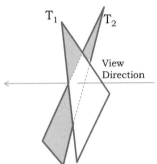

Figure 13.5. The shows two intersecting triangles viewed from a the shown view direction. The orange regions shows the visible parts of each triangle from this view direction which should be rendered after occlusion resolution.

Let us now consider the future stage of scan conversion where we are interpolating the attributes of a point half-way between s_0 and s_1 in screen coordinates, i.e. $\frac{s_0+s_1}{2}$. Therefore, we will compute the depth of this point using the same interpolation coordinates as $\frac{Z_0+Z_1}{2}$. However, the depth of the object point C that is projected at the screen coordinate $\frac{s_0+s_1}{2}$ is not $\frac{Z_0+Z_1}{2}$. The green curve in the object space in Figure 13.6 plots the depth computed through linear interpolation in the screen space. e shows the difference between the actual depth and the linearly interpolated depth in the screen space. Of course, the shape of the green curve and the amount of error will change based on the exact positions of A and B.

Therefore, the question is what kind of interpolation would yield the correct result? For this, let us consider the point (X_t, Z_t) between A and B defined by the interpolation coefficient t in 3D. Therefore,

$$X_t = X_0 + t(X_1 - X_0) \tag{13.13}$$
$$Z_t = Z_0 + t(Z_1 - Z_0) \tag{13.14}$$

Let their projection on the image plane be between s_0 and s_1 defined by the 2D interpolation coefficient u, s_u. Therefore,

$$s_u = s_0 + u(s_1 - s_0) = \frac{X_0}{Z_0} + u\left(\frac{X_1}{Z_1} - \frac{X_0}{Z_0}\right) \tag{13.15}$$

Since s_u is the image of (X_t, Z_t), we can derive the following.

$$s_u = \frac{X_t}{Z_t} \tag{13.16}$$

$$Or, \quad \frac{X_0}{Z_0} + u\left(\frac{X_1}{Z_1} - \frac{X_0}{Z_0}\right) = \frac{X_0 + t(X_1 - X_0)}{Z_0 + t(Z_1 - Z_0)} \tag{13.17}$$

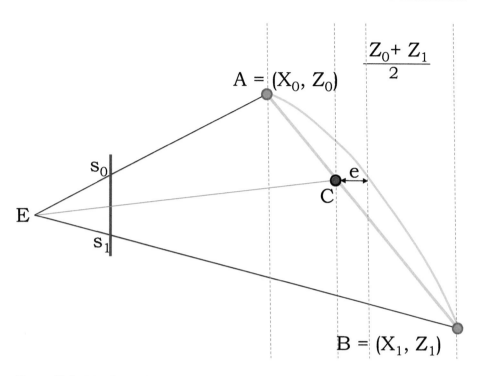

Figure 13.6. This figure illustrates the error that occurs if depth (z) is interpolated in screen space to find the depth of points internal to the triangles from the depth of the vertices.

Solving the above equation, we can find u as

$$u = \frac{Z_1 t}{Z_0 + t(Z_1 - Z_0)}. \tag{13.18}$$

Rearranging terms in the above equation, we can find t as

$$t = \frac{u Z_0}{Z_1 - u(Z_1 - Z_0)}. \tag{13.19}$$

Using the above equations, we can now compute Z_t, the accurate 3D depth of the point as

$$Z_t = Z_0 + t(Z_1 - Z_0) \tag{13.20}$$

$$= Z_0 + \frac{uZ_0}{Z_1 - u(Z_1 - Z_0)}(Z_1 - Z_0) \tag{13.21}$$

$$= \frac{Z_0 Z_1}{Z_1 - u(Z_1 - Z_0)} \tag{13.22}$$

$$= \frac{1}{\frac{1}{Z_0} + u\left(\frac{1}{Z_1} - \frac{1}{Z_0}\right)}. \tag{13.23}$$

The above derivation shows that

$$\frac{1}{Z_t} = \frac{1}{Z_0} + u\left(\frac{1}{Z_1} - \frac{1}{Z_0}\right), \tag{13.24}$$

i.e. the reciprocal of Z_t can be linearly interpolated from the reciprocal of depth at A and B using the screen space interpolation parameter u. Hence, linearly interpolating the reciprocal of Z instead of Z would yield the correct answer.

Therefore, the depth of a point C half-way in 3D between A and B, Z_c can be computed as

$$\frac{1}{Z_c} = \frac{1}{2Z_0} + \frac{1}{2Z_1} \tag{13.25}$$

Therefore, when retaining the depth in the third coordinate after applying L, we should retain $\frac{1}{Z}$ instead of Z so that we can readily apply linear interpolation of depth during scan conversion. Intuitively, this is due to the fact that perspective projection is not directly proportional to the depth but inversely proportional to depth.

Next, we will deduce the transformations required to retain $\frac{1}{Z}$ as the third coordinate. From equation 13.12, $L(n, r, l, t, b) = D(n)S(\frac{2}{r-l}, \frac{2}{t-b}, 1)Sh_z$ $(-\frac{l+r}{2n}, -\frac{t+b}{2n})$ and $LP_v = P_l$. Let $P_v = (X_v, Y_v, Z_v)$. Replacing L with multiplication of the matrices in Equation 13.12 we get

$$P_l = LP_v = L \begin{pmatrix} X_v \\ Y_v \\ Z_v \\ 1 \end{pmatrix} = \begin{pmatrix} \frac{2X_v}{r-l} - \frac{(l+r)Z_v}{n(r-l)} \\ \frac{2Y_v}{t-b} - \frac{(t+b)Z_v}{n(t-b)} \\ Z_v \\ \frac{Z_v}{n} \end{pmatrix} = \begin{pmatrix} \frac{2X_v n}{Z_v(r-l)} - \frac{l+r}{r-l} \\ \frac{2Y_v n}{Z_v(t-b)} - \frac{t+b}{t-b} \\ n \\ 1 \end{pmatrix} \tag{13.26}$$

In the above equation, the third coordinate of P_l is n. To retain the depth in this coordinates, we want the third coordinate of P_l to be $\frac{1}{Z_v}$. Not only so, we need the third coordinate to be normalized between -1 and $+1$ as $\frac{1}{Z_v}$ goes between $\frac{1}{n}$ to $\frac{1}{f}$ to transform the truncated pyramid to a cuboid as shown in Figure 13.4.

To achieve this, we have to map $\frac{1}{n}$ to -1, $\frac{1}{f}$ to $+1$, and the center of the range $[\frac{1}{n}, \frac{1}{f}]$ to 0. The center is given by

$$\frac{\frac{1}{f} + \frac{1}{n}}{2} = \frac{f + n}{2nf}. \tag{13.27}$$

This movement of the center is achieved by a translation by $-\frac{f+n}{2nf}$. Then, the extent between $\frac{1}{n}$ to $\frac{1}{f}$ is scaled to 2 via a scale factor of

$$\frac{2}{\frac{1}{f} - \frac{1}{n}} = \frac{2nf}{n - f}. \tag{13.28}$$

Therefore the final expression for the third coordinate of P_l is given by

$$\left(\frac{1}{Z_v} - \frac{f + n}{2nf}\right)\frac{2nf}{n - f} = \frac{2nf}{(n - f)Z_v} + \frac{f + n}{f - n} = \frac{\frac{2nf}{n-f} - \frac{n+f}{n-f}Z_v}{Z_v} \tag{13.29}$$

Therefore, P_l that we would like to achieve is given by

$$\begin{pmatrix} \frac{2X_v n}{Z_v(r-l)} - \frac{l+r}{r-l} \\ \frac{2Y_v n}{Z_v(t-b)} - \frac{t+b}{t-b} \\ \frac{\frac{2nf}{n-f} - \frac{n+f}{n-f}Z_v}{Z_v} \\ 1 \end{pmatrix} = \begin{pmatrix} \frac{2X_v n}{r-l} - \frac{l+r}{r-l}Z_v \\ \frac{2Y_v n}{t-b} - \frac{t+b}{t-b}Z_v \\ \frac{2nf}{n-f} - \frac{n+f}{n-f}Z_v \\ Z_v \end{pmatrix} \tag{13.30}$$

This can be achieved by making $D(n)$ into a matrix that depends on both n and f and is given by

$$D(n, f) = \begin{pmatrix} n & 0 & 0 & 0 \\ 0 & n & 0 & 0 \\ 0 & 0 & -\frac{n+f}{n-f} & \frac{2fn}{n-f} \\ 0 & 0 & 1 & 0 \end{pmatrix} \tag{13.31}$$

Therefore, the entire perspective projection matrix L, now dependent on f also, is given by

$$D(n,f)S(\frac{2}{r-l},\frac{2}{t-b},1)Sh_z(-\frac{l+r}{2n},-\frac{t+b}{2n}) \tag{13.32}$$

$$= \begin{pmatrix} n & 0 & 0 & 0 \\ 0 & n & 0 & 0 \\ 0 & 0 & -\frac{n+f}{n-f} & \frac{2nf}{n-f} \\ 0 & 0 & 1 & 0 \end{pmatrix} \begin{pmatrix} \frac{2}{r-l} & 0 & 0 & 0 \\ 0 & \frac{2}{t-b} & 0 & 0 \\ 0 & 0 & 1 & 0 \\ 0 & 0 & 0 & 1 \end{pmatrix} \begin{pmatrix} 1 & 0 & -\frac{r+l}{2n} & 0 \\ 0 & 1 & -\frac{t+b}{2n} & 0 \\ 0 & 0 & 1 & 0 \\ 0 & 0 & 0 & 1 \end{pmatrix} \tag{13.33}$$

$$= \begin{pmatrix} \frac{2n}{r-l} & 0 & 0 & 0 \\ 0 & \frac{2n}{t-b} & 0 & 0 \\ 0 & 0 & -\frac{n+f}{n-f} & \frac{2nf}{n-f} \\ 0 & 0 & 1 & 0 \end{pmatrix} \begin{pmatrix} 1 & 0 & -\frac{r+l}{2n} & 0 \\ 0 & 1 & -\frac{t+b}{2n} & 0 \\ 0 & 0 & 1 & 0 \\ 0 & 0 & 0 & 1 \end{pmatrix} \tag{13.34}$$

$$= \begin{pmatrix} \frac{2n}{r-l} & 0 & -\frac{r+l}{r-l} & 0 \\ 0 & \frac{2n}{t-b} & -\frac{t+b}{t-b} & 0 \\ 0 & 0 & -\frac{n+f}{n-f} & \frac{2nf}{n-f} \\ 0 & 0 & 1 & 0 \end{pmatrix} \tag{13.35}$$

This matrix $L(n,f,r,l,t,b)$ is often called the frustum transformation matrix since it is dependent on the parameters that define the view frustum.

The top-left 3×3 submatrix of Equation 13.35 looks exactly like the intrinsic parameter matrix in Equation 7.10 where the focal length is n, the horizontal and vertical scale factor are $\frac{2}{r-l}$ and $\frac{2}{r-l}$ respectively and the horizontal and vertical offsets are $-\frac{l+r}{2}$ and $-\frac{t+b}{2}$ respectively. In other words, while the view transformation is essentially the extrinsic parameter matrix, the frustum transformation matrix L is essentially the intrinsic parameter matrix. The camera model in the synthesis pipeline is essentially the same as the pinhole camera model, but we arrive at the same equations from different directions and use it differently.

13.1.5 Window Coordinate Transformation

The perspective transformation normalizes each of the three coordinates to be between -1 to $+1$. However, the final drawing in any image synthesis has to be done on a window on the display screen which is usually defined by integer coordinates of the top left and bottom right corner of the window. Therefore, this is exactly similar to providing the viewport. Let the coordinates of these top, left, right and bottom window boundaries be t_w, l_w, b_w and r_w respectively. Therefore, the center of the window is given by $(\frac{l_w+r_w}{2}, \frac{t_w+b_w}{2})$ and the length and height of the window is given by $r_w - l_w$ and $t_w - b_w$ respectively. The transformation to the window coordinates involve a translation by $(\frac{l_w+r_w}{2}, \frac{t_w+b_w}{2})$ and a scaling by

$\frac{r_w - l_w}{2}$ and $\frac{t_w - b_w}{2}$ in horizontal and vertical directions respectively while keeping the z-coordinate unaffected. This is achieved by the transformation

$$W(t_w, l_w, b_w, r_w) = \begin{pmatrix} \frac{r_w - l_w}{2} & 0 & 0 & \frac{l_w + r_w}{2} \\ 0 & \frac{t_w - b_w}{2} & 0 & \frac{t_w + b_w}{2} \\ 0 & 0 & 1 & 0 \\ 0 & 0 & 0 & 1 \end{pmatrix} \tag{13.36}$$

13.1.6   The Final Transformation

Therefore the complete transformation G of a point P is given by

$$G = W(l_w, r_w, t_w, b_w)L(n, f, r, l, t, b)R(N, V)T(-E)M \tag{13.37}$$

The above transformation projects the point from the object coordinate system to the window coordinates. This is exactly how the vertices a, b, c and d are generated in Figure 13.1 from the 3D vertices A, B, C and D.

13.2   Clipping and Vertex Interpolation of Attributes

Clipping is usually done in the graphics hardware and the application programmer does not need to worry about. Yet, we provide a very short overview here. Clipping is done in 2D following the projection of points. We will discuss some 3D clipping methods in later chapters. Any 2D clipping algorithm fundamentally depends on finding intersections of primitive (lines or polygons) with the edges of the window. Therefore, the mathematics behind these algorithms are straight forward. However, what makes it challenging in the context of the interactive graphics pipeline is its efficiency. Every primitive or triangle needs to go through the process of clipping and when the scene consists of millions of triangles efficiency of the algorithm becomes important even if it is done in the hardware. Following are some of the ways to increase the efficiency; they are most likely to be deployed one after the other in sequence.

Fun Facts

In computer graphics you will come across the *revered Utah teapot* model which has become synonymous to CG innovation ever since Martin Newall, a graduate student at the University of Utah, modeled and introduced the object to the computer graphics community. The actual teapot that Newall used to create the digital model resides at the Computer History Museum in Mountain View, California. So why a teapot? It is said that Newall's wife suggested the object while the two were having tea. But her idea was perfect technically due to a large number of reasons cited over the years. It is round, has saddle points, has a non-zero genus due to the hole in the handle, can project a shadow on itself, can have a self-reflection, and looks reasonably aesthetic even when rendered without a texture. It is amazing that such a simple object provided computer graphics researchers with so much complexity so as to become the benchmark geometric model. In 2006, Professor Peter Shirley of the University of Utah paid homage to this model through his Siggraph Talk "The Teapot Through the Ages". Each year at SIGGRAPH (the biggest conference for computer graphics academicians, industry and enthusiasts) Pixar hands out hundreds of tiny wind-up teapot to collectors.

Performing a number of floating point intersection operations for every primitive is definitely not the most efficient way to achieve clipping, especially when a large majority of triangles can fall either completely outside the window or completely inside it and only a few will actually intersect the window boundaries. One way to improve performance is to make sure that intersection computations are only performed when there is a high probability that the primitive actually intersects the window boundaries. Therefore, a fast acceptance or rejection test for primitives completely inside or outside the window is critical. Such tests can be achieved in multiple ways.

Using Bounding Boxes: We can compute the axis aligned bounding box of each triangle and see if it lies completely inside or outside the window. An axis aligned bounding box is the smallest box enclosing the primitive, with sides parallel to the axes of the window. If this bounding box is completely outside, the primitive is rejected. If it is completely inside, it is accepted. An axis-aligned bounding box for each primitive can be computed by just finding the minimum and maximum extent of the vertices in the horizontal and vertical direction. Testing of this bounding box is also easily achieved without any intersection computation by checking for intersection between the extents of the bounding box and window. Only primitives whose bounding box intersects the window go through the intersection computation. In Figure 13.7, both the horizontal and vertical ranges of the bounding box of A are completely within the respective ranges of the window. So A is inside the window. In case of D, while the vertical range of the bounding box intersects that of the window, the horizontal range of the bounding box is completely outside that of the window. So D lies outside the window. For the other two cases, both the horizontal and vertical ranges of the bounding box and that of the window partially intersect leading to possible intersection of the primitive with the window. The primitive B intersects the window and will be clipped via intersection computation. However, there can be cases like C where the bounding box intersects though the triangle does not. The actual intersection computation in such cases will yield negative results.

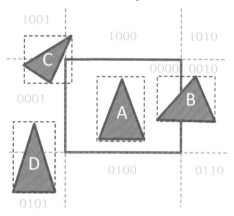

Using Logic Operations: Another technique to expedite acceptance/rejection tests is to divide the 2D image plane into regions and assign binary codes to the regions. For example, we can have four bits, $b_1 b_2 b_3 b_4$ associated with each projected vertex (x, y) such that $b_1 = y < t_w$, $b_2 = y > b_w$, $b_3 = x > r_w$, and $b_4 = x < l_w$. The four-bit code divides the image plane into nine different regions, each with an unique code (shown by green codes in Figure 13.7). Consider the four bit codes of two end points of a line segment. If the bitwise AND of these two codes is not zero, then both end points are outside the same window boundary and the line is com-

Figure 13.7. This figure illustrates some of the efficiency improvements in clipping algorithms using bounding boxes or binary codes.

pletely outside the window and therefore rejected. If each of the two codes is zero then the line segment is inside the window and therefore accepted. If at least one of the two codes is not zero, but the AND operation of the two codes is zero, then the line intersects with the boundary of the window and therefore an accurate intersection test has to be performed. Such logic operations can be extended to triangles also to efficiently clip them. An acceptance or rejection test using logic operations is equivalent to, yet simpler than, the test using a bounding box.

Using Integer Operations: Intersection of the lines and triangles that are not trivially accepted/rejected with the window boundaries have to be computed. For this we first find the window boundary that intersects the primitive and then find the exact intersection. For both of these steps, it is far more efficient if they can be achieved primarily via integer computation rather than floating point computation. Let us consider the red and green lines shown in Figure 13.8. The green line intersects the left window boundary before the top. This indicates that the line is entering the window at the left boundary. But, the red line meets the top boundary before left that can only happen when the line is completely outside the window and should be rejected. Therefore, to figure out which boundary intersection computation needs to be performed, it may be useful to find the parametric value α of the line intersection with different window boundaries. Denoting the alpha boundaries by α_l, α_r, α_t and α_b for the left, right, top and bottom boundary and simply ordering these parametric values we can find the portion of the line that is inside the window. However, computing these parametric values involve floating point operation. So, the next question is how can we make this operations more efficient?

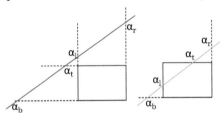

Consider a line given by two points defined by integers (x_1, y_1) and (x_2, y_2) where $x_1 < x_2$ and $y_1 < y_2$ (assuming the bottom left corner of the screen to be origin). We know

$$\alpha_t = \frac{t - y_1}{y_2 - y_1} \qquad (13.38)$$

$$\alpha_l = \frac{l - x_1}{x_2 - x_1} \qquad (13.39)$$

Figure 13.8. This figure shows the computation of which boundaries to intersect with using integer computations.

If $\alpha_t < \alpha_l$, i.e.

$$\frac{t - y_1}{y_2 - y_1} << \frac{l - x_1}{x_2 - x_1}, \qquad (13.40)$$

the line should be rejected. Instead of carrying this test on the floating point numbers α_l and α_t, the same results can be obtained if we derive the decision factor from Equation 13.40 to be

$$(l - x_1)(y_2 - y_1) < (t - y_1)(x_2 - x_1). \qquad (13.41)$$

The advantage of Equation 13.41 is that it is completely in integers and does not involve any floating point computation. The methods of clipping are fraught with such techniques to avoid floating point computation thereby making the pipeline extremely efficient.

Put a Face to the Name

Z-buffer (also called depth buffer) is considered one of the milestone concepts of interactive computer graphics. Prior to that primitives had to be sorted in 3D and rendered from back to front to resolve occlusion and there was no easy way to handle intersecting primitives other than to split them. Edwin Catmull, president of Pixar and Disney Animation Studios, was the first to invent this concept though it was invented independently by Wolfgang Straber. Catmull was also the inventor of the concept of texture mapping which brought in  an unforeseen realism in interactive graphics. Born in 1945 in West Virginia, he was raised in a Mormon family in Utah. Though from very early in life he dreamed of becoming a feature animator, instead of pursuing a career in the movie industry he pursued his talent in math and science to study physics and computer science at the University of Utah where he returned as a graduate student in the 1970s to pursue his PhD under Ivan Sutherland. His discoveries of texture mapping, bicubic patches (also called Clark-Catmull patches), subdivision surfaces and anti-aliasing methods changed the face of graphics forever. His first contribution to the movie industry was in 1972 via an animated version of his left hand which was picked up by a Hollywood producer to be used in the 1976 movie *Futureworld* and its sequel *Westworld* which were the first films to use 3D computer graphics. This sequence, simply known as the Computer Animated Hand was chosen for preservation by the National Film Registry in 2011. He started the computer graphics division in Lucasfilm in 1979 which was later bought by Steve Jobs in 1986 to be called Pixar. Popular among peers as Ed Catmull, he developed the first complete rendering system to be used in movies, Renderman, while at Pixar for which he received the Academy Award in 1993. Since Disney's acquisition of Pixar in 2006, Ed Catmull has been the president of both Pixar and Disney Animation studios. He has won many awards since then for his pioneering contributions to modeling animation, and rendering including another Academy Award in 1996, the IEEE John von Neumann Medal in 2006, and the Gordon E. Sawyer Award in 2008.

Using Pipelining: Finally, one more technique that is often used for efficiency is pipelining. For example, once we have detected that intersection computations need to be done, the polygon, represented as a list of vertices, can pass through the four stages of clipping against left, top, right and bottom edges of the window, in a pipelined fashion. Clipping against an edge of the window clips out the part

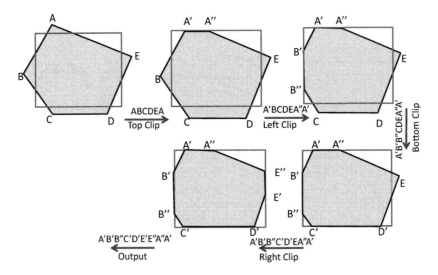

Figure 13.9. This figure shows the Sutherland Hodgeman algorithm for clipping attained via clipping the orange polygon against the top, left, bottom and right edges of the blue window successively. The list of vertices input to each of these steps is also shown above the red arrows.

of the polygon that lies in the half plane formed by the edge that does not contain the window. This half plane is denoted by OUT while the other one that contains the window is denoted by IN. In the Sutherland-Hodgeman method (Figure 13.9), a polygon clipping is attained by such successive clipping of the polygon against the top, left, bottom and right edges of the window. The input to each of these steps is a cyclic list of vertices (i.e. starting and ending with the same vertex) defining the polygon.

Let us first consider the list of vertices of the polygon $ABCDEA$ in Figure 13.9 going through clipping against the top edge of the window. The clipping algorithm parses this list of vertices in sequence from left to right and outputs one existing or a new vertex per parsed vertex based on transitions in the locations of the vertices in terms of the IN and OUT half planes as follows.

1. If first vertex is IN output the same, or else nothing;

2. Loop through the rest of the vertices testing transitions.

    (a) If IN-TO-OUT, output intersection with edge;

    (b) If IN-TO-IN, output the vertex;

    (c) If OUT-TO-IN, output intersection with edge and the vertex ;

    (d) If OUT-TO-OUT, output nothing;

We will now execute the top clip with the cyclic list $ABCDEA$. The first vertex A is OUT resulting in no output. The next vertex is B and the transition from A to B is that of OUT-TO-IN resulting in the output of the intersection of edge AB with the top edge, A' and B. The next transitions from B to C, C to D, and D to E are all IN-TO-IN leading to the output for the vertices C, D and E respectively. Finally, the transition from E to A is IN-TO-OUT resulting in the output of the intersection of EA with the top edge, A''. Therefore, the output vertex list is $A'BCDEA''$ which is made into a cyclic list by repeating the first vertex at the end resulting in the list $A'BCDEA''A'$ which acts as an input for clipping against the next window edge. This process continues for all four edges as shown in Figure 13.9 finally creating the clipped polygon $A'B'B''C'D'E'E''A''A'$.

The pipelining is possible due to the fact that each stage of the clipping against a window edge can output a vertex as soon as it reads vertex without waiting for the entire list of vertices to be parsed. Further, the output vertex can be pushed as input to the next stage before the entire input list is created from the previous step. This improves throughput tremendously since each step hands over partial results to the next which can work with it.

There are several clipping method that use one or more of the above techniques. The Cohen-Sutherland method use logic operations, the Liang-Barsky method uses integer operations and the Sutherland-Hodgeman method uses pipelining. However, they can be combined in multiple ways to create more efficient methods, some variant of which is probably being implemented by the current graphics hardware.

13.3   Rasterization and Pixel Interpolation of Attributes

Rasterization is the last step of the interactive graphics pipeline where all the pixels inside the clipped polygons (triangles may not remain triangles after clipping) have to be computed, and colors and other attributes interpolated from the those of the vertices of the polygon. During the clipping operation, the attributes at the edge-window intersection points are themselves computed using interpolation of colors at the vertices of the given triangle. The process of rasterization is performed in the graphics hardware. We only provide very basic methods and some key insights of how such methods are made efficient. The buffer in which we draw the color is called the *framebuffer* and the buffer in which we handle the depth is called the *z-buffer* or *depth buffer*. Both of these buffers are the size of the window defined by the API. We start with a clear framebuffer (all pixels initialized to black) and the depth-buffer set to 0. Since we will deal with reciprocal of depth in the Z-buffer, initializing it to 0 means the depth is at ∞.

The rasterization process is applied to each primitive and it proceeds line by line from top left of the window to the bottom right. For every scanline, the intersection of the scanline is computed with all the edges of the polygon and the

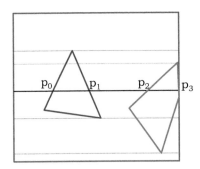 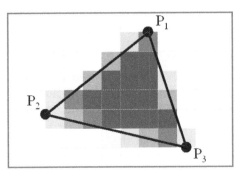

Figure 13.10. This figure illustrates the process of polygon rasterization (left) and shows a rasterized triangle with interpolated gray scale values (right).

intersections sorted in the increasing order of their x (note all of them have the same y since we are dealing with a horizontal scanline). Consider the two triangles in Figure 13.10 and the black scanline. The intersection points when ordered will be p_0, p_1, p_2, p_3. Next, the pixels within pairs of their intersections are filled up. Therefore, p_0 to p_1 and p_2 to p_3 is filled up. When filling up these pixels, their color and depth are also interpolated. For every pixel on a scanline that has been detected to be inside the triangle, first its reciprocal of depth is interpolated from the reciprocal of the depths stored at the intersection points of the scanline and the edges. If the interpolated value is larger than the existing value at that pixel in the Z-buffer (i.e. depth is smaller), only then the framebuffer is updated at that pixel with the interpolated color. Otherwise, this pixel is occluded and is not drawn in the framebuffer.

The polygon rasterization is also made efficient by several measures like incrementally updating the intersection of a polygon edge with a scan line by using the results with the previous scanline and the slope of the edge thereby avoiding computation of the intersections anew for every scanline. Several data structures are used to reduce computation. For example, the extent of the scanlines that a triangle spans (showed using red and orange dotted lines in Figure 13.10) can be maintained so that only the triangles whose span include the scanline under consideration are included in the processing for that scan line. Other more complicated data structure like edge table (that is a bucket sort of edges bucketed by scanlines) are used. Several improvements are also achieved by using integer computation for as much of the process as possible. Details of such processes are available in most traditional computer graphics books. The final rasterized polygon painted with interpolated color or gray values is shown in Figure 13.10.

13.4   Conclusion

We learnt about the interactive graphics pipeline in this chapter. We have deliberately kept this treatise API independent and given you the fundamental concepts. We hope that following this you can adapt API specific aspects into the pipeline easily. For example, OpenGL assumes the normal to the image plane in the view set-up to be negative Z. This means the view transformation and the perspective transformation will change slightly and we hope you can work through it. In this chapter we have not given you details about clipping and rasterization methods assuming they will be done in the graphics hardware. To learn more about such techniques, please look up [Foley et al. 90, Shirley and Marschner 09, Watt 99].

Bibliography

[Foley et al. 90] James D. Foley, Andries van Dam, Steven K. Feiner, and John F. Hughes. *Computer Graphics: Principles and Practice (2nd Ed.)*. Boston, MA, USA: Addison-Wesley Longman Publishing Co., Inc., 1990.

[Shirley and Marschner 09] Peter Shirley and Steve Marschner. *Fundamentals of Computer Graphics*. A. K. Peters, Ltd., 2009.

[Watt 99] Alan Watt. *3D Computer Graphics*. Addison Wesley, 1999.

Summary: Do you know these concepts?

- Model Tranformation
- View Transformation
- Perspective Transformation
- View Frustum
- Window Coordinate Transformation
- Framebuffer
- Depth Buffer
- Clipping
- Scan Conversion
- Rasterization
- Interpolation of Attributes

Exercises

1. Consider a 2D square on the XY plane with side 2 units, the center at the origin and four sides parallel or perpendicular to the coordinate axes.

   (a) Draw the picture of the transformed square after performing the following sequence operations: rotation of 45 degrees counter clockwise about Z-axis, translation by $(\sqrt{2}, 0, 0)$, and again a rotation of 45 degrees counter clockwise about Z-axis. Can you reduce the transformations thus giving the new sequence of transformation to achieve the same result?

   (b) Draw the picture of the transformed square after performing the following sequence of operations: translation by $(2, 2, 0)$, scaling by $(3, 2, 1)$.

   (c) Draw the picture of the square if these two operations in the previous question were swapped.

   (d) We would like to achieve the result of the transformations of the previous questions where scaling is followed by translation by applying a translation followed by scaling. How would the parameters of the translation and the scaling change?

2. A viewer is defined by the following. (a) Eye position: $(0, 0, 0)$, (b) View up vector: $(0, 2, 0)$, (c) Equation of the image plane: $x + y + z = 6$. Find the view transformation matrix generated for this view-setup. Let the left, right, top and bottom planes be at -2, +2, 4, and 8 respectively. Let the far plane be at 10. Find the perspective projection matrix given by L. Find what would be projected coordinates of a point $P = (10, 4, 6)$ for this viewer.

3. The model transformation for a scene is a rotation R about the Y axis in the counter clockwise direction by 90 degrees, followed by a translation T in the positive X direction by 20 units. What is the resulting transformation?

4. Consider a default view with the near plane (or image plane) at a distance 5. The gaze direction is at $(2, 1)$ and the size of the window in X and Y direction in which it is centered are 10 and 6 respectively.

   (a) Provide the r, l, t, b for the view frustum?

   (b) Provide the transformation that would make the gaze direction coincident with the normal to the image plane?

   (c) Following this, find the transformation to normalize X and Y coordinates between -1 to $+1$.

5. Provide the window coordinate transformation for a window whose center is located at $(200, 400)$ and whose width and height are given by 800 and 600 respectively?

6. We say that interpolation of Z in screen space is mathematically wrong and we should interpolate the reciprocal of Z to correct for the effect Yet, we interpolate colors using the same screen space interpolation. Is this mathematically correct? Justify your answer.

14

Realism and Performance

In the last chapter, we discussed the geometric fundamentals of the interactive graphics pipeline. However, a scene rendered using this basic pipeline with no lighting effects (e.g. specular highlights or shadows) or finer details or some patterns or bumps on objects would not look realistic. In this chapter, we will study a number of techniques that will allow us to render more realism in the scene. However, these do not come for free, rather with a risk of degrading performance (e.g. frame rate). Therefore, we will also discuss some techniques to enhance realism without compromising performance.

14.1   Illumination

Computing illumination of a scene is an extremely complex problem. The total amount of illumination at any surface point is due to both *direct* and *indirect* illumination. Direct illumination accounts for the light coming directly from a light source and reaching a surface point on the object. In addition, light reflected off, transmitted through or refracted by other surfaces can also reach the same surface point after multiple bounces and is called the indirect illumination. Thus, in order to compute that total amount of illumination at any surface point on an object we need to compute all the indirect illumination resulting from multiple bounces across multiple surfaces in addition to the direct illumination from the light source, as summarized in Equation 9.23 of Chapter 9. Such compute-intensive complex light models can be extremely time-consuming and therefore not suitable for interactive graphics. Therefore, much simpler illumination models are used to meet the interactive rates performance criterion.

The first simplification comes from assuming point light sources where we will start our discussion. Second, only the direct illumination (i.e. the light that comes at a point on the object directly from the point light source) is modeled while all the indirect illumination (i.e. light that reaches a point after bouncing multiple times from multiple objects) is combined under a single term called *ambient illumination*.

Let us consider a single light source illuminating a surface point P with normal N from the direction $\mathbf{L}$ and the eye looking at this point from the direction $\mathbf{V}$ as in Figure 14.1. Let R be the vector formed by reflecting L about the normal N. Note the similarity between the figures 14.1 and 9.1. Let the intensity of the light be I. The ambient illumination I_a is modeled very simplistically as

$$I_a = c_a I \tag{14.1}$$

where c_a is called the coefficient for ambient illumination.

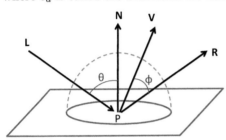

Figure 14.1. This illustrates the parameters for the simple ambient, diffused and specular illumination models at P.

The direct illumination is modeled in two parts – the view-independent component (that remains constant with change in view point and direction) called the *diffused component* and the view-dependent component (that changes with change in view point and direction) called the *specular component*. Different illumination models differ in the way they compute the specular component. We will discuss the most commonly used Phong Illumination model named after Bui Tong Phong. But more complicated models (e.g. Cook Torrance model) can also be employed for this purpose at the cost of performance.

The view-independent diffused illumination I_d is given by

$$I_d = c_d I (N \cdot L) = c_d I \cos\theta \tag{14.2}$$

where c_d is called the coefficient of diffused reflection. Note the similarity of this equation with Equation 9.22 in Chapter 9. c_d in Equation 14.2 is equivalent to ρ in Equation 9.22. However, since the amount of light reflected in the direction of the viewer is independent of his location, c_d does not have any dependency on V.

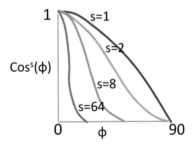

Figure 14.2. This figure shows the function $cos^s(\phi)$ for different values of s.

A specular component of the Phong illumination model is given by

$$I_s = c_s I \cos(R.V)^s = c_s I \left(\cos(\phi) \right)^s \tag{14.3}$$

where c_s is the coefficient for specular reflection and s is a parameter that controls the size of the view-dependent specular highlight. Figure 14.2 shows how the cosine fall off becomes steeper as s increases to achieve this effect. Since R depends on the incident direction L with respect to the normal vector N, and

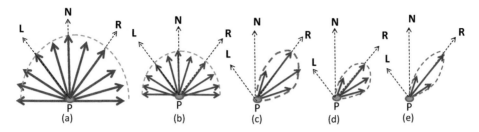

Figure 14.3. This illustrates effect of the parameters c_d and c_s of the simple illumination model. (a) and (b) show diffused illumination where c_d for latter is smaller than that of the former. (c), (d) and (e) show the specular illumination where (c) and (d) has smaller s than (e) while (d) has smaller c_s than (c) and (e).

the measurement of light is done in the outgoing direction V, the term $c_s(R.V)$ in Equation 14.3 is equivalent to $\rho(k_i, k_o)$ in Equation 9.22.

Figure 14.3 shows the effect of these parameters on the illumination. Assuming that light direction, intensity and distance from the surface point remians that same (given by L), we consider a sampling of the light vectors, indicating the amount of light seen in that direction, in red. (a) and (b) shows diffused reflection where equal amount of light is reflected in all directions illustrated by the equal length of the red vectors. Therefore, the amount of light received by the viewer V is the same irrespective of the angle between R and V. However, (b) has a smaller c_d illustrated by the fact that the vectors in (b) are shorter than those in (a). (c), (d) and (e) all show specular reflection where the length of the reflected vectors, indicating the amount of light seen in that direction, change based on how much the vector V deviate from the reflected vector R. If V is aligned with R, the light reaching V would be the maximum and would diminish as the angle between R and V reduce. Hence, specular illumination is view-dependent. (d) has smaller c_s than (c) while the directions in which they are reflected remain the same. However, (e) has a sharper view dependency shown by a sharper lobe in the outgoing light direction. Therefore, s controls the sharpness of the view-dependency as shown in Figure 14.2.

Figure 14.4 shows the effect of different parameters of the aforementioned Phong illumination model. Note that with the increase in c_a, the entire object looks brighter and shows no dependence on the direction of light, as is expected. On the contrary, when c_d is increased the shadow effect becomes more prominent since the directionality of the lighting plays a role. You can also see the effect of c_s and s on specular lighting. While c_s changes the amount of light reflected without changing the size of the highlight, changing s changes the size of the highlight.

Finally, Figure 14.5 shows the combined effect of the ambient, diffused and specular illumination under a point light source. Athough the Phong model is

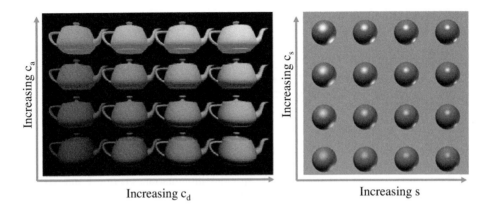

Figure 14.4. This figure shows the effect of our simple ambient, diffused and specular illumination on an object.

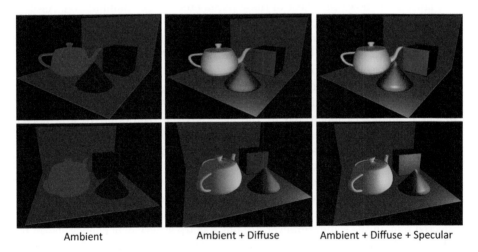

Ambient   Ambient + Diffuse   Ambient + Diffuse + Specular

Figure 14.5. This figure shows the effect of our simple ambient, diffused and specular illumination on an object from two different viewpoints.

restrictive and may not be able to provide the visual effect of a large number of materials, it is quite effective in interactive applications due to its simplicity.

For both diffused and specular lighting, often an attenuation parameter is used to model the attenuation of intensity of light as the distance from the source increases. Therefore, instead of I we use $\frac{I}{f}$ where f is a function dependent on the distance d of the point light source from the surface point. Physically, $f \propto d^2$. But in order to provide more control parameters to the application, f is defined

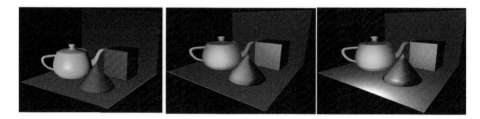

Figure 14.6. This figure shows the effect of the same lights with flat (left), Gouraud (middle) and Phong shading (right). Note how Gouraud shading misses the specular highlight on the pyramid captured by the Phong shading.

as $f = ad^2 + bd + c$ where a, b and c are parameters that can be set by the application.

Recall that the aforementioned model assumed point light sources. Other kinds of light sources are directional or area light sources. Directional light sources are lights that shine in a single uniform direction, i.e. all the rays originating from the source are parallel to each other. Area light sources are more like light panels instead of a point of light. Directional light sources are commonly used in computer graphics to mimic strong distant light sources like the sun. Directional light sources can be modeled by point light sources that are infinitely far away. Therefore, only the light direction vector L is used and the attenuation factor is ignored (by assigning it to 1). Area or extended light sources are modeled by a set of closely place point light sources. Another kind of light source that is often useful in computer graphics is a spotlight. This is modeled as a light source whose angular extent is restricted. The angle is given by the angle of the cone defined by the point light source and circle on the surface that defines the spot to be lighted by the spotlight.

14.2   Shading

Once we have computed the illumination at every vertex, we need to compute the illumination of a point inside the triangle during rasterization. This process of painting the interior of the triangle based on the illumination at the vertices is called *shading*. There are three shading algorithms used in interactive graphics.

Flat Shading: Here we compute the illumination once for each triangle. This can be done by averaging the normal vectors at the three vertices of the triangle. Then, we can compute the color using an illumination model once for the entire triangle and apply the same color to every pixel of the triangle during rasterization. The advantage of flat shading is its simplicity. However, an edge can get two vastly different colors from the two triangles incident on it. Therefore, it creates a gradient discontinuity in the shading of the surface creating visible artifacts, such as Mach bands.

Gouraud Shading: This technique is named after its inventor, Henri Gouraud. Here, the RGB color is computed at each of the vertices using an illumination model. The color inside the triangle is computed from the color at its vertices using screen space interpolation. Therefore, an edge will always get the same color from multiple incident triangles. This makes the shading continuous, but cannot still guarantee gradient continuity. Hence, the Mach band artifacts are less than in flat shading, but still exist.

A bigger problem of the Gouraud shading is diffusion or missing of specular highlights. For a piecewise linear interpolation of a smooth surface, the interior of the triangle represents a small smooth surface patch with smoothly varying normal. One way to compensate for the shading artifacts in the piecewise linear (or triangular) approximation of curved surfaces is to use accurate normals at the vertices. Gouraud shading computes the illumination only at the vertices of the triangle then using then normal at the vertices, and interpolates the color in the interior of the triangle, without reconstructing the normals in the interior of the triangles. Gouraud shading, therefore, cannot capture a specular highlight that exists in the interior of the triangle but not at its vertices. This is essentially the problem of not sampling the normals adequately when reconstructing the shading function.

Phong Shading: To alleviate this problem, Phong proposed a shading model in which the normal is interpolated across the triangle using screen space interpolation of its normals at its vertices during rasterization and then the color is computed at each pixel using an illumination model. Note that this does address the issue of inadequate sampling – per pixel sampling is the best one can do. However, this still does not guarantee that the shading will have no gradient discontinuities. So, though Mach bands are greatly reduced, they are not entirely non-existant with Phong shading. The differences among these three shading techniques are illustrated in Figure 14.6. Also, note that Phong illumination is a model for illumination, while Phong shading is an entirely different technique for shading. Therefore, be careful not to confuse these two, just because they are invented by the same person.

14.3   Shadows

Though we have discussed illumination models, we have not yet discussed rendering shadows. However, shadows can completely change our perception of a scene as shown in Figure 14.7. Note that the location of the spheres with respect to the checkered ground is exactly the same in the image, but only with the shadows can we perceive their correct height with respect to the ground.

In the context of interactive rendering, we use a very simple definition of shadows. If a point is not visible from the light source, it is in shadow with respect to that light source. Also, we would not attempt to compute the exact attenuation of each pixel in shadow for its accurate physical representation in the form of umbra or penumbra. Instead we will focus on a relative attenuation of the pixel color that would help provide the missing cues that location of shadows provide (e.g. depth).

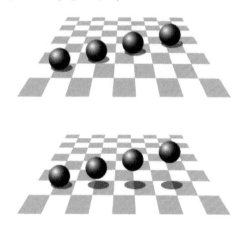

The primary concept behind shadows is to detect if a particular screen pixel is visible from the light or not. A pixel is in shadow if the depth of the 3D point corresponding to that pixel from the light is more than the value in the z-buffer of the corresponding reprojected rendering from the light. The shadow pixels thus detected are marked and stored as an image. This means some other object with smaller depth is in front of the 3D point in front of the visible pixel as seen from the light and therefore the 3D point is in shadow. It is evident that to make this decision, we have to render the scene multiple times, once from the light position and another time from the viewer position. Such methods are often called *multi-pass rendering* methods which we will explore next in greater detail.

Figure 14.7. This figure shows the perceptual effect of shadows on our perception of depth. The position of the spheres with respect to the checkered ground is identical in both images, but the height thereof only becomes clear with the presence of shadows.

Let the first pass render the scene from the light's viewpoint. Let Z_l denote the depth of a closest 3D point at a pixel, which is taken from the depth buffer from the first rendering pass. Therefore, any 3D point that projects on the same pixel from the light and has greater depth than Z_l at that point will be in shadow. The depth map consisting Z_l for all pixels is called the *shadow map* and is stored to be used in the later rendering pass. The 3D to 2D projection matrix post view transformation, M_l, is stored and used in this pass. There will be one shadow map associated with each light. Since we will only consider the depth buffer in this stage, we do not need to render illumination or shading or any other complexities during this first rendering pass.

Put a Face to the Name

Bui Tuong Phong is considered one of the stalwart figures in the advancement of interactive computer graphics due to his work on a computationally inexpensive simple illumination model that enabled lighting at video rates of 30 frames per sec. He is known for his famous quote "We do not expect to be able to display the object exactly as it would appear in reality, with texture, overcast shadows, etc. We hope only to display an image that approximates the real object closely enough to provide a certain degree of realism" which summarizes the philosophy behind interactive graphics. Bui Tuong Phong  was born in 1942 in Hanoi, part of French Indo China. He later moved to Saigon and then to France where he received his Licences Sciences from Grenoble Institute of Technology in 1966 and his Diplome d'Ingnieur from the ENSEEIHT Toulouse, in 1968. In 1968, he joined the Institut de Recherche en Informatique et en Automatique (then INRIA) as a researcher in Computer Science, working in the development of operating systems for digital computers. He went to the University of Utah College of Engineering in September 1971 for his Ph.D. and graduated in 1973 thereafter joining Stanford as a professor. Phong knew that he was terminally ill with leukemia while he was a student and died not long after finishing his dissertation. Though he lived for only 33 years, he has left a long lasting impression in the domain of interactive computer graphics.

Let us now consider the second pass of the rendering from the viewer. The depth buffer, Z_v, created in this process records the depths of only the 3D points visible to the viewer. The rest of the 3D points are irrelevant in our context of deciding whether it is in shadow or not. For these visible points from the viewer, we need to find their depth from the light. Let us consider a 3D point $P = (X, Y, Z_v)$ that is the final rendered point at (x_v, y_v) after 3D to 2D projection (post view transformation) using matrix M_v and occlusion resolution. We know that

$$M_v^{-1} \begin{pmatrix} ix_v \\ y_v \\ \frac{1}{Z_v} \\ 1 \end{pmatrix} = \begin{pmatrix} X \\ Y \\ Z_v \end{pmatrix} \tag{14.4}$$

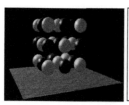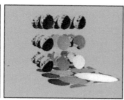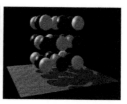

Figure 14.8. This figure shows pipeline for rendering shadows. From left to right: The 3D scene rendered without shadows, the shadow map after the first rendering pass, the points in the framebuffer which are in shadow denoted by non-green values, these non-green pixel colors are attenuated to create the effect of shadow. Note that balls cast shadow on each other also.

The depth of the same point from the light, Z_v^l, can therefore be found using

$$
\begin{pmatrix} x_l \\ y_l \\ \frac{1}{Z_v^l} \end{pmatrix} = M_l \begin{pmatrix} X \\ Y \\ Z_v \end{pmatrix} = M_l M_v^{-1} \begin{pmatrix} x_v \\ y_v \\ \frac{1}{Z_v} \\ 1 \end{pmatrix}
\tag{14.5}
$$

Multiplication by a 4×4 matrix $M_l M_v^{-1}$ matrix yields the projection and the depth of the same point P from the light. If $Z_l(x_l, y_l) < Z_v^l$, then the point P is in shadow and the framebuffer at (x_v, y_v) is attenuated by a factor less than 1.0 to create the effect of shadow. The whole process is illustrated in Figure 14.8.

14.4 Texture Mapping

Texture mapping is the process of pasting a 2D image on an object in order to increase the richness and visual detail of a digital scene. Texture mapping uses three coordinate systems: 2D texture space, 3D object space, and 2D screen space. Texture image's 2D coordinates are defined in the *2D texture space*. The 3D coordinates of the vertices of the object in the scene are defined in the *3D object space*. Finally, the pixel coordinates of the interior of the projected primitives of the object are defined in the *2D screen space*. Each 3D vertex coordinate in the object space is assigned a texture coordinate in the texture space. Screen space is used during rasterization to map the image to the interior of the primitives of the projected object during the texture mapping process.

14.4.1 Texture to Object Space Mapping

In this step a rectangular 2D image gets mapped onto an arbitrary 3D shape. Informally, this is akin to gift wrapping a complex 3D object (e.g. vase, fruit bowl, a tray). The more complex the shape, the more difficult is this mapping (e.g. a book is easy to gift wrap while a globe is not). Therefore, when mapping 2D images on complex shapes, different amounts of stretching or wrinkles in different places can be seen which are completely dependent on the underlying local geometry and how we choose to wrap the texture around it locally. Let us define the texture space with two coordinates s and t where $0 \leq s, t \leq 1$. Let (x, y, z) represent the 3D coordinate of the vertex in object space for which we need to assign the 2D texture coordinate. In this section we will describe ways to compute the 2D texture coordinates for 3D object coordinates. There are two ways to compute such a mapping.

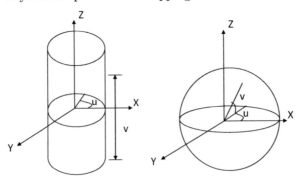

Parametric Shapes: There are shapes that have 2D parametric representations, such as a sphere or a cylinder. In such cases, we will map the two coordinates of the texture space to the two parameters used for the parametric surface representation. Consider an example of a cylinder.

Figure 14.9. This figure shows the parametric representation of the object or surface to be texture mapped using two parameters for a cylinder (left) and sphere (right).

A 3D point (x, y, z) on a cylinder of radius r can be described using two parameters: u, $-180 \leq u \leq 180$, defining the angle around the axis and v, $0 \leq v \leq 1$, defining the height of the cylinder (Figure 14.9). Therefore, a 3D point (x, y, z) on the surface can be expressed by the two parameters (u, v) as follows.

$$x = r cos(u); \tag{14.6}$$

$$y = r sin(u); \tag{14.7}$$

$$z = v. \tag{14.8}$$

By solving the above equations, we can compute the 2D parameters (u, v) associated with any vertex (x, y, z) on the cylinder. Next, we can relate the (u, v) to the normalized texture coordinates (s, t) as

$$s = \frac{u + 180}{360}; \tag{14.9}$$

$$t = v. \tag{14.10}$$

Similarly, a sphere can be parametrized by two angles u, $-180 \leq u \leq 180$, and v, $-90 \leq u \leq 90$ (Figure 14.9). Again, we can define the 2D parametrization

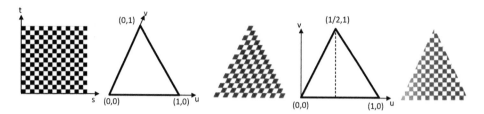

Figure 14.10. This shows the effect of parametrization on texture mapping. The two different parametrization for the same triangle is shown along with the mapped (s,t) coordinates at each vertex. Note that the appearance changes dramatically with two different parametrization.

for a point (x,y,z) on the sphere as follows.

$$x = rcos(v)cos(u); \qquad (14.11)$$
$$y = rcos(v)sin(u); \qquad (14.12)$$
$$z = rsin(v). \qquad (14.13)$$

The (u,v) coordinates related to a 3D point (x,y,z) can then be mapped to texture coordinates (s,t) as

$$s = \frac{u+180}{360}; \qquad (14.14)$$

$$t = \frac{v+90}{180}. \qquad (14.15)$$

The mappings from the (s,t) to (x,y,z) coordinates thus achieved in the above two cases are called cylindrical and spherical mapping respectively. Once the texture coordinates have been defined as above for every 3D vertex, the vertex is colored based on the color at the mapped texture image coordinate (s,t). Note that color is defined only at integer values of (s,t), often called texels. However, after mapping, there is no guarantee that (s,t) will be integers. Therefore, if (s,t) falls between integer values, the color value can be interpolated from the nearest texels either by picking the color of the nearest texel or interpolating the colors from the a few nearest texels in the texture.

An important point to note in the context of texture mapping is that the appearance of the textured objects depends completely on the parametrization. Let us consider a black and white checker texture on a triangle to illustrate the importance of parametrization as shown in Figure 14.10. A planar triangle is parameterized differently to create two different mappings of the texture coordinates at its vertices creating two different appearances.

More Complex Shapes: For more complex shapes, it is difficult to find an easy 2D parametrization. In such cases, we enclose or project the complex shape to a

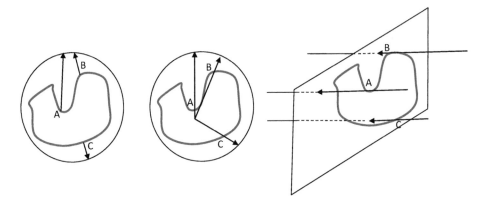

Figure 14.11. This figure shows a complex shape being texture mapped using a spherical (top) and an orthogonal (bottom) mapping. The spherical mapping uses the normals (left) or rays going out from the center of the object to pick the texture coordinates.

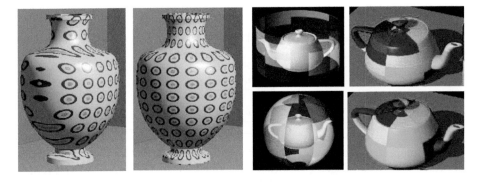

Figure 14.12. This figure shows orthogonal (left) and cylindrical (right) mapping on a vase and a cylindrical mapping (top) and a spherical mapping (bottom) on a teapot.

simpler shape that can be easily parametrized and for which texture coordinates can be assigned using the above method. Then we find a way to map vertices of the complex shape to the vertices of the simple shape and assign the texture coordinates to the corresponding vertex in the complex shape. Many methods can be designed to achieve this mapping from the complex to the simpler shape and we consider a few examples here.

The object can be enclosed in a simple geometry like a sphere (Figure 14.11). At any vertex of the complex shape, the normal can be extended and the texture coordinates at the point where this extended normal meets the enclosing simpler geometry can be used as the texture coordinates at the vertex. Or, a ray can be drawn from the center of the complex object through the vertex at which the texture coordinate needs to be assigned. The texture coordinate at the intersection of this ray with the enclosing simple geometry can be used to define the texture coordinate at the vertex of the complex geometry. Similarly, a cylindrical mapping can use a cylinder as an intermediate geometry. Far simpler mappings can also be done. For example, the texture coordinates can be assigned by an orthogonal projection of the vertices on a textured plane. This is called orthogonal mapping. Perspective projection mapping can also be used where the texture is treated as the image plane in perspective projection. The texture coordinate at a 3D point is defined by the intersection of a ray that connects the single center of projection to the 3D point. This is also called *projective textures*. Such textures are often used to simulate the effect of projections in large theaters or virtual reality environments.

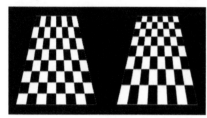

Figure 14.13. A texture mapped polygon with depth going from the front to back. The texture coordinate assignment on the left does not account for perspective projection and therefore the shrinkage of the checkerboard squares is only in the horizontal direction due to the trapezoidal shape of the polygon. The size squares of the checkerboard on the right change based on the depth due to the perspectively correct texture coordinate assignment.

However, the closer the geometry of the object to the enclosing geometry, the better the results achieved. Figure 14.12 illustrates this. The vase is closer to a cylinder in shape and therefore the orthogonal mapping shows severe and unrealistic distortion than in cylindrical mapping. Even in cylindrical mapping distortions are high only in places that deviate from the cylindrical structure like the neck and the base of the vase. However, for some objects like the teapot, it maybe hard to choose an enclosing geometry since it is close to both a sphere and a cylinder. In such cases, note that the distortions do not differ too much other than the colors that land in specific regions of the object and it is entirely up to the user's discretion to choose the one that is best for their applications.

14.4.2 Object to Screen Space Mapping

After assigning texture coordinates to vertices, it is treated like any other attribute such as color. The texture coordinates in the interior of the triangle are interpolated and computed from those at the vertices. This interpolation is computed during rasterization.

We have learned in the last chapter that correct interpolation of depth in screen space is achieved by interpolating the reciprocal of the depth. Let us consider two points P_1 and P_2 in 3D with depth Z_1 and Z_2. Let us consider the depth Z_t of a point P_t on the line connecting these two points given by the parameter q, $0 \leq q \leq 1$ as

$$Z_q = Z_1 + q(Z_2 - Z_1). \tag{14.16}$$

We know from Chapter 13 that if the screen space parameter for the same point is given by p, $0 \leq q \leq 1$, then

$$\frac{1}{Z_q} = \frac{1}{Z_1} + p\left(\frac{1}{Z_2} - \frac{1}{Z_1}\right) \tag{14.17}$$

From the above two equations we can find the relationship between p and q as

$$q = \frac{pZ_1}{pZ_1 + (1-p)Z_2} \tag{14.18}$$

When interpolating the texture coordinates we would like to achieve the correct coordinate based on their correct depth. Therefore, if the mapping from the texture space to object space assigns texture coordinates T_1 and T_2 to P_1 and P_2 respectively, then the texture coordinate at P_q is given by

$$T_q = T_1 + q(T_2 - T_1) = \frac{\left(\frac{T_1}{Z_1} + p\left(\frac{T_2}{Z_2} - \frac{T_1}{Z_1}\right)\right)}{\frac{1}{Z_q}} \tag{14.19}$$

Figure 14.13 shows the effect of perspectively correct texture coordinate assignment.

(a)   (b)   (c)   (d)

Figure 14.14. This figure shows the spherical mapping of the texture (a) on the 3D sphere (b) and 3D cylinder (c). The texture mapped sphere in (b) is now illuminated in (d).

Instead of coloring a vertex from the object color, texture coordinates provide a color that is picked up from a texture. Therefore, texture mapped objects can be illuminated just as a colored object is illuminated. One can compute a diffused or phong illumination based on the color picked from the texture to have an illuminated textured object as shown in Figure 14.14.

14.4.3 Mipmapping

The final rendered primitive gets assigned a texture coordinate at every pixel in it. Therefore, one can think of the pixels in a triangle as samples of the texture image. For example, if one side of a triangle is rasterized to have 5 pixels and is mapped along one side of the texture image that is 180 texels in size, we are expecting these 5 pixels to sample a function of 180 pixels and provide an accurate representation. As we know from the Nyquist sampling criterion that this leads to undersampling and therefore incorrect reconstruction of the signal.

This problem is illustrated in Figure 14.15. Let us consider the rasterized triangle on the right shown by the gray pixels. The center of these pixels are shown with different colored dots and the corresponding interpolated texture coordinates during rasterization are shown with similar colored dots on the texture. Note that since these coordinates sample the texture at much lower frequency than is desired to capture the stripes, though the triangle is supposed to get an appearance of green and white striping via texture mapping, it will end up appearing a flat green. This is the artifact of aliasing due to insufficient sampling of the texture via the pixels of the rendered triangle.

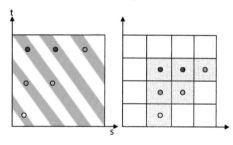

Figure 14.15. The colors picked by the pixels of a rasterized triangle (right) from the texture (left). Note that though the texture is that of green stripes, the triangle will only get painted a flat green.

The best way to avoid this problem is to keep a Gaussian pyramid of textures where the texture is filtered to have different frequency cut offs. Based on the number of pixels in the rendered triangle an appropriate level of the Gaussian pyramid is chosen such that the pixels are more than double the size of the texture at that level and can therefore capture all the different frequencies in it adequately. Let us consider the texture to be of size $2^N \times 2^N$ organized in a Gaussian pyramid with $ln(N)$ levels where the image at level i, $1 \leq ln(N)$, is of size $2^{N-i+1} \times 2^{N-i+1}$. Mipmapping offers a compact way to store this Gaussian pyramid. The size of the mipmapped RGB texture is $4 \times 2^N \times 2^N$ bytes. The image is divided into four quadrants, each of size $2^N \times 2^N$. Three of these are used to store the R, G and B channels of the first level of the Gaussian pyramid while the fourth one is used to store the next level RGB image of size $2^{N-1} x 2^{N-1}$. The fourth quadrant is recursively divided into four quandrants as before to store the R, G, and B components separately, and the fourth quadrant is again used to store the next level of the RGB image of size $2^{N-2} x 2^{N-2}$. This continues until the original image is filtered down to a single pixel image. The sequence of images is also called a Gaussian pyramid representation. During run-time the appropriate level of the pyramid is accessed based on the instruction provided in the application program. Figure 14.16 shows the mipmap organization and a scene rendered with and without mipmapping.

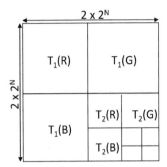

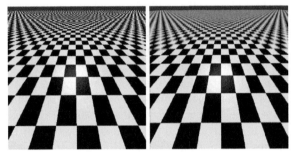

Figure 14.16. This figure shows the Gaussian pyramid organized in a mipmap (left) and rendering a ground scene viewed from an oblique perspective texture mapped with a checkerboard without (middle) and with (right) mipmapping being used. Note the strong aliasing artifacts without mipmapping is removed once mipmapping is applied. The gray that is visible at the distance area of the ground is exactly how our brain will perceive this scene.

14.5 Bump Mapping

Bump mapping is a technique by which we can simulate the effects of small bumps on the surface of an object without changing the number of primitives, as shown in Figure 14.17. The two tori in this figure have the same number of triangles. But one looks much richer geometrically than the other due to the bumps. Note that both objects are also texture mapped with a blue and yellow texture. The bumps are simulated by perturbing the normal vectors in a predefined way so that the lighting changes in a manner that is consistent with the presence of the bump. This makes us perceive the bumps even if they are absent in the mesh.

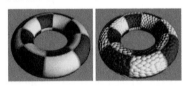

Figure 14.17. The same object rendered without bump mapping (left) and with bump mapping (right).

Let us consider a surface parametrized with two parameters (u, v). This is akin to what we did for texture mapping. Let us consider the point $P(u, v)$ with normal N. Let P_u and P_v denote the tangents at P in u and v direction respectively. In bump mapping, we want to perturb the normals at the vertices based on a scalar bump function $B(u, v)$. Consider B to be a gray scale image where white indicates the maximum bump and black the minimum. Therefore, we would like to move the point $P(u, v)$ to $P'(u, v)$ in the direction of its normal such that

$$P'(u, v) = P(u, v) + B(u, v)N. \tag{14.20}$$

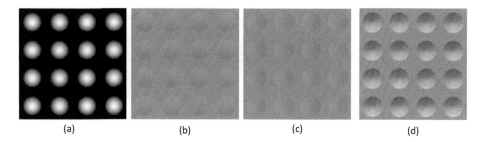

(a) (b) (c) (d)

Figure 14.18. This figure shows the bump image $B(u, v)$ (a), its derivative in u, B_u, direction found by subtracting every pixel from its right neighbor (b), its derivative in v direction, B_v, found by subtracting every pixel from its bottom neighbor (c), and the normal map (d).

Note that the above addition is a vector addition. Given this displacement, we would like to find the perturbed normal so that we can use this normal in illumination computation instead of N. For this, we find the tangent vectors P_u' and P_v' in u and v direction respectively at the displaced point $P'(u, v)$. These are given by partial derivatives of Equation 14.20 in u and v directions as

$$P_u' = P_u + B_u N + B N_u = P_u + B_u N, \tag{14.21}$$
$$P_v' = P_v + B_v N + B N_v = P_v + B_v N, \tag{14.22}$$

where B_u and B_v are the partial derivatives of the bump function in the horizontal and vertical derivation. Although N_u and N_v represent directional curvatures along u and v, assuming a locally planar surface we consider them to be zero, i.e. $N_u = N_v = 0$. Figure 14.18 shows an example bump image and its derivative, B_u and B_v, obtained by subtracting the value at a pixel from its right (or left) and bottom (or top) neighbor respectively.

Therefore, the perturbed normal N' at P' is given by

$$N' = P_u' \times P_v' \tag{14.23}$$
$$= P_u \times P_v + B_v(P_u \times N) + B_u(P_v \times N) + B_v B_u(N \times N) \tag{14.24}$$
$$= N + B_v(P_u \times N) + B_u(P_v \times N) \tag{14.25}$$

since $N \times N = 0$. Note that N, $P_u \times N$ and $P_v \times N$ are unit vectors that are orthogonal to each other. Therefore, they define a local coordinate system at P with $P_v \times N$, $P_u \times N$ and N denoting the X, Y and Z coordinates respectively. Therefore, the coordinates of N' in this coordinate system are given by $(B_u, B_v, 1)$. Therefore, we can store these perturbed normals as an image whose RGB value at location (u, v) is $(B_u, B_v, 1)$ and denotes the perturbed normal at (u, v). This image will be bluish in color and is called a normal map, denoted by $n(u, v)$ (Figure 14.18).

Figure 14.19. This figure shows some results of bump mapping: Original object on the left, the bump image in the middle and the bump mapped object on the right.

We can now use this normal map to achieve the bump mapping using the following steps.

1. Define a local coordinate system at parameter (u, v) on a surface using its normal and tangent vectors.

2. Find the transformation from the global coordinate system to this local coordinate system.

3. Transform the light and view vectors to this local coordinate system.

4. Find the perturbed normal $n(u, v)$.

5. Compute the lighting using the perturbed normal and the transformed light and view vectors.

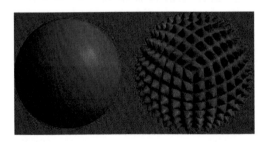

Figure 14.20. This shows an example of displacement map of a sphere.

Figure 14.19 shows some results. Can you see what is missing? First, note that no bumps show up at the silhouette. This is because there is no real displacement of pixels and the silhouette where geometry is most easily visible fails to fool the eye. Also, since geometry is not changed, self shadowing effects are not visible giving the trick away.

Another technique, more commonly called a *displacement map* actually perturbs the geometry from P to P' guided by an image. In this case, microgeometry needs to be created and rendered, something that the bump-map avoids. With displacement map, therefore, we can see the bumps at the silhouettes, self-occlusions and self-shadowing. Figure 14.20 shows an example.

14.6 Environment Mapping

In the real world, we often see objects that are extremely shiny (Figure 14.21). Environment mapping, also referred to as reflection mapping, is a technique used in computer graphics to render such shiny objects interactively. The key feature of a very shiny object is that you can find the entire environment reflected off it. Sometimes the environment in which people are (e.g. home, cafe) are not inside the field of view of rendering, but can be deduced from the reflection of this environment off the rendered shiny object. However, to create such an effect can be quite expensive via accurate tracing of light rays between the object and the environment. Such accurate rendering will also show the effects of self-reflection (e.g. the handle of the teapot reflected off its main body as in Figure 14.21). As we already know, this is rather complicated to achieve in interactive graphics.

We can achieve a rudimentary approximation of the accurate reflection of the environment off a shiny object by first creating, what we call an environment map. This is a simple geometry (e.g. a cube or a sphere) on which

Figure 14.21. Examples of real world shiny objects.

the environment is mapped. In real world, this can be achieved by taking an image using a fish eye lens. This image has 180 and 90 degrees horizontal and vertical field of view respectively. Two such images (left and right or top and bottom hemispheres) can create a *spherical* environment map. For a digital scene, one can employ a multi-pass rendering of a digital scene, where each pass generates a face of a cube of a *cubic* environment map. Six passes will be required to capture the entire field of view from a viewpoint placed at an appropriate loca-

Figure 14.22. This shows a cubic and spherical environment map. The cubic map is shown unfolded (middle) and is generated by using a 6-pass rendering each rendering a face of a cube (left) seen from the center of the cube. The spherical map (right) is generated from images of a cafe taken using a fish-eye lens camera.

tion, maybe the center of the object on which the environment will be mapped. Alternatively, a spherical environment map can also be created by a ray tracing process where rays are traced from the center of the sphere out to the environment. The color of the point in the environment where the ray hits first is used to color the point where the ray meets the sphere. Figure 14.22 shows an example of a cubic and spherical environment map.

Fun Facts

Reflection mapping have been used in movies for a long time, especially on robots in science fiction movies, even before it became a common computer graphics technique. The technique was developed independently by Gene Miller working with Ken Perlin, and also by Michael Chou working with Lance Williams, around 1982 or 1983. The first two instances in which reflection mapping was used to place objects into scenes were of a synthetic shiny robot standing next to Michael Chou in a garden, and of a reflective blobby dog floating over a parking lot. In 1985, Lance Williams was part of a team at the New York Institute of Technology who used reflection mapping in a moving scene with an animated CG element in a piece called "Interface" that featured a young woman kissing a shiny robot. In reality, she was filmed kissing a 10-inch shiny ball, and the reflection map was taken from the reflection of the ball. The first feature film to use the technique was Randal Kleiser's Flight of the Navigator in 1986 to render a shiny morphing spaceship flying over and reflecting fields, cities, and oceans. Its ground breaking appearance as an instrument concept exhibited through a movie was in films by James Cameron, *The Abyss* and *Terminator 2*.

Once the environment map is generated, it is mapped onto an arbitrary shaped object in a fashion which is very similar to texture mapping on complex surfaces, but is guided by the location of the viewer with respect to the object. In Figure 14.23, let us consider the arbitrary blue shape to be environment mapped. Let us consider a spherical map enclosing it. Let P be a vertex on the object whose environment map coordinates (i.e. the coordinates on the spherical map whose color will be picked to color P) we would like to compute. Let V be the view vector usually achieved by connecting the viewpoint to P. Let N be the normal at P. Reflecting V about N creates the vector R. Note that if the blue shape were mirror like shiny object, the 3D point where R hits the environment will be the color seen by the viewer after getting reflected off P. This point is captured by the point Q where a ray R' parallel to R passing through the center of the spherical map intersects the environment map. Therefore, the point Q will be mapped on to P thereby imparting its color to P.

Figure 14.23. Left: The figure illustrates the environment mapping process for mapping the blue geometry enclosed in the spherical environment map. Q is the point mapped at P on the object. Middle: This shows a torus mapped using the cubic environment map in Figure 14.22. Right: A wine glass mapped using the spherical map in Figure 14.22.

However, since we do not do an accurate environment map computation we do miss out on some effects. For example, we cannot see self-reflections, a very common phenomenon in reflective objects as in the reflection of a spout or a handle of a teapot off the surface of the teapot. So, environment mapping creates a compelling realism via only a rudimentary approximation of the real phenomena. But such anomalies may go unnoticed to a few, such as a gamer in interactive gaming applications.

14.7 Transparency

So far, we have only considered rendering opaque objects. However, we encounter a large number of materials in the real world which are transparent or translucent (e.g. glass, liquids). In order to achieve interactive rendering of such objects, we introduce the concept of *alpha blending*. For this, we introduce a new channel of attributes in addition to the 3-channel RGB color. This is called the alpha channel, A. The alpha value A allows a rendering application to have a fractional contribution of the color from a source pixel S (i.e. the color of the pixel that is being rendered) blended with a fractional contribution from the destination pixel D (i.e. the color already existing in the framebuffer at that pixel). Let the color at S and D be (s_r, s_g, s_b, s_a) and (d_r, d_g, d_b, d_a) respectively where s_a and d_a are values of the alpha channel. Note that prior to the introduction of the alpha channel, we assumed that the new value at the destination pixel $D' = S$, i.e. the new source color replaces the destination color.

However, in alpha blending, we adopt a more general way to achieve D' as a combination of D and S given by

$$D' = f_s(s_a, d_a)S + f_d(s_a, d_a)D \qquad (14.26)$$

where f_s and f_d provide fractional values between 0 and 1. Note that the above equation is general and can be used to achieve a large variety of effects by different choices of f_s and f_d. For the particular case of transparency or translucency, the functions we use are

$$D' = s_a S + (1 - s_a)D. \qquad (14.27)$$

Therefore, if the pixel being rendered is transparent, $s_a = 0$ achieving $D' = D$, i.e. the destination pixel will not change color at all since the source pixel is transparent. If the source pixel is opaque, as we have been assuming so long, then $s_a = 1$. Therefore, $D' = S$. i.e. the destination pixel gets overwritten by the color of the source pixel. If s_a is any other fraction between 0 and 1, we

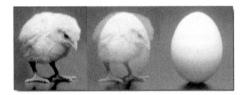

Figure 14.24. This figure shows the effect of alpha channel blending. Left: Chicken=1, Egg=0; Middle: Chicken = Egg = 0.5; Right: Chicken=0, Egg=1.

would get a combination of the source and destination colors to create the effect of translucency. Figure 14.7 shows the concept. Assume the egg to be the source and chicken to be the destination. The images from left to right are achieved with the above functions for transparency using $s_a = 0, 0.5$ and 1 respectively. It should also be noted that the order in which the translucent objects are drawn also determines the final color. Let O_1 and O_2 be two objects with alpha values s_1 and s_2 respectively. Let B be the background color to start with. If O_1 is drawn first and then O_2, then the final color will be $O_2 s_2 + (1 - s_2)(O_1 s_1 + (1 - s_1)B)$. If the order is reversed, then the final color using the same blending function will be $O_1 s_1 + (1 - s_1)(O_2 s_2 + (1 - s_2)B)$. Obviously, these two might result in different colors.

However, to achieve translucency correctly, there is depth to be considered. Only alpha channel manipulation will not work. Let us consider the following scenarios in Figure 14.25 to illustrate this. Consider the image plane shown by the horizontal line and the line of sight shown by the vertical dashed line. Three objects are shown. A is opaque while B and C are translucent. Their order of rendering is shown by the numbers on the right. In the first case (left), B will be rendered first and then its color will get attenuated by the alpha value of C, s_c. However, next the attenuated color will be completely replaced by the color of opaque A. However, physically since B and C are in front of A, we should see a combination of colors of all of A, B and C. Therefore, the result is wrong. In the second case (right), A will be rendered. Following this, during rendering of C, the framebuffer will be attenuated based on the alpha of C, s_c, and then again attenuated by the alpha value of B, s_b. Therefore, the final color will be the color of A attenuated by $s_b s_c$. However, note that since B is behind opaque A, physically s_b should not attenuate the color of A in the final rendering. Therefore, this result is also wrong.

These examples are designed to drive the point that the depth of a primitive is very important for transparency or translucency and we need to account for that. In order to achieve that without compromising performance, we assume that the number of transparent objects in the scene is relatively small. Therefore, first all the opaque objects are rendered which resolves occlusion via the depth buffer. The depth buffer is then

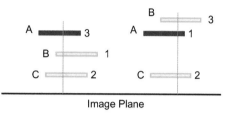

Image Plane

Figure 14.25. This shows different scenarios of depth order arrangement of translucent and opaque objects to evaluate their rendering based on alpha blending.

set to read-only which allows it to retain the depth of the rendered opaque objects. Then the pixel of a translucent object is rendered only if the pixel passes the z-buffer test, i.e. if no opaque object is in front of it. Further, the translucent/-transparent objects are drawn from back to front in order to get the composition of the colors correct. 14.26 shows some renderings using this technique.

14.8   Accumulation Buffer

An accumulation buffer is a higher precision frame buffer that is used to accumulate multiple images in real time rendering. The higher precision of the accumulation buffer allows higher precision sum, multiplication or division of images. It can be used to achieve several effects like blending, depth of field (simulating the

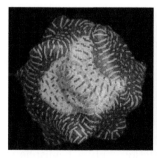  

Figure 14.26. The figure shows rendering of transparency using alpha blending. Additional effects of lighting, shadows and texture mapping are included in the renderings.

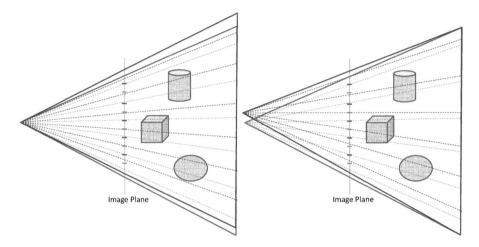

Figure 14.27. The figure shows rendering anti-aliasing via jittering view settings using an example in 2D. The red and blue show the two different view frustums along with the rays they sample. Left: The view frustum is jittered. Right: The view point is jittered.

effect of human eye where only objects at a certain depth appear focused while others are blurred) or anti-aliasing. Here we will see how an accumulation buffer can be used for anti-aliasing. One way to achieve anti-aliasing is to sample each pixel more than once and then average the samples. This achieves the effect of rendering the image at a higher resolution and then low-pass filtering it.

The process starts by clearing the accumulation buffer. One way to sample multiple values for the same pixels is to jitter the view settings so that the center of the rendering pixels moves slightly but remains within the pixel. This can be easily achieved by moving the image plane slightly. The scene is then rendered multiple times, each time with different jittered view settings given by a slightly moved image plane or slightly moving the view frustum or the viewer position (Figure 14.27). Each of these renderings is weighted by a fraction and accumulated in the accumulation buffer to achieve a low pass filtering. The final anti-aliased result is moved to the frame-buffer for rendering.

14.9   Back Face Culling

When we render closed abjects, there are certain parts of the object which are *back facing* and therefore occluded by its *front facing parts*. Consider a sphere. Anytime we look at it from any conceivable view direction, only a hemisphere can be seen while the other one will be at the back occluded by this visible hemisphere. We can improve the rendering performance considerably if we can prevent the back facing primitives from going through the rendering pipeline. Back face culling is a technique by which we detect such back facing polygons and remove them from using up computational resources to go through the model, view and projection transformations.

Put a Face to the Name

Jim Blinn, a retired scientist, educator and industrial legend, is considered as one of the father figures of computer graphics (CG), in particular in light-matter interaction. He is the first person to introduce concepts of bump and reflection mapping which provided a very powerful tool to early computer graphics animators. Though the shading model using normal interpolation goes by Phong  shading model, it should be more accurately called Blinn-Phong shading model since Blinn worked together with Phong on this model. Blinn was born in 1949 and received his bachelors degree from University of Michigan in 1970. He received his PhD from University of Utah in 1978. He first became widely known for his work as a computer graphics expert at NASA's Jet Propulsion Laboratory (JPL), particularly for his work on computer graphics animations for various space missions to Jupiter, Saturn and Uranus, especially the Voyager project. These animations were shown on many news broadcasts as part of the press coverage of the missions and were the first exposure to computer animation for many people in the industry today. He is also known in the computer graphics community for his enthusiastic and inspirational role as an educator, mentor and a visionary. His columns "Jim Blinn's Corner" (today published as a book by Morgan Kaufman) has inspired many to take computer graphics as their calling. These were articles, covering math, graphics pipelines and a wealth of tips and tricks which always kept graduate students motivated to work on the next big thing. He is well known for creating animation for three television education series: Carl Sagan's *Cosmos: A Personal Voyage*; *Project MATHEMATICS!*; and the pioneering instructional graphics in *The Mechanical Universe*. His talks in CG venues are still very popular due to his reputation of throwing a challenge to the community. In 1998, in a keynote talk in SIGGRAPH (the premier CG conference) he asked the CG community "to figure out to drop a piece of spaghetti onto the plate and how it squiggles up and model the sauce on there for the frictional coefficients and so forth". This led to a large amount of research finally resulting in accurate CG simulations of protein foldings.

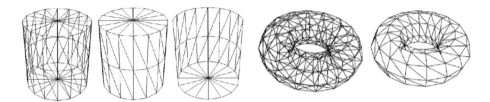

Figure 14.28. The figure shows back face culling in action in wire frame rendering of a cylinder (left) and torus (right). For each, you see the model rendering without back face culling on the left and with back face culling from the same or different viewpoint on the right.

To achieve this we can apply a very simple test. Note that any front facing primitive will have a normal vector whose angle with the view direction (from the vertex to the view point) will be within -90 to $+90$ degrees i.e. its cosine is positive. The sign of the cosine can be computed using $N.V$ where N is the normal to the plane of the triangle and V is the view direction. Therefore, if this dot product is greater than 0, then these primitives should be rendered, otherwise they are discarded. This process is called back-face culling. Obviously, back-face culling cannot be used with transparent and translucent objects. Figure 14.28 illustrates the process of back face culling.

14.10   Visibility Culling

When we navigate a scene, usually the view frustum has a limited horizontal and vertical field of view. Therefore, objects which are not within this view frustum (e.g. objects behind us) are invisible and we should not spend resources to render those objects. Therefore, instead of making these objects go through the entire pipeline and be culled away in the last stage of clipping, the performance can be improved tremendously if objects *which are not inside the view frustum* are culled away very early in the pipeline. This process is called view frustum culling and is illustrated in Figure 14.29. Only the objects within the frustum or which intersect the frustum are rendered. The purple torus intersects the view frustum though it is only partially inside the frustum. Such objects are special cases and are handled appropriately. In the next few sections, we will introduce methods for achieving view-frustum culling with wireframe objects that shows exactly the triangles that are rendered and those that are culled.

14.10.1   Bounding Volumes

The first method consists of defining a simple bounding volume around each object in the scene - e.g. a cube or a sphere. This bounding volume should be

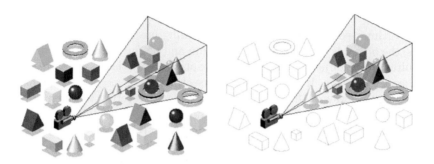

Figure 14.29. The figure shows the view frustum and all the objects in the scene (left) and then shows the objects which has been culled away by the view frustum culling by (right) using no color on them.

the smallest possible volume that encloses the object. Then, instead of checking if every triangle of the object is inside the view frustum, we can first check if the bounding primitive is completely inside or outside the view frustum. If it is completely inside, the whole object has to be rendered. If it is completely outside, the whole object is culled. These two are the most common cases and lead to quick culling of all the objects that are clearly outside the view frustum. The small cases of objects whose bounding volume intersect the view frustum needs to be treated differently. The easiest way to deal with them is to retain them for rendering and let the screen space clipping of the rendering pipeline clip away the part outside the view frustum. A more complex approach is to subdivide the object into multiple smaller objects and test their bounding volumes against the view frustum and go down the hierarchy of only those smaller objects whose bounding volume intersects the view frustum. The choice of the geometry used as bounding volumes are typically simple ones such as spheres and cuboids, whose intersection with the view volume can be efficiently computed. Bounding volumes should also bound the given object tight in order to reduce false positives during intersection computations.

Bounding Box: The first bounding volume that comes to mind is often a bounding box. A simpler bounding box is an axis-aligned bounding box that can be computed using the minimum and maximum extents in the X, Y and Z values of the vertices of the object. The box thus defined is the smallest box enclosing the object such that the the edges of the box are parallel to the X, Y, Z axis of the world coordinate system. Note that if the model is rotated, this box may not be axis aligned anymore and hence a new axis aligned bounding box has to be computed for the transformed object.

Each plane of the view frustum divides the 3D space into two half-spaces — one that is inside the frustum and the other which is outside the frustum.

The bounding box is tested against each of these planes to see if it is inside, outside or intersecting the plane. If the bounding box is outside any one of the planes, the object is outside the view frustum and no further testing with planes is required. If the object is inside all the six planes, only then is it completely inside the view frustum. If the object is intersecting one or more of the planes and is not outside of any of the six planes, then it is intersecting the view frustum. Figure 14.30 shows the axis aligned bounding boxes for the different objects in dotted blue lines. Therefore, the next step is how to compute the intersection of a bounding box with a plane that comprises testing all the eight points of the bounding box. If they are all outside or all inside, the object is completely outside or on the same side of the frustum respectively. If not, then it is intersecting the frustum.

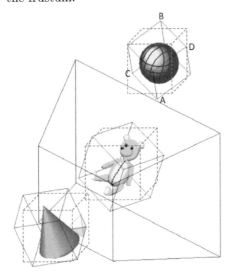

Figure 14.30. This shows the axis aligned bounding box (in dotted blue) and axis aligned bounding box (in solid blue) for different objects and the trapezoidal view frustum (in solid blue) during view frustum culling. The ball is culled out, the teddy is accepted for rendering while the pyramid is considered to be intersecting with the view frustum during the process.

Note that more often than not, an axis-aligned bounding box has a large amount of empty space and therefore is a rather inaccurate approximation of the volume occupied by the object. A more accurate approximation can be achieved via an oriented bounding box (OBB), as shown in solid blue lines in Figure 14.30. The three directions of the OBB are computed via a principal component analysis of the vertices of the objects and then finding the maximum extent of the object in those directions. The advantage of the OBB is that the OBB need not be recomputed with transformations – rotations, scaling and translations. The same transformation applied to the object when applied to the OBB generates the OBB for the transformed object. To learn about OBBs in detail, refer to [Gottschalk et al. 96].

However, the intersection computation of an oriented bounding box with a plane of the view frustum is more complicated. In this case, first the two diagonals that pass through the center of the oriented bounding box are computed. Next, one of these two diagonals which has a closer alignment with the normal of the plane is chosen. The endpoints of this diagonal form the closest and farthest points in the OBB from the plane. If both the endpoints of this diagonal are inside or outside the plane, the object is completely inside or outside the plane respectively. If not, it is intersecting the plane. This is illustrated in Figure 14.30. Let us consider the bounding box for the red ball. When considering the far plane of the frustum, AB is the diagonal which is closer in alignment to the normal to the far lane. Therefore, the inclusion test has to be run on the nearest and farthest point, A and B, respectively. However, when considering the left plane, the closest and the farthest points are C and D respectively.

Bounding Sphere: Bounding spheres can also be used as bounding volumes. In this case, the intersection computation becomes even simpler. First it is detected whether the center of the sphere is inside or outside the plane. Next the distance of the center of the sphere from the plane is computed. If this is smaller than the radius, then the object is intersecting with the view frustum. If the distance is bigger than the radius, then the object is accepted or culled based on its center being inside or outside the plane respectively. Bounding spheres are not affected by the rotation of the enclosed object, and if the object is translated, the bounding sphere is also translated by the same amount. So it is easy to update the bounding sphere with rigid transformations of the enclosed object.

14.10.2 Spatial Subdivision

Object space subdivision using bounding volumes as seen in the previous section can adapt to unique shapes of the objects and are effective in applications such as collision detection and view frustum culling. However, in applications that requires computation of relative positioning of objects, for example, from a view point in a particular direction, spatial subdivision of the scene becomes more useful than object level subdivision. A few spatial subdivision techniques in 3D include octree, k-d tree, and binary space partitioning. We will discuss octree subdivision in this section. For an in-depth treatise on other kinds of spatial partitioning techniques, refer to [Jimenez et al. 01].

The octree is a tree data structure where each node has eight children nodes. The root node corresponds to the axis-aligned bounding box of the entire scene defined by minimum and maximum coordinates in the X, Y and Z directions. This bounding box is subdivided into half in each of the X, Y and Z directions to partition the space into eight *equal sized* bounding boxes, each associated with a child node of the parent. Therefore, the space associate with each child is completely contained in the space associated with its parent and the union of the spaces of all the children creates the space of its parent. This continues in a hierarchical manner for each child thus creating a tree in which every node has eight children, as shown in Figure 14.31. Note that there is no need to associate a bounding box with every node since the bounding box at the root node provides a predefined subdivision that defines the extent of every box in the tree which can be computed very easily during the tree traversal.

For spatial subdivision in graphics applications, each node stores a list of indices of primitives contained in its corresponding bounding box. If any box has only one primitive in it, that node will not be subdivided any further and it becomes a leaf node. If a primitive intersects more than one sibling node, it can be handled in two different ways. It can be split across the boundary to have different parts of it contained in the different boxes. Or, it can be repeated in all the boxes in which it partially belongs. Octree construction involves populating the tree nodes with an index list and is performed as a pre-processing step.

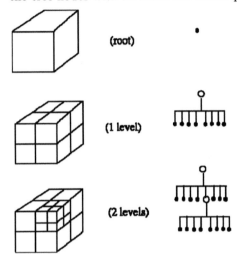

Figure 14.31. This figure illustrates how the octree is build by hierarchical spatial sub-division.

Let us discuss the use of octree in view frustum culling application. During runtime, the view frustum will be cutting through the bounding box defined at the root. The scene is rendered by the following algorithm starting at the root. If the bounding box at the node is completely inside the view frustum render the triangles associated with it. If the bounding box at the node is completely outside the view frustum reject it (do nothing). If the bounding box at the node intersects with the view frustum apply the process recursively to all its children. This algorithm achieves a depth first traversal of the octree and a cut of the tree for each view frustum that is rendered. This is illustrated using a quadtree for a 2D curve in Figure 14.32. In the 2D case, each box will be subdivided into four equal sized bounding boxes.

14.10.3 Other Uses

Bounding volumes and spatial subdivision techniques are used in many applications other than view frustum culling. One common application is collision detection widely used in games and scientific simulations. Examples are a digital pool game or a simulation of pistons in an engine. In such applications, objects move based on some rules and if they collide, it should be detected and an appropriate action should be taken. For example, in pool if two balls collide they should be reflected in opposite directions. A collision is detected when one or more of the triangles in an object intersect with one or more of the triangles in another object or with itself in case of self intersections in non-rigid objects. The brute force way to compute this is to intersect every triangle in one object with every primitive in the other. As is evident, this leads to a tremendous amount

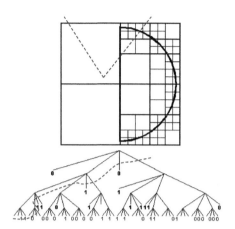
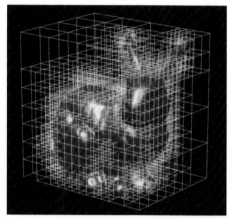

Figure 14.32. Left: We consider a quadtree for a 2D curve (instead of octree for 3D surface) to illustrate the concept. The tree created has four children listed from left to right as the top-left, top-right, bottom-left and bottom-right boxes contained in the node. The red dotted line shows 2D the view frustum and the corresponding cut in the quadtree. Right: We show the spatial subdivision of a 3D bunny model into an octree.

of computation. For example, for objects with around 1 million traingles, it will lead to 10^{12} intersection computations that can hardly be achieved in interactive rates, a mandatory requirement in such applications. Therefore, bounding volumes are used for fast rejection of non-collision and spatial subdivision is used for fast detection of candidate collisions.

Each object maintains a bounding volume data structure using hierarchical spatial subdivision by modifying the octree-based spatial subdivision slightly. This is called hierarchical bounding volumes where, unlike octree-based spatial subdivision, the bounding box at each node is not a pre-determined half-way subdivision of the parent bounding box. Instead, it is the smallest bounding box that fits all the triangles in the bounding volume created by the half-way subdivision of the parent bounding box. But, the list of triangles associated with each node of the hierarchical bounding volumes is identical to the octree-based spatial subdivision. Therefore, unlike octree-based spatial subdivision where the bounding box at each node need not be stored but can be easily derived from the bounding volume at the root of the tree, the bounding volume has to be explicitly stored at each node of the hierarchical bounding volumes. Figure 14.33 illustrates this difference using an example in 2D. The green, red and magenta show these tightest fitting bounding boxes at levels 0(root), 1 and 2 of the tree respectively.

To detect collision between two objects, bounding volume intersection tests are first performed at the level 0 of the hierarchical bounding volume representation of the objects. No collision between the bounding volumes implies no collision between the enclosed objects. If the bounding volumes collide, there is a possibility that the enclosed objects will collide. Note that the bounding volumes can collide in the empty regions of its volume. Therefore, a collision of bounding volumes does not always imply a collision of objects. If the bounding volumes intersect, pairwise intersection tests between the bounding volumes of the children nodes are performed. Therefore, the above process is repeated on the bounding volumes of the children nodes recursively. The trees are thus traversed in depth first search and a collision is detected when one primitive remains in each bounding volume and their intersection computation is essentially a triangle-triangle intersection computation to detect the point of collision.

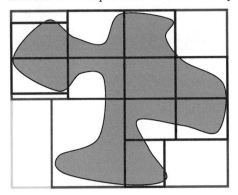

Figure 14.33. This shows the hierarchical bounding volumes for a 2D object. Unlike octree-based spatial subdivision, at every node the tightest bounding box enclosing the part of the object subdivided bounding box is stored.

We have discussed the use of bounding boxes for collision detection and how hierarchical methods can be used for fast collision detection. However, it is important to note that different types of bounding volumes can be used for collision detection like spheres or a spherical shell. In fact, three criteria dictate the choice of the type of bounding volumes to be used during collision detection. First, how tightly does the bounding volume hug the object so that the empty space in the bounding volume is minimized? For example, when considering contemporary objects (e.g. table, chair, room, house etc) which have many straight edges, a box-like volume is probably most appropriate. However, when dealing with cellular, biological or astronomical simulations where objects have closed curved contours, a spherical volume is probably more conducive. A tightly fit bounding volume will reduce false positives in collision detection and will minimize going down the hierarchy for collision resolution. The second criterion is the complexity of computing and updating the bounding box. This criterion would dictate if the system can be used for collision detection in dynamic environments. An axis aligned bounding box is easy to construct but more expense to update under a few transformations such as rotation of the enclosed object. An oriented bounding box is more difficult to construct, but easier to update. Even for static scenes, since the collision computations have to be done many times down the hierarchy, the third criterion is the complexity of computing the collision test between two bounding volumes. The simplest collision test is between two spheres - if the distance between the centers of the sphere is more than the sum of the radii, then the spheres do not collide, otherwise, they

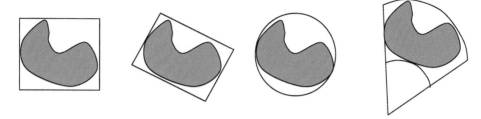

Figure 14.34. The figure illustrates the concepts of bounding volume of an object (in blue) in 2D. From left to right: axis aligned bounding box, oriented bounding box, a sphere, and a spherical shell.

collide. More complex to compute an axis aligned bounding box. If at least one of the X, Y, or Z ranges of the bounding boxes do not overlap, then the boxes do not intersect, otherwise, they intersect. Finally, intersection computation of an oriented bounding box requires us to find a separating plane and is more complex than for the other two primitives [Gottschalk et al. 96]. There are other bounding volumes such as spherical shell [Krishnan et al. 98] which is the region enclosed between two concentric spheres and a cone. Spherical shells can provide a tight fit for higher order surfaces. They are relatively more complex to compute, easier to update, and require moderate computation for detecting collision between two spherical shells. An illustration of all these bounding volumes is shown in Figure 14.34.

14.11   Conclusion

In this chapter we introduced you to the most common ways to enhance realism in interactive computer graphics through rudimentary approximations of reality. Again, in this chapter concepts are explained in an API independent manner. We hope these fundamental concepts help the readers to code up using any suitable API. We should also acknowledge that enhancing realism does not come free – and almost all the time trades off with performance. For example, bump maps come at no cost with respect to increased geometry but require Phong illumination computation during rasterization and cannot show realism at the silhouettes. While a displacement map can alleviate this problem, it comes at the cost of lower rendering speed due to the significant increase in geometry. The challenge is to make the right choices that are suitable for specific applications.

Bibliography

[Gottschalk et al. 96] Stefan Gottschalk, Ming Lin, and Dinesh Manocha. "OBB-Tree: A Hierarchical Structure for Rapid Interference Detection." *Computer Graphics (SIGGRAPH 1996 Proceedings)*, pp. 171–180.

[Jimenez et al. 01] P. Jimenez, F. Thomas, and C. Torras. "3D Collision Detection: a Survey." *Computers and Graphics*, 25:2 (2001), 269–285.

[Krishnan et al. 98] Shankar Krishnan, Amol Pattekar, Ming Lin, and Dinesh Manocha. "Spherical Shell: A Higher Order Bounding Volume for Fast Proximity Queries." *Robotics: The Algorithmic Perspective: WAFR*, pp. 177–190.

Summary: Do you know these concepts?

- Ambient, Diffuse and Specular Illumination

- Flat, Gouraud, Phong Shading

- Shadow Map

- Texture Mapping and Mipmap

- Environment Map

- Bump and Displacement Map

- Alpha Blending

- Accumulation Buffer

- Anti-aliasing

- Backface Culling

- Visibility Culling

- Spatial Subdivision and Octree

- Bounding Volumes

- Collision Detection

Exercises

1. Consider a gray world with no ambient and specular lighting (only diffuse lighting). The light is at infinity and its direction and color are $(1,1,1)$ and 1.0 respectively. The coefficient of diffuse reflection is $1/2$. The normals at points P_1, P_2 and P_3, are $N_1 = (0,0,1)$, $N_2 = (1,0,0)$ and $N_3 = (0,1,0)$ respectively. Find the illumination at the points P_1, P_2 and P_3.

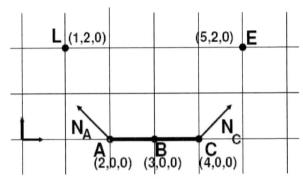

2. In the above figure, the light and the eye are denoted by L and E respectively. On the surface AC the normals at A and C are $N_A = (-1,1,0)$ and $N_C = (1,1,0)$ respectively. Everything is drawn to scale. Use the illumination model given by $I = I_a k_a + I_L k_d (N.L) + I_L k_s (R.V)^n$ where R denotes the reflected light vector at the surface point, $I_a = 0.8$, $I_L = 1.0$, $k_a = 0.2$, $k_d = 0.9$, $k_s = 0.5$, $n = 2$. Find the illumination at A and B. (Hint: Treat negative dot products as 0.)

3. You are rendering a black and white checkered tiled floor using a single texture mapped polygon. The view is simulating a person standing on the floor and looking at a point far away from him on the floor. (1)Artifacts at the distant end of the floor can be seen. How would you remove these artifacts? (2) How can you explain why this method works using the sampling theorem?

4. One artifact of Gouraud shading is that it can miss specular highlights in the interior of the triangles. How can this be explained as an aliasing artifact?

5. Consider five objects in the line of sight from the eye. Object i is behind Object $i-1$. Object 1, 3 and 5 are opaque while the others are translucent. In what order would you render the objects to get the correct effect of translucency? Justify your answer.

6. Consider the above 2D gray world and the primitive AB in it (shown by the red line). The blue vectors show the normal at A and B. L and E are the

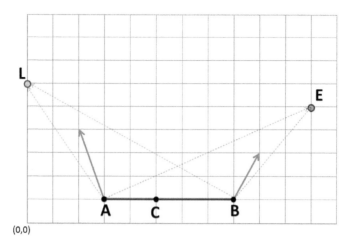

position of the light and the eye respectively. Let the coefficient of diffused illumination be 0.5 respectively. Let the intensity of light be 0.5.

(a) What are the coefficients of A and B respectively for bilinear interpolation of C?

(b) What are the normals at A, B and C?

(c) Find the diffused illumination at A and B.

(d) Find the diffused illumination at C using Gouraud shading.

(e) Find the diffused illumination at C using Phong shading.

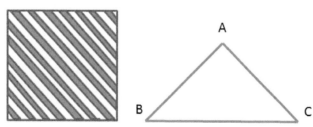

7. Consider the above striped texture on the left and the triangle ABC which we would like to texture map using this texture. Consider the bottom left corner of the texture to be $(0, 0)$ and the top right to be $(1, 1)$. Find the texture coordinates assigned to A, B and C respectively to create the appearance of stripes in each of the following directions: (a) horizontal, (b) vertical, (c) diagonal in the same orientation as the texture, and (d) diagonal in the perpendicular orientation to the texture.

8. You are seeing an object which is either texture mapped, bump mapped or displacement mapped but you don't know which one. However, You have the liberty to move the light and the viewpoint of an object and see it from different angles and for different positions of the light. How will you figure out which technique was used?

9.

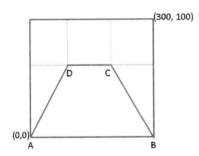

Consider the framebuffer of size 300×100. ABCD is a rectangle in 3D space which has been projected as a trapezium in the 2D. AB is projected on the bottom scanline. CD is projected on a scanline (shown in brown) that is $\frac{3}{5}$ way above and has a projection length $\frac{1}{3}$ of AB. The depth of side AB and CD are 60 and 30 respectively. Consider a 512×512 checkerboard texture T that will be used to texture map ABCD. T is stored in different resolutions using mipmapping.

(a) On which scanline is CD projected?

(b) Consider a scanline S that is half way in screen space between AB and CD. Find the depth of S. What level of the T will be used to texture map AB and CD respectively?

(c) Find the length of S contained in the trapezium. What level of T will be used to texture map S?

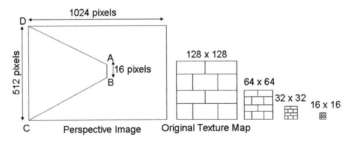

10. Consider the figure above. Suppose we have a brick wall that forms the left hand side of a corridor in a maze game as shown. The image is drawn to scale. This wall is defined in world coordinates by points ABCD, the projection of which are shown in the image. Assume that the brick wall is 16 bricks high.

(a) If we assume the brick wall to be 16 bricks high, how many times do we have to repeat the texture in the vertical direction?

(b) What level of mipmapped image pyramid on the right will be used for texture mapping the near end CD?

(c) What is the minimum number of pixels each texel should cover to avoid aliasing?

11.

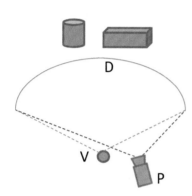

Let us consider the building of a virtual reality environment. Its top view is shown in the above figure. D is the cylindrical display whose digital representation (shape and position) is known. The scene will be projected from the projector P. The digital objects are placed behind the screen. The user will be tracked. The view seen from the location V should be projected on D from P. How can you use projected texture to generate the image that P should project? How many passes would this rendering take?

12. How can you create the effect of mirror in a scene using some of the techniques you have learnt in this chapter? How many rendering passes would it take?

13. Consider a scene in 2D. The scene comprises a parallelogram whose four vertices are (1,0), (-1,0), (0,-1) and (0,1).

(a) What would be four vertices of the 2D axis aligned bounding box for this parallelogram?

(b) If we want to translate this parallelogram by (2,2), how would you use the transformation parameters to find the new axis-aligned bounding box?

(c) Would you be able to use a similar approach if we have to rotate the parallelogram by 45 degree (instead of translation)?

(d) Can you think of any other bounding shape that would allow you to recompute of the bounding shape for all different kinds of rigid body transformations in the same manner as you did in (b)?

14. Silhouette edges are the edges in the manifold that have one back-facing polygon AND one front facing polygon incident on it.

(a) How do you compute the silhouette edges of a manifold?

(b) In OpenGL you can draw only back-facing polygons, or only front facing polygons. If you render the manifold (front facing polygons), then clear the frame-buffer but not the depth buffer, then again render only the back facing polygons. What do expect to see?

(c) Assume that the thickness of a line is an attribute of a line. Thickness of three means that the line would be drawn three pixels thick. If the thickness of the line was one and now is increased to three only for the second rendering (rendering of back faces), what do you expect to see?

15

Graphics Programming

Graphics programming requires navigating through several APIs and libraries while using specialized graphics hardware. We first present the history and development of the modern graphics processing unit (GPU) that provides a perspective on its existing form and functionalities in the context of the different interactive techniques discussed in the previous chapters. Next, the fundamental aspects of the modern graphics hardware and existing APIs and libraries that facilitate programming it for both graphics and general purpose computation is presented.

15.1 Development of Graphics Processing Unit

The graphics processing unit (GPU) is a specialized hardware unit designed to off-load and accelerate 2D or 3D processing from the central processing unit (CPU) to assure interactive performance. Today, almost all desktops, laptops and mobile devices come equipped with some kind of GPU. GPUs have undergone revolutionary transformations in recent years and it is important to know how the GPU aids in the computations of the effects we render in the interactive graphics pipeline.

Graphics hardware first came to use in 1980s, though they were called GPUs much later. GPU is a term introduced by nVidia in 1999. However, we will refer to such hardware as GPU in the rest of this chapter for the sake of consistency. The early GPUs were essentially integrated framebuffers that could only achieve line rasterizations, also termed as wireframe rasterizations, thereby off-loading to the GPU, the rendering of the edges of the polygonal primitives. The first hardware dedicated to the graphics pipeline was the IBM professional graphics controller (PGA) that used a microprocessor hardware to off-load the simple tasks of rendering, like drawing and coloring filled polygons, from the CPU opening up CPU cycles for other general purpose processing while the graphics processing was done in parallel on the PGA card. A separate PGA card onboard marked an important step in the evolution of a separate GPU. By 1987 more features were added to the GPUs including shaded solids, vertex lighting, rasterization of filled polygons, depth buffer, and alpha blending. However, there was still a huge reliance on the CPUs where most of the computations used to happen and the data transfer from the CPU to the GPU was a major bottleneck.

Pipeline Functions	1996	1997	1999
Application specific tasks (Moving objects, setting up camera, object level culling)	CPU	CPU	CPU
Transformation	CPU	CPU	GPU
Lighting	CPU	CPU	GPU
Clipping	CPU	GPU	GPU
Rasterization	GPU	GPU	GPU

Figure 15.1. Evolution of early graphics pipeline leading to a fixed function pipeline in 1999.

The GPU evolution got a boost from the release of the graphics industry's most widely used application programming interface (API) of SGI-GL by SGI (Silicon Graphics Inc.) in 1989, which later gained popularity as OpenGL. In 1993, SGI released its first graphics cards for workstations while the first 3D consumer graphics hardware was offered by companies like Matrox, nVidia, 3DFX and ATI. However, the distinction between GPU and CPU was not that clear in this hardware. While much of the later stages of the graphics pipeline were instrumented in the GPU hardware, there was still a significant reliance on the CPU, especially for the first part of the pipeline involving transformations. Games like *Quake* and *Doom* drove the fast adoption of the graphics cards in the gaming industry.

The first graphics processing unit close to its current form was introduced in 1996 by 3DFX and was called the Voodoo card. The CPU still did the vertex transformation and lighting while Voodoo provided shading, texture mapping, z-buffering and rasterization. It was still not possible to evaluate the lighting model at every pixel and therefore effects of Phong shading or bump mapping were still not possible at interactive rates. In 1999, the first present-day GPU hardware became a reality via the introduction of nVidia's Geforce 256 and ATI's Radeon 7500 (Figure 15.1) where the vertex transformation and lighting computations were also moved to the GPU. Four parallel pipelines aided faster rendering with new features of multi-texturing and bump-mapping. A faster communication channel between CPU and GPU allowed even higher performance. However, this hardware still followed a *fixed function pipeline* since once the data was sent to the GPU pipeline, it could not be modified. Fixed functions were achieved via feature sets defined by APIs like OpenGL and DirectX. Therefore, if newer features were added to the graphics API, the fixed function hardware could not take advantage of them. Figure 15.2 shows the different stages of such a fixed pipeline. The vertex control receives the triangle data from the CPU. The VS/T & L (Vertex Shader/Texture and Lighting) stage transforms the vertices and assigns attributes to each vertex (e.g. color, texture coordinates, tangents). Lighting computation can also take place in the VS/T & L stage to assign colors to the vertices. Clipping and interpolation of attributes at the clipped vertices occur in the next stage. Following this, rasterization happens to mark pixels covered by the clipped triangles. The shader achieves the interpolation of attributes (e.g. texture coordinates, colors, normals) for every pixel touched by the triangle.

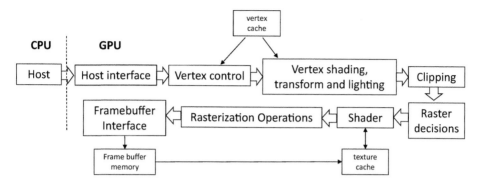

Figure 15.2. A fixed graphics hardware pipeline.

Finally, the raster operations are performed to blend colors for anti-aliasing or translucency effects. Depth buffer tests are performed to resolve occlusion. The frame buffer interface manages the reads and writes out of the framebuffer.

In 2001, we saw the advent of the first *programmable graphics pipeline* via ATI Radeon 8500 and Microsoft X-box, where unlike ever before, parts of the GPU could be programmed. Instead of sending all the data to the GPU and simply letting it flow through the fixed pipeline, programmers could now send this data along with vertex programs (commonly called shaders) that would be operating on the data while it was passing through the GPU. These shaders were small kernels written in assembly-like specific shader language giving a limited amount of programmability in the vertex processing stage of the pipeline. The programmable graphics pipeline was followed by the advent of the first fully programmable GPU in 2002 via nVidia GeForce FX and Radeon 9700. These graphics card allowed for per-pixel operations with dedicated hardware for programming both the vertex and the pixel(fragment) shaders. By 2003, full floating point support and advanced texture processing started to appear in the cards enabling the first wave of applications that started using GPUs for non-graphics computing as well.

By 2006, the graphics hardware started to capitalize on the tremendous data independence provided by the graphics pipeline. The goal was to push the flexibility to the shader programs. Early high level shader languages started to appear to provide easier programming interface. The GeForce 6 was the first GPU that streamlined the data independence to create a pipeline that has multiple parallel multi-core stages with fixed stages in between. The first parallel stage is that of a vertex shader that reads a vertex position and computes its position on the framebuffer. Multiple threads process different vertices independently. A fragment shader processed the floating point RGBA color contributing to every pixel. Similarly, multiple threads process different pixels independently as well.

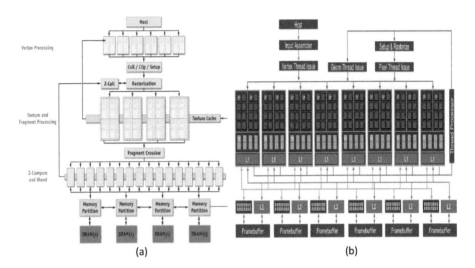

Figure 15.3. This shows the programmable GPU via the abstraction of the GeForce 6 card (a) and GeForce 8 card (b).

In between these two parallel stages is the fixed stage of clipping, and rasterization. Similarly, the operations of blending and processing the depth buffer can also be performed in parallel in units called raster operations processors (ROP). In between the parallel fragment shaders and the ROP processors, there is a fixed operation that assembles the fragments from the different threads together. Therefore, such alternating parallel and fixed stages turn the GPUs into massively parallel and programmable processors. Figure 15.3 shows an abstraction of such a pipeline in nVidia GeForce 6. Note that the vertex shader has 6 threads, the pixel shaders have 4 threads and the ROP opertaions have 16 threads in this case. Also, the partitioning of the framebuffer memory allows much higher resolution graphics without reducing the frame-rate. Therefore, from this time GPUs have been viewed as a powerful programmable floating point computational and storage unit that can be exploited for compute intensive applications that need not have anything to do with graphics.

At this stage, the graphics hardware still had specialized shaders for vertices, pixels and ROP operations. The GeForce 8 changed that by unifying the shaders and making them into a fully programmable unified processors which are called streamlined processors or SP (Figure 15.3). With this change, the graphics pipeline model became a purely software abstraction. To harness the GPU power, new programming languages were devised. CUDA is such a language provided by nVidia for nVidia cards. Similarly, there is ATI Stream for ATI cards and DirectX 10 for either cards.

15.2 Development of Graphics APIs and Libraries

Computer graphics has become very popular, especially in the video games and simulation community. Therefore, there are several specialized API to ease programming of different stages of the graphics pipeline and application requirements. These APIs provide a way to access the hardware in an abstract manner while taking advantage of the special hardware capabilities of a specific graphics card. However, since even a few years back a fixed pipeline was in vogue, several APIs, especially those which evolved in the age of the fixed graphics pipeline, have evolved to adapt to the programmable pipeline.

OpenGL is one of the oldest cross-language cross-platform interfaces for 3D graphics rendering providing a way to interact with GPUs. GLUT is the OpenGL Utility toolkit for writing OpenGL programs independent of the window system used for rendering the scene. It implements a simple windowing API for OpenGL making it much easier to learn OpenGL. GLUT also provides a portable API across multiple OS and PCs. OpenGL is defined as a set of functions which can be called by the client program. The functions are similar superficially to C, but are language independent. OpenGL's popularity is primarily due to its quality of official documentation which are known by the colors of their covers (the red, orange, green and blue books are the first to fourth edition of the OpenGL Programming Guide). Often accompanying libraries like GLU, GLEE or GLEW bind with OpenGL to support useful features that may not be supported in contemporary hardware like mipmapping or tessellation. OpenGL Shading language (GLSL) is a high level shading language based on the syntax of C, first designed to allow OpenGL to access the programmable GPUs with using assembly level or hardware specific languages. OpenGL ES is an extension of OpenGL API for programming for embedded devices. WebGL is a Javascript API for rendering 3D graphics. Direct3D is a similar API offered by Microsoft which promises better performance on Windows OS while Metal is an API that debuted for Apple's iOS8.

Vulkan is a more recent cross-platform API. It was initially referred to as a "next generation openGL initiative". It was build upon AMD's API called Mantle. In addition to optimizing performance on GPUs like OpenGL, Vulkan is also optimized to reduce CPU use and distribute whatever is needed from the CPU across multiple cores. It works for both high end-graphics cards and mobile devices. Unlike OpenGL, it provides a unified management of compute kernels and graphics shaders removing the need to use a shader API in conjunction with a graphics API.

The modern GPU offers tremendous potential to solve largely parallelizable general purpose problems using the GPU. However, this means that the programmer must know the graphics API and GPU hardware well to map general purpose problems onto the vertex, textures and shader programs. To alleviate this problem, nVidia has developed CUDA, a programming language that offers a parallel computing platform and API for programmers to use a CUDA enabled nVidia GPU for general purpose computing. CUDA gives direct access to GPU's instruction set and parallel computing elements to general purpose programmers.

Fun Facts

Today's generation won't know much about SGI (Silicon Graphics Inc), the company which pioneered graphics workstations in the early days of computer graphics and to whom we owe much of the advancement of the graphics hardware. SGI introduced the concept of the geometry engine resulting in the first very large scale integration (VLSI) of the graphics pipeline with specialized hardware to accelerate the geometric computations needed to display three dimensional imagery. SGI was founded by yet another father figure of computer graphics who also got his PhD from the University of Utah, Jim Clark (top left). Jim Clark was born in Texas and had a difficult childhood and dropped out of high school after being suspended. However, his life turned with his 4 year tenure in the Navy where he was introduced to electronics and fell in love with it. He worked hard with night courses at Tulane University which opened up the doors of the University of New Orleans for a B.S. and M.S. in physics. After his PhD from University of Utah in 1974 Jim Clark was a faculty member at UC-Santa Cruz before moving to Stanford University in 1979. Jim Clark founded SGI in 1982 with seven of his graduate students, one of whom, Kurt Akeley, played a major role in recent years in bringing the light field camera, Lytro, to the market. SGI spearheaded the development of several graphics workstations including the indigo, prism, onyx, crimson and finally the high performance computing multi-core room-size machine called infinite reality engine (from top right in counter clockwise order). SGI's core market was impacted adversely by the advent of the consumer graphics card (e.g. nVidia, ATI) and the company moved its thrust area to high performance computing in 1999.

15.3   The Modern GPU and CUDA

Modern GPUs are no longer tied to their graphics ancestry but have proven themselves to be the most successful desktop supercomputing architecture for general purpose computing. Once thought to be for video games, today's GPUs find their place in solving varied problems in varied areas from astrophysics and arts to seismology and surgery [Luebke 09]. In this chapter we will describe the modern GPU architecture in brief. More details are available in [Azad 16]. We specifically focus on nVidia's GPU architecture and CUDA programming language.

15.3.1   GPU Architecture

The CUDA programming model is a parallel programming model that provides an abstract view of how the processes can be run on underlying GPU architectures. The evolution of GPU architecture and the CUDA programming language have been quite parallel and interdependent. While the CUDA programming model has stabilized over time, the architecture is still evolving in its capabilities and functionality. GPU architecture has also grown in terms of number of transistors, and number of computing units over years, while yet supporting the CUDA programming model. The CUDA programming model has been used to implement many other algorithms and applications other than graphics and this explosion of use and permeability of CUDA in hitherto unknown applications has catapulted the GPU's near ubiquitous use in many domains of science and technology. Since then all the GPUs designed are CUDA capable. It should be noted that before CUDA was released, there were attempts to create high level langues and template libraries such as Glift [Lefohn et al. 06] and Scout [McCormick et al. 07]. But such efforts tapered down with the introduction of CUDA, and more effort was spent on refining CUDA and building libraries using its constructs.

One of the conceptual differences between CPU and GPU is that CPU is defined for minimum latency so that context switch time is minimum, while GPU is primarily designed for maximum throughput through fine grain pipelining (and hence more latency than CPU). In other words, in the CPU design, there is plenty of cache memory and control logic that would reduce the time taken to bring the data to the ALU and thus reduce the wait and latency. On the other hand, GPU has a lot of ALUs and may wait for the data to be fetched from external DRAM to its local cache. The fundamental optimization of GPU programming hence focusses on hiding this latency by providing enough work for the ALUs while the data is fetched from DRAM.

The two main components of the GPU board that go into the PCI express bus in the PC are the global memory (around 12 GB currently) and the actual streaming multiprocessor chip along with associated circuitry. The basic GPU processing flow consists of three steps: (1) moving data from the host's (main CPU, memory, etc.) main memory to the device's (GPU board) global memory, (2) the CPU issuing instructions to the GPU while the data for this computation is taken from the GPU's global memory and the results are put back in that global memory, and finally (3) the results transferring from the GPU's global memory back to the host's main memory through the PCI express bus.

In latest GPU architectures starting from Kepler, it is possible to communicate between multiple GPUs directly from one GPU's global memory to another's through MPI calls, without going through the intermediate host's memory. Further, not all jobs that the GPU is doing need to be instructed by the CPU – the GPUs can launch their own jobs. The latter feature is also called CUDA dynamic parallelism and can be useful in several ways: It would reduce the communication required between GPU and CPU through the slow PCI bus; it can be used to program recursive parallel algorithms and dynamic load balancing; features like adaptive hierarchical spatial subdivision and computational fluid dynamic grid simulation can effectively be done for efficient and accurate simulation. Conceptually, dynamic parallelism moves the GPU from being a co-processor to an autonomous, dynamic parallel processor.

Each multiprocessing chip has many processors. Each processor can handle thousands of threads of processes. Each basic hardware unit that handles one thread is called a core, also called a CUDA core. For example, the Kepler streaming multiprocessor chip has 15 processors, and each processor can manage 2048 threads – it has 2048 CUDA cores. Each of these 15 processors have plenty of registers (over 64K 32-bit registers) and also shared memory (around 48KB) that are accessible to all the threads running in a processor. The threads running in the same processor can cooperate and share data using the registers and shared memory.

15.3.2   CUDA Programming Model

CUDA is basically an extension of C++. The goal of CUDA design is to let the programmer focus on parallel algorithms rather than the underlying multiprocessor architecture. It is both a programming model as well as a memory model. A typical CUDA application has a mixture of serial code and parallel code. The serial segments of the code run on the host (CPU) while the parallel code, also called kernels, run on the device (GPU) across multiple processing elements. While the parallel code is run on the GPU, the serial code can continue to work on the CPU.

A kernel is a piece of code for one thread. Many instances of the kernel are executed in parallel with potentially different data for each thread under the Single Instruction Multiple Data (SIMD) model. All the threads that run in the same processor are grouped together and are called a block. Each block is run on different processor in the multiprocessor chip, and potentially at different times. A group of these blocks is called a grid. In other words, a grid consists of all the instances of the kernel partitioned as thread blocks. The hardware takes care of scheduling these thread blocks on each of the cores in the processor and there is no cost associated with switching between threads. When the number of blocks exceed the number of available processors, multiple thread blocks may be scheduled to the same streaming processor and in any arbitrary order. Hence there is no simple way to communicate between threads in different blocks as they may be separated both in space (different processors) and time. The goal of efficient CUDA code is to make sure that the task is partitioned into sufficiently fine grained threads such that the latency of data transfer from the DRAM to the multiprocessor chip is well hidden through overlap of computation tasks of the threads.

Each block has a unique id, and each thread within a block also has a unique id. They are referred using the built-in variables threadIdx and blockIdx. These ids can be one, two or three dimensional entities. The number of threads within a block can be read back from the variable blockDim and the number of blocks in a grid is stored in the variable gridDim. The linear id of a thread among all the threads spawned by the kernel is given by blockDim.x * blockIdx.x + threadIdx.x. The .x refers to the first dimension of the three dimensions assuming that the other two dimensions have values 1 each, thus representing just a linear array of one dimensional threads. (Note that the maximum number of the block index value is 64K. So if you need more than 64K blocks, you may need to fold that vector into two dimensional array of blocks in which the index of each dimension can go upto 64K.) The 3D representation of block and thread indices is just to give flexibility in representation that might implicitly align with the problem description. For example, a kernel operating on each element of a dense matrix might need to refer to each thread using a 2D index. Linearization of the index of a thread in a 2D array of threads is done as follows:

```
int iy = blockDim.y*bloxkIdx.y + threadIdx.y;
int ix = blockDim.x*blockIdx.x + threadIdx.x;
int idx = iy*w + ix;
```

Kepler architecture can keep track of 2048 threads or 64 warps (number of threads that can run in lock-step at the same time in a processor), or 16 blocks per stream processor. In other words, each block should have at least 128 threads to keep the GPU compute-busy. It is good to have a number of threads per block that is a multiple of 32 since that is the warp size. For each device there are 14 stream processors. So we need to have at least 224 blocks to keep the GPU busy, and typically will have 1000 or more blocks in a grid. This will also make the code future GPU ready.

15.3.3   CUDA Memory Model

There is a memory hierarchy used by CUDA and supported by the streaming multiprocessor and GPU architectures. Within each processor inside the chip, we noted that there are registers that are accessible per thread and this space is valid until that thread is alive. If a thread uses more registers than are available, the system automatically uses "Local memory" which is actually the off-chip memory on the GPU card (device). So, although the data can be transparently fetched from the local memory as if it is in the register, the latency of this data fetch is as high as the data fetched from the global memory, for a simple reason that "local" memory is just a part of allocated global memory. The "shared" memory is an on-chip memory like registers, but is allocated per-block, and the data in the shared memory is valid until the block is being executed by the processor. Global memory, as mentioned earlier, is off-chip, but on the GPU card. This memory is accessible by all threads of all kernels, as well as the host (CPU). Data sharing between threads in different blocks of the same kernel or even different kernels can be done using the global memory. The host (CPU) memory, which is the slowest from the GPU perspective is not directly accessible by CUDA threads, but the data has to be explicitly transferred from the host memory to the device memory (global memory). However, CUDA 6 introduces unified memory using which the data in the host memory can be directly indexed from the GPU side without explicitly transfering data between the host and the device. Finally, communication between different GPUs have to go through the PCI express bus and through the host memory. This is clearly the most expensive communication. However, the latest NVLink a power-efficient high-speed bus between the CPU and GPU, and between multiple GPUs, allows much higher transfer speeds than those achievable by using PCI Express.

15.4   Conclusion

This chapter gives you a very brief overview of the graphics hardware and API to bootstrap your process of learning a graphics API for programming. Several books exist that teach graphics using APIs, mostly using OpenGL [Angel 02,Hill and Kelly 06,Hearn and Baker 03] or WebGL [Angel and Shreiner 14]. These can be a great starting point for graphics programming. The famous redbook [Kessenich et al. 16] is a great handbook for any questions about the OpenGL API. Several books exist to get an in depth knowledge about CUDA programming [Sanders 10,Cheng and Grossman 14] which can help you get the maximum mileage out of your GPU.

Bibliography

[Angel and Shreiner 14] Edward Angel and Dave Shreiner. *Interactive Computer Graphics: A Top-Down Approach with WebGL*, 7th edition. Pearson, 2014.

[Angel 02] Edward Angel. *Interactive Computer Graphics*, Third edition. Addison-Wesley Longman Publishing Co. Inc., 2002.

[Azad 16] Hamid Azad. *Advances in GPU Research and Practice*, First edition. Morgan Kaufman, 2016.

[Cheng and Grossman 14] John Cheng and Max Grossman. *Professional CUDA C Programming*. Wrox, 2014.

[Hearn and Baker 03] Donald D. Hearn and M. Pauline Baker. *Computer Graphics with OpenGL*, Third edition. Prentice Hall Professional Technical Reference, 2003.

[Hill and Kelly 06] F.S. Hill and Stephen M. Kelly. *Computer Graphics using OpenGL*. Prentice Hall, 2006.

[Kessenich et al. 16] John Kessenich, Graham Sellers, and Dave Shreiner. *OpenGL Programming Guide: The Official Guide to Learning OpenGL, Version 4.5*, 9th edition. Addison-Wesley Longman Publishing Co., Inc., 2016.

[Lefohn et al. 06] Aaron E. Lefohn, Shubhabrata Sengupta, Joe Kniss, Robert Strzodka, and John D. Owens. "Glift: Generic, efficient, random-access GPU data structures." *ACM Trans. Graph.* 25:1 (2006), 60–99. Available online (http://doi.acm.org/10.1145/1122501.1122505).

[Luebke 09] David P. Luebke. "Graphics hardware & GPU computing: past, present, and future." In *Proceedings of the Graphics Interface 2009 Conference, May 25-27, 2009, Kelowna, British Columbia, Canada*, p. 6, 2009. Available online (http://doi.acm.org/10.1145/1555880.1555888).

[McCormick et al. 07] Patrick McCormick, Jeff Inman, James Ahrens, Jamaludin Mohd-Yusof, Greg Roth, and Sharen Cummins. "Scout: A Data-parallel Programming Language for Graphics Processors." *Parallel Comput.* 33:10-11 (2007), 648–662. Available online (http://dx.doi.org/10.1016/j.parco.2007.09.001).

[Sanders 10] Jason Sanders. *CUDA by Example: An Introduction to General-Purpose GPU Programming*, First edition. Addison Wesley, 2010.

Summary: Do you know these concepts?

- Graphics Processing Unit (GPU)
- Vertex Shaders
- Fragment Shaders
- Unified Shaders
- Rasterizing Operations (ROP)
- OpenGL
- CUDA

Index

Printed and bound by CPI Group (UK) Ltd, Croydon, CR0 4YY

24/10/2024

01778290-0010